Wave Propagation in Electromagnetic Media

Julian L. Davis

Wave Propagation in Electromagnetic Media

With 30 Illustrations

Springer-Verlag
New York Berlin Heidelberg
London Paris Tokyo

Julian L. Davis
22 Libby Avenue
Pompton Plains, NJ 07444
U.S.A.

Library of Congress Cataloging in Publication Data
Davis, Julian L.
    Wave propagation in electromagnetic media / Julian L. Davis.
        p.   cm.
    Includes bibliographical references.
    ISBN-13:978-1-4612-7950-1
        1. Electromagnetic Waves—Transmission.   2. Differential equations, Hyperbolic.
3. Hamilton-Jacobi equations.   I. Title.
    QC665.T7D38   1989
    530.1'41—dc20   89-21864
                                                                                  CIP

Printed on acid-free paper.

© 1990 by Springer-Verlag New York, Inc.
Softcover reprint of the hardcover 1st edition 1990

Typeset by Asco Trade Typesetting Ltd., Hong Kong.

9 8 7 6 5 4 3 2 1

ISBN-13:978-1-4612-7950-1   e-ISBN-13:978-1-4612-3284-1
DOI: 10.1007/978-1-4612-3284-1

To Dorothy

Preface

This is the second work of a set of two volumes on the phenomena of wave propagation in nonreacting and reacting media. The first, entitled *Wave Propagation in Solids and Fluids* (published by Springer-Verlag in 1988), deals with wave phenomena in nonreacting media (solids and fluids). This book is concerned with wave propagation in reacting media—specifically, in electromagnetic materials. Since these volumes were designed to be relatively self-contained, we have taken the liberty of adapting some of the pertinent material, especially in the theory of hyperbolic partial differential equations (concerned with electromagnetic wave propagation), variational methods, and Hamilton–Jacobi theory, to the phenomena of electromagnetic waves.

The purpose of this volume is similar to that of the first, except that here we are dealing with electromagnetic waves. We attempt to present a clear and systematic account of the mathematical methods of wave phenomena in electromagnetic materials that will be readily accessible to physicists and engineers. The emphasis is on developing the necessary mathematical techniques, and on showing how these methods of mathematical physics can be effective in unifying the physics of wave propagation in electromagnetic media.

Chapter 1 presents the theory of time-varying electromagnetic fields, which involves a discussion of Faraday's laws, Maxwell's equations, and their applications to electromagnetic wave propagation under a variety of conditions. Chapter 7 is concerned with a discussion of magnetohydrodynamics and plasma physics. Since modern ideas in electromagnetic phenomena involve quantum mechanics and the special theory of relativity, we present a chapter on a survey of quantum mechanics including cosmology (Chapter 6) and a chapter on the special theory of relativity (Chapter 8). The mathematical foundation of electromagnetic waves, vis-à-vis partial differential equations, is contained in Chapters 2 and 3. They present a rather complete mathematical discussion of the hyperbolic partial differential equations, which give the mathematical basis for a proper understanding of propagating waves in electromagnetic media. It is the author's experience, in teaching physicists and engineers, that they need a more thorough grounding in the mathematical structure of these partial differential equations in order to gain a better insight into electromagnetic wave propagation. Chapters 4 and 5 present variational methods and Hamilton–Jacobi theory, which are essential to a deeper understanding of modern concepts in wave propagation and quantum mechanics.

The phenomenon of electromagnetic wave propagation is an important branch of physics. It would not be amiss at this stage to quote Michael Faraday on the definition of physics: "Physics is to make experiments and publish them". He goes on to say that physics is rooted in experiment, in active, inquisitive, and skillful intercourse with nature. Obviously, such a profound thinker as Faraday was jesting when he considered publication as the sole contribution to physics. He certainly realized that all experiments are blind if they are not interpreted by theories. Maxwell understood this very well when he developed the mathematical structure for the interpretation of Faraday's fundamental laws of electricity and magnetism. Maxwell's great contribution was the wedding of electricity and magnetism into a unified electromagnetic theory. One of the important consequences of this theory was the deduction that light waves are electromagnetic waves. This led Hertz to invent and develop wireless radio transmission. It is rather interesting that mathematicians, such as Riemann and Jacobi who developed modern concepts in partial differential equations (somewhat independent of electromagnetic theory), gave the physicist the necessary mathematical tools for the study of electromagnetic waves. In fact, Riemann's great contributions to the theory of partial differential equations stemmed from his thesis on fluid mechanics and shock waves in nonreacting gases. In any case, mathematical physics supplies the conceptual thought in physics, for it treats, in a quantitative way, the relations amongst the various measurable quantities of nature. Thus experimental and mathematical physics both work together to attempt to give us a more profound insight into natural phenomena.

It is clear that we have omitted many topics of interest to some of our readers. It appears that the prerogative of an author is to write profusely on what interests him or her, and leave out material that he or she thinks is of secondary interest. However, we hope that the reader will get a better understanding, not only of the phenomenon of electromagnetic wave propagation, but of its place in the grand scope of physics which includes quantum mechanics and special relativity.

Julian L. Davis
Spring, 1989

Contents

Preface   vii

CHAPTER 1
Time-Varying Electromagnetic Fields   1
     Introduction   1
1.1.   Maxwell's Equations   1
1.2.   Conservation Laws   4
1.3.   Scalar and Vector Potentials   8
1.4.   Plane Electromagnetic Waves in a Nonconducting Medium   10
1.5.   Plane Waves in a Conducting Medium   28

CHAPTER 2
Hyperbolic Partial Differential Equations in
Two Independent Variables   32
     Introduction   32
2.1.   General Solution of the Wave Equation   33
2.2.   D'Alembert's Solution of the Cauchy Initial Value Problem   35
2.3.   Method of Characteristics for a Single First-Order Equation   40
2.4.   Method of Characteristics for a First-Order System   48
2.5.   Second-Order Quasilinear Partial Differential Equation   50
2.6.   Domain of Dependence and Range of Influence   55
2.7.   Some Basic Mathematical and Physical Principles   56
2.8.   Propagation of Discontinuities   60
2.9.   Weak Solutions and the Conservation Laws   63
2.10. Normal Forms for Second-Order Partial Differential Equations   69
2.11. Riemann's Method   74
2.12. Nonlinear Hyperbolic Equations in Two Independent Variables   82

CHAPTER 3
Hyperbolic Partial Differential Equations in
More Than Two Independent Variables   91
     Introduction   91
3.1.   First-Order Quasilinear Equations in n Independent Variables   91
3.2.   First-Order Fully Nonlinear Equations in n Independent Variables   94

3.3.   Directional Derivatives in n Dimensions   95
3.4.   Characteristic Surfaces in n Dimensions   99
3.5.   Maxwell's Equations   105
3.6.   Second-Order Quasilinear Equation in n Independent Variables   107
3.7.   Geometry of Characteristics for Second-Order Systems   112
3.8.   Ray Cone, Normal Cone, Duality   118
3.9.   Wave Equation in n Dimensions   120

Appendix: Similarity Transformations and Canonical Forms   127
       Introduction   127
3A.1. Geometric Considerations   128
3A.2. Orthogonal Transformations and Eigenvectors in Relation
       to Similarity Transformations   130
3A.3. Diagonalization of A' 133

CHAPTER 4
Variational Methods   135
       Introduction   135
4.1.   Principle of Least Time   136
4.2.   One-Dimensional Calculus of Variations, Euler's Equation   137
4.3.   Generalization to Functionals with More Than
       One Dependent Variable   144
4.4.   Special Case   145
4.5.   Hamilton's Variational Principle and Configuration Space   146
4.6.   Lagrange's Equations of Motion   148
4.7.   D'Alembert's Principle, Constraints, and Lagrange's Equations   151
4.8.   Nonconservative Force Field, Velocity-Dependent Potential   158
4.9.   Constraints Revisited, Undetermined Multipliers   165
4.10. Hamilton's Equations of Motion   167
4.11. Cyclic Coordinates   169
4.12. Principle of Least Action   171
4.13. Lagrange's Equations of Motion for a Continuum   174
4.14. Hamilton's Equations of Motion for a Continuum   178

CHAPTER 5
Canonical Transformations and Hamilton–Jacobi Theory   181
       Introduction   181

Part I.   Canonical Transformations   181
5.1.   Equations of Canonical Transformations and Generating Functions   181
5.2.   Some Examples of Canonical Transformations   186

Part II.   Hamilton–Jacobi Theory   187
       Introduction   187
5.3.   Derivation of the Hamilton–Jacobi Equation for
       Hamilton's Principle Function   188
5.4. S Related to a Variational Principle   189
5.5.   Application to Harmonic Oscillator   190
5.6.   Hamilton's Characteristic Function   191

5.7. Application to n Harmonic Oscillators   193
5.8. Hamilton–Jacobi Theory Related to Characteristic Theory   193
5.9. Hamilton–Jacobi Theory and Wave Propagation   197
5.10. Hamilton–Jacobi Theory and Quantum Mechanics   203

CHAPTER 6
Quantum Mechanics—A Survey   206
      Introduction   206

CHAPTER 7
Plasma Physics and Magnetohydrodynamics   231
      Introduction   231
7.1. Fluid Dynamics Equations—General Treatment   232
7.2. Application of Fluid Dynamics Equations to Magnetohydrodynamics   235
7.3. Application of Characteristic Theory to Magnetohydrodynamics   239
7.4. Linearization of the Field Equations   243

CHAPTER 8
The Special Theory of Relativity   254
      Introduction   254
8.1. Collapse of the Ether Theory   256
8.2. The Lorentz Transformation   258
8.3. Maxwell's Equations with Respect to a Lorentz Transformation   262
8.4. Contraction of Rods and Time Dilation   263
8.5. Addition of Velocities   265
8.6. World Lines and Light Cones   267
8.7. Covariant Formulation of the Laws of Physics in
      Minkowski Space   270
8.8. Covariance of the Electromagnetic Equations   274
8.9. Force and Energy Equations in Relativistic Mechanics   278
8.10. Lagrangian Formulation of Equations of Motion in
      Relativistic Mechanics   283
8.11. Covariant Lagrangian   285

Bibliography   289

Index   291

CHAPTER 1

Time-Varying Electromagnetic Fields

Introduction

In this chapter we discuss time-varying or nonsteady state electromagnetic
fields by way of the powerful unifying theory of the great Scottish mathematical physicist, James Clerk Maxwell (1831–1879). The reader may want
to refer to the original work of Maxwell [32]. In electrostatics and magneto-
statics the steady state electric and magnetic fields are treated separately.†
Although similar mathematical techniques were used, electric and magnetic
phenomena were treated as being independent of each other. The only link
is that an electrical current produces a magnetic field. But when we consider
time-dependent electric and magnetic fields, Maxwell taught us that these
fields are interdependent, being bound together by a unified electromagnetic
theory. It is this theory that we consider in detail in this chapter. In particular,
we shall investigate Maxwell's equations and their solutions in a physical
setting, as a basis for a discussion of the mathematical properties of these
equations vis-à-vis the partial differential equations (PDEs) of wave propagation in an electromagnetic medium. The mathematical theory of these PDEs
will be treated in detail in subsequent chapters.

1.1. Maxwell's Equations

The fundamental equations of electricity and magnetism can be summarized
in Maxwell's equations and the conservation laws of physics. The equations
of Maxwell can be represented as the following set of four PDEs:

$$\nabla \cdot \mathbf{D} = 4\pi\rho, \tag{1.1}$$

$$\nabla \times \mathbf{E} = -\frac{1}{c}\frac{\partial \mathbf{B}}{\partial t}, \tag{1.2}$$

---

† The reader is assumed to have a working knowledge of this subject. Reference is
made to standard works on electricity and magnetism, such as [18], [27], [36], and
[42].

$$\nabla \cdot \mathbf{B} = 0, \tag{1.3}$$

$$\dagger \nabla \times \mathbf{H} = \left(\frac{4\pi}{c}\right)\mathbf{J} + \frac{1}{c}\frac{\partial \mathbf{D}}{\partial t}, \tag{1.4}$$

where $\mathbf{D}$ is the displacement vector, ρ is the charge density, $\mathbf{E}$ is the electric field, $\mathbf{B}$ is the magnetic-flux density or magnetic induction, $\mathbf{H}$ is the magnetic field, and $\mathbf{J}$ is the current density (current per unit cross-sectional area). These equations are in Gaussian units and are written in macroscopic form (molecular structure is ignored). (Note: Boldface type refers to vectors.) We have the following relations:‡

$$\mathbf{D} = \mathbf{E} + 4\pi\mathbf{P} = \varepsilon\mathbf{F},$$

$$\mathbf{P} = \chi_e\mathbf{E},$$

$$\mathbf{H} = \mathbf{B} - 4\pi\mathbf{M},$$

$$\mathbf{B} = \mu\mathbf{H},$$

$$\mathbf{J} = \sigma\mathbf{E}, \qquad \text{Ohm's law,}$$

where $\mathbf{P}$ is the polarization (electric dipole moment per unit volume), ε is the dielectric constant, χ_e is the electric susceptibility, $\mathbf{M}$ is the magnetization (magnetic moment per unit volume), μ is the permeability, and σ is the conductivity. The fundamental electric and magnetic fields are $\mathbf{E}$ and $\mathbf{B}$, respectively. The derived electric and magnetic fields are $\mathbf{D}$ and $\mathbf{H}$, respectively. They are introduced as a matter of convenience. We take into account, in an average way, the effects of electric and magnetic polarization. Equation (1.1) is Gauss's law. It tells us that the flux of $\mathbf{D}$ through a closed surface S surrounding a volume V equals 4π times the total charge inside V. The integral representation is

$$\oint_S \mathbf{D} \cdot \mathbf{n} \, dS = \int_V \nabla \cdot \mathbf{D} \, dV = 4\pi \int_V \rho \, dV,$$

upon using the divergence theorem. $\mathbf{n}$ is the unit normal to the surface element dS and dV is the volume element. The integral representation of (1.2) is

$$\oint_C \mathbf{E} \cdot d\mathbf{s} = \int_S \nabla \times \mathbf{E} \cdot \mathbf{n} \, dS = -\frac{1}{c}\frac{d}{dt}\int_S \mathbf{B} \cdot \mathbf{n} \, dS,$$

where the first integral is a line integral around a closed loop C on the closed surface S and $d\mathbf{s}$ is an element of arc length on C. Stokes's theorem is used to transform this line integral to the surface integral of the normal component

---

† The ∇ operator is defined by

$$\nabla = \mathbf{i}\frac{\partial}{\partial x} + \mathbf{j}\frac{\partial}{\partial y} + \mathbf{k}\frac{\partial}{\partial z}, \qquad \nabla \cdot \mathbf{D} \equiv \text{div } \mathbf{D}, \qquad \nabla \times \mathbf{E} = \text{curl } \mathbf{E}$$

‡ These are called the *constitutive equations*.

of **curl E** across S. Equating this surface integral to the surface integral of the normal component of $d\mathbf{B}/dt$ across S gives us the integral formulation of *Faraday's law* (1.2), which expresses the experimental fact that the induced electromotive force is proportional to the total time derivative of the magnetic flux. Equation (1.3) expresses the fact that the flux of **B** through the closed surface S vanishes. This is easily seen by the integral representation, upon using the divergence theorem.

Equation (1.4) requires further explanation. For an electrostatic field it reduces to

$$\nabla \times \mathbf{H} = \left(\frac{4\pi}{c}\right)\mathbf{J}, \tag{1.5}$$

which is *Ampere's law*. Its integral representation is obtained by applying Stokes's theorem to the normal component of the above equation over a surface S bounded by a closed curve C. We obtain

$$\int_S \nabla \times \mathbf{B} \cdot \mathbf{n}\, dS = \oint_C \mathbf{B} \cdot d\mathbf{s} = \frac{4\pi}{c} \int_S \mathbf{J} \cdot \mathbf{n}\, dS = \left(\frac{4\pi}{c}\right) I,$$

where we have used the fact that the surface integral of the normal component of the current density **J** is equal to the total current I passing through the closed curve C. Now it is easily shown that **J**, as given by (1.5) for a static field, does not satisfy the conservation of charge law for a dynamic or time-varying field. This is seen as follows: The law of conservation of charge for the dynamic case is an example of the continuity equation which is one of the three conservation laws of physics (the other two being the conservation of momentum or equation of motion and conservation of energy). The differential form of the continuity equation (conservation of mass) applied to time-varying charges is

$$\nabla \cdot \mathbf{J} = -\frac{\partial \rho}{\partial t}. \tag{1.6}$$

The integral representation of (1.6) tells us that the flux of **J** across the surface S enclosing the volume V equals the divergence of **J** out of V which is equal to the total decrease of current out of V. Now, if we take the divergence of both sides of (1.5) (by multiplying each term by the operator ∇), we obtain

$$\nabla \cdot \mathbf{J} = 0,$$

which means that the charge is time invariant so that there is no current flow—the electrostatic case.

Maxwell recognized that (1.1), (1.2), and (1.3) are valid for static and time-varying fields, since neither of these equations involves differentiation with respect to time. These equations coupled with (1.5) represent a complete system for the static electric and magnetic fields. But, as shown, Ampere's law as given by (1.5) breaks down for time-varying fields because it does not satisfy the continuity equation (which is a fundamental law of physics and must be satisfied). Maxwell must have reasoned that if he could only generalize Ampere's law to satisfy the continuity equation, he would then have four equations which would unify the electric and magnetic fields—this would be a tremendous breakthrough! How could he do this? He tried the inspired idea of adding a fictitious current term in the form $\partial \mathbf{D}/\partial t$† to the right-hand side of (1.5), so that he replaced (1.5) by the more general form

$$\nabla \times \mathbf{H} = \left(\frac{4\pi}{c}\right)\mathbf{J} + k\left(\frac{\partial \mathbf{D}}{\partial t}\right),$$

where k is a positive constant adjusted for the appropriate units. He now took the divergence of each term of this equation and used Gauss's law (1.1) obtaining

$$\nabla \cdot \nabla \times \mathbf{H} = 0 = \left(\frac{4\pi}{c}\right)\nabla \cdot \mathbf{J} + k\left(\frac{\partial}{\partial t}\right)\nabla \cdot \mathbf{D} = \left(\frac{4\pi}{c}\right)\nabla \cdot \mathbf{J} + 4\pi k\left(\frac{\partial \rho}{\partial t}\right),$$

which is indeed (1.4), so that the system $(1.1)\cdots(1.4)$ represents the complete formulation of Maxwell's equations which gives us a unified electromagnetic theory. This system forms the basis of all electromagnetic phenomena. When combined with the three conservation laws, the four Maxwell equations provide a complete description of the classical dynamics of interacting charged particles and electromagnetic fields. Maxwell called the added term $\partial \mathbf{D}/\partial t$ the *displacement current*. We thus see that this necessary addition to Ampere's law, yielding the generalized form of Ampere's law (1.4), is of crucial importance for time-varying fields. Without (1.4) there would be no electromagnetic radiation (and this book could not be written!). Maxwell went further by predicting that light waves were electromagnetic waves, and that electromagnetic waves of a great range of frequencies could be produced. This captured the attention of the nineteenth-century physicists and stimulated much theoretical and experimental research into electromagnetic phenomena.

1.2. Conservation Laws

It was mentioned above that Maxwell's equations, along with the three conservation laws of physics, give a complete description of the classical dynamics of interacting charged particles. It is therefore worthwhile, in this section, to state these laws in the setting of electromagnetic theory.

The *continuity equation* applied to moving charges is called the *conservation of charge* and was given as (1.6).

The *conservation of linear momentum* or *equation of motion* for a current involves the *Lorentz force* $\mathbf{F}$. The force on a charge q in an external electric field $\mathbf{E}$ is $q\mathbf{E}$. Suppose that in addition to E we have an external magnetic field

---

† Apparently the reason for using $\mathbf{D}$ was to involve Gauss's law (1.1).

B. Since current involves moving charges, it can be shown that the magnetic force acting on a charge moving with a velocity **v** due to **B** is (q/c)**v** × **B**. Thus the total electromagnetic force on a charged particle is

$$F = q\left(E + \frac{v}{c} \times B\right). \tag{1.7}$$

Consider a continuous distribution of charged particles having a density ρ and momentum **p**, contained in a volume V. Then, applying Newton's second law to the Lorentz force given by (1.7), we obtain

$$\frac{d\mathbf{p}}{dt} = \int_V \left[\rho E + \left(\frac{1}{c}\right) J \times B\right] dV. \tag{1.8}$$

Note: In this treatment of the conservation laws† we shall treat the charges in terms of the microscopic equations which involve the electric and magnetic fields **E** and **B**, rather than **D** and **H** as given in Maxwell's equations. We now solve for ρ in (1.1) and for **J** in (1.4), and insert the resulting expressions into the integrand of the right-hand side of (1.8). The integrand becomes

$$\rho E + \left(\frac{1}{c}\right) J \times B = \frac{1}{4\pi}\left[E(\nabla \cdot E) + \left(\frac{1}{c}\right) B \times \frac{\partial E}{\partial t} - B \times (\nabla \times B)\right].$$

Since **B** × $\partial E/\partial t = -(\partial/\partial t)(E \times B) + E \times \partial B/\partial t$ and **B**$(\nabla \cdot B) = 0$, the above expression for the integrand becomes

$$\rho E + \left(\frac{1}{c}\right) J \times B = \frac{1}{4\pi}[E(\nabla \cdot E) + B(\nabla \cdot B) - E \times (\nabla \times E) - B \times (\nabla \times B)]$$

$$- \frac{1}{4\pi c}\frac{\partial}{\partial t}(E \times B). \tag{1.9}$$

The total rate of change of momentum becomes

$$\frac{d\mathbf{p}}{dt} + \frac{d}{dt}\int_V \frac{1}{4\pi c}(E \times B)\, dV = \frac{1}{4\pi}\int_V [E(\nabla \cdot E) - E \times (\nabla \times E)]\, dV$$

$$+ \frac{1}{4\pi}\int_V [B(\nabla \cdot B) - B \times (\nabla \times B)]\, dV. \tag{1.10}$$

The first term of the left-hand side of (1.10) is the rate of change of the "mechanical momentum" **p**, while the second term on the left is the rate of change of the "electromagnetic momentum" $d\mathbf{p}_{em}/dt$ where $\mathbf{p}_{em}$ is defined by

$$\mathbf{p}_{em} = \frac{1}{4\pi c}\int_V (E \times B)\, dV, \tag{1.11}$$

---

† Refer to [15], for example, for a treatment of the conservation laws for wave propagation in a nonconducting medium.

where the integrand in (1.11) may be interpreted as the "electromagnetic momentum". We observe that the integrand of the first integral on the right-hand side of (1.10) is the same function of **E** as the second integrand is of **B**. Therefore the same vector identity that we shall use for the first integrand also holds for the second. For any vector **v**, the identity is

$$\tfrac{1}{2}\nabla(\mathbf{v}\cdot\mathbf{v}) = (\mathbf{v}\cdot\nabla)\mathbf{v} + \mathbf{v}\times(\nabla\times\mathbf{v}).$$

Letting **v** stand for **E** or **B**, we use this identity to write the integrands of the right-hand side of (1.10) as

$$\mathbf{v}(\nabla\cdot\mathbf{v}) + (\mathbf{v}\cdot\nabla)\mathbf{v} - \tfrac{1}{2}\nabla v^2 = \nabla\cdot(\mathbf{vv} - \tfrac{1}{2}\mathbf{I}v^2), \qquad (1.12)$$

where the expression **vv** is a *dyadic* or tensor of rank two and the *identity dyadic* is defined by $\mathbf{I} = \mathbf{ii} + \mathbf{jj} + \mathbf{kk}$, $(\mathbf{i}, \mathbf{j}, \mathbf{k})$ are unit vectors in the (x, y, z) directions, respectively. For a more symmetric notation the coordinate system **x** is replaced by (x_1, x_2, x_3) and $(\mathbf{i}, \mathbf{j}, \mathbf{k})$ by $(\mathbf{e}_1, \mathbf{e}_2, \mathbf{e}_3)$ so that $\mathbf{I} = \mathbf{e}_i\mathbf{e}_i$, where we use the *tensor summation convention* (sum over the repeated index i from 1 to 3—for three space). We use the definition of $\mathbf{P}_{em}$ given by (1.11) and apply (1.12) to the right-hand side of (1.10), obtaining

$$\frac{d(\mathbf{p} + \mathbf{p}_{em})}{dt} = \int_V \nabla\cdot\mathbf{T}\, dV = \oint_S \mathbf{n}\cdot\mathbf{T}\, dS, \qquad (1.13)$$

where we used the divergence theorem to convert the volume integral over the volume V to the surface integral over the closed surface S surrounding V. The tensor **T** is called *Maxwell's stress tensor*. It is defined by

$$\mathbf{T} = \frac{1}{4\pi}[\mathbf{EE} + \mathbf{BB} - \tfrac{1}{2}\mathbf{I}(E^2 + B^2)]. \qquad (1.14)$$

Let T_{ij} be the ijth component of the tensor **T**. Then **T** can be represented by

$$\mathbf{T} = \mathbf{e}_i T_{ij}\mathbf{e}_j,$$

where we sum over i and j from 1 to 3. Conversely, given the tensor **T** and the set of unit vectors $(\mathbf{e}_1, \mathbf{e}_2, \mathbf{e}_3)$, we can obtain the ijth element of **T** by taking the appropriate scalar products

$$T_{ij} = \mathbf{e}_i\cdot\mathbf{T}\cdot\mathbf{e}_j.$$

The ijth element of **T** is obtained from (1.14) as

$$T_{ij} = \frac{1}{4\pi}[E_i E_j + B_i B_j - \tfrac{1}{2}\delta_{ij}(E^2 + B^2)],\dagger \qquad i, j = 1, 2, 3. \qquad (1.15)$$

Note that the purpose of the above exercise with the vector identity is to transform the integrands on the right-hand side of (1.10) to the divergence of

---

$\dagger\ \delta_{ij} = \begin{cases} 1, & i = j, \\ 0, & i \neq j. \end{cases}$ $\quad \delta_{ij}$ is the Kronecker delta.

a tensor (which turns out to be Maxwell's stress tensor **T**), so that we can thereby transform the volume integral of **div T** to the surface integral of the normal component of **T** across S. This powerful idea allows us to physically interpret (1.13) as follows: The rate of change of the mechanical plus the electromagnetic momentum equals the force per unit area transmitted across S (normal to S) bounding V (which is the flux of the Maxwell tensor **T** across S). This equation allows us to calculate the forces acting on a body in an electromagnetic field by enclosing it with a surface S and computing the electromagnetic forces by using the right-hand side of (1.15).

The *conservation of energy* is given by *Poynting's theorem*, which we shall now discuss. We start, as usual, with a continuous distribution of charge and current in an electromagnetic material occupying a finite volume V. Since the magnetic field does no work (being normal to the direction of current flow given by the current density **J**), the total rate of work or power dW/dt due to the electromagnetic fields in V is

$$\frac{dW}{dt} = \int_V \mathbf{J} \cdot \mathbf{E} \, dV. \tag{1.16}$$

We now appeal to Faraday's law given by (1.2), the Ampere–Maxwell law (1.4), and the following vector identity:

$$\mathbf{E} \cdot (\nabla \times \mathbf{H}) = \mathbf{H} \cdot (\nabla \times \mathbf{E}) - \nabla \cdot (\mathbf{E} \times \mathbf{H}).$$

Equation (1.16) is then transformed into

$$\frac{dW}{dt} = \int_V \mathbf{J} \cdot \mathbf{E} \, dV = -\frac{1}{4\pi} \int_V \left[c\nabla \cdot (\mathbf{E} \times \mathbf{H}) + \mathbf{E} \cdot \frac{\partial \mathbf{D}}{\partial t} + \mathbf{H} \cdot \frac{\partial \mathbf{B}}{\partial t} \right] dV. \tag{1.17}$$

Let w be the **energy density** due to the **E** and **H** fields so that the total power due to these fields is

$$\frac{dW}{dt} = \int_V \frac{dw}{dt} \, dV.$$

It can be shown, for example, [26, Chap. 6], that w is given by

$$w = \frac{1}{8\pi} (\mathbf{E} \cdot \mathbf{D} + \mathbf{B} \cdot \mathbf{H}), \tag{1.18}$$

and for a linear electromagnetic material (1.17) is approximated by

$$\frac{dW}{dt} = \int_V \mathbf{J} \cdot \mathbf{E} \, dV = -\int_V \left[\left(\frac{c}{4\pi} \right) \nabla \cdot (\mathbf{E} \times \mathbf{H}) + \frac{\partial w}{\partial t} \right] dV. \tag{1.19}$$

Since (1.19) is valid for an arbitrary small volume dV, we obtain the differential form of the law of conservation of energy

$$\frac{\partial w}{\partial t} + \nabla \cdot \mathbf{S} = -\mathbf{J} \cdot \mathbf{E}, \tag{1.20}$$

where **S** is called *Poynting's vector*. It represents the energy flow, has the dimensions of energy per unit area per unit time, and is given by

$$\mathbf{S} = \frac{c}{4\pi}(\mathbf{E} \times \mathbf{H}). \tag{1.21}$$

The physical meaning of the law of conservation of energy, given in integral form by (1.19) and in differential form by (1.20), is: The rate of increase of electromagnetic energy per unit time (electromagnetic power) within V, plus the energy flowing out through S bounding V, is equal to the negative of the total work done by the electromagnetic fields on the sources within V. This statement of the conservation of energy does not take into account any nonlinear effects such as hystersis in ferromagnetic materials. In this case, the energy law must be supplemented by terms representing the power loss due to hysteresis.

1.3. Scalar and Vector Potentials

We return to Maxwell's equations given by $(1.1) \cdots (1.4)$, which is a system of linear first-order PDEs for the electromagnetic fields and the current density $\mathbf{J}$ to be solved for as functions of space and time. They can be solved in this form for simple situations. However, it is convenient to transform them to a smaller number of second-order PDEs by introducing a vector potential $\mathbf{A}$ and a scalar potential Φ.

To define the vector potential we observe from (1.3) that $\mathbf{B}$ can be given as the curl of a vector, since vector analysis tells us that the diverence of the curl of a vector vanishes. We therefore define the vector potential $\mathbf{A}$ by the following equation:

$$\mathbf{B} = \nabla \times \mathbf{A}. \tag{1.22}$$

By using (1.22), Faraday's law (1.2), which is the other homogeneous equation of Maxwell, can be put in the form

$$\nabla \times \left(\mathbf{E} + \frac{1}{c}\frac{\partial \mathbf{A}}{\partial t} \right) = 0. \tag{1.23}$$

Equation (1.23) tells us that the curl of some vector vanishes. This means that the vector is the gradient of some scalar Φ, according to a theorem in vector analysis.† We therefore define the scalar potential Φ by the equation

$$\mathbf{E} + \left(\frac{1}{c}\right)\frac{\partial \mathbf{A}}{\partial t} = -\nabla\Phi \quad \text{or} \quad \mathbf{E} = -\nabla\Phi - \left(\frac{1}{c}\right)\frac{\partial \mathbf{A}}{\partial t}. \tag{1.24}$$

We now observe that, having defined $\mathbf{B}$ in terms of $\mathbf{A}$ and $\mathbf{E}$ in terms of Φ, the two homogeneous Maxwell equations (1.2) and (1.3) are satisfied identically.

---

† See, for example, [25].

Since $\mathbf{D} = \varepsilon\mathbf{E}$, (1.1) becomes

$$\nabla^2\Phi + \left(\frac{1}{c}\right)\left(\frac{\partial}{\partial t}\right)(\nabla \cdot \mathbf{A}) = -\left(\frac{4\pi}{\varepsilon}\right)\rho. \tag{1.25}$$

Using the vector identity

$$\nabla \times (\nabla \times \mathbf{A}) = \nabla(\nabla \cdot \mathbf{A}) - \nabla^2\mathbf{A},$$

and the relations $\mathbf{H} = (1/\mu)\mathbf{B}$ and $\mathbf{D} = \varepsilon\mathbf{E}$, (1.4) becomes

$$\nabla^2\mathbf{A} - \left(\frac{\varepsilon\mu}{c^2}\right)\frac{\partial^2\mathbf{A}}{\partial t^2} - \nabla\left(\nabla \cdot \mathbf{A} + \left(\frac{\varepsilon\mu}{c}\right)\frac{\partial\Phi}{\partial t}\right) = -\left(\frac{4\pi\mu}{c}\right)\mathbf{J}. \tag{1.26}$$

We have thus reduced the four first-order Maxwell equations to the two coupled second-order equations (1.25) and (1.26) for Φ and $\mathbf{A}$.

We can now uncouple these equations by exploiting the fact that the vector potential $\mathbf{A}$ is arbitrary in the sense that the gradient of any scalar function ψ can be added without changing $\mathbf{B}$; there is also an arbitrariness in Φ which we will subsequently use. The plan is to find the transformation $\mathbf{A} \to \acute{\mathbf{A}}$ and $\Phi \to \acute{\Phi}$ such that (1.25) uncouples into a nonhomogeneous wave equation for $\acute{\Phi}$ and (1.25) becomes a nonhomogeneous wave equation for $\acute{\mathbf{A}}$. We must also satisfy the constraints that $\mathbf{E}$ and $\mathbf{B}$ must be invariant with respect to these transformations. To this end, we set

$$\mathbf{A} = \acute{\mathbf{A}} + \nabla\psi, \qquad \Phi = \acute{\Phi} + f, \tag{1.27}$$

where ψ and f are scalar functions of space and time which must satisfy the constraining conditions of the transformation. Clearly, in order for (1.26) to be uncoupled with respect to the transformed vector potential $\acute{\mathbf{A}}$ we must have

$$\nabla \cdot \acute{\mathbf{A}} + \left(\frac{\varepsilon\mu}{c}\right)\frac{\partial\acute{\Phi}}{\partial t} = 0. \tag{1.28}$$

This tells us that (1.26) is transformed to

$$\nabla^2\acute{\mathbf{A}} - \left(\frac{\varepsilon\mu}{c^2}\right)\frac{\partial^2\acute{\mathbf{A}}}{\partial t^2} = -\left(\frac{4\pi\mu}{c}\right)\mathbf{J}, \tag{1.29}$$

which is the uncoupled nonhomogeneous wave equation for the transformed vector potential $\acute{\mathbf{A}}$. Applying the transformation equation (1.27) to (1.24) gives

$$\mathbf{E} = -\nabla\acute{\Phi} - \nabla f - \left(\frac{1}{c}\right)\frac{\partial\acute{\mathbf{A}}}{\partial t} - \left(\frac{1}{c}\right)\frac{\partial\nabla\psi}{\partial t} = -\nabla\acute{\Phi} - \left(\frac{1}{c}\right)\frac{\partial\acute{\mathbf{A}}}{\partial t},$$

in order for $\mathbf{E}$ to be invariant with respect to the transformation. This means that we have determined f as

$$f = -\left(\frac{1}{c}\right)\frac{\partial\psi}{\partial t}. \tag{1.30}$$

Inserting (1.30) into (1.27) tells us that the transformation into the primed potentials involves one arbitrary scalar potential ψ. Inserting (1.27) and (1.30) into (1.25) and using (1.28) yields

$$\nabla^2 \acute{\Phi} - \left(\frac{\varepsilon\mu}{c^2}\right)\frac{\partial^2 \acute{\Phi}}{\partial t^2} = -\left(\frac{4\pi}{\varepsilon}\right)\rho, \tag{1.31}$$

Equations (1.29) and (1.31) are the uncoupled nonhomogeneous wave equations for $\acute{A}$ and Φ, respectively, in accordance with the transformations (1.27) and (1.30) which are called *gauge transformations*. The condition given by (1.28) is called the *Lorentz condition*.

1.4. Plane Electromagnetic Waves in a Nonconducting Medium

For a nonconducting medium the current density $\mathbf{J} = 0$. Upon using $\mathbf{D} = \varepsilon\mathbf{E}$, Maxwell's equations (1.1)$\cdots$(1.4) then become

$$\nabla \cdot \mathbf{E} = 0, \qquad \nabla \times \mathbf{E} + \left(\frac{1}{c}\right)\frac{\partial \mathbf{B}}{\partial t} = 0,$$

$$\nabla \cdot \mathbf{B} = 0, \qquad \nabla \times \mathbf{B} - \left(\frac{\varepsilon\mu}{c}\right)\frac{\partial \mathbf{E}}{\partial t} = 0. \tag{1.32}$$

We now eliminate $\mathbf{B}$ from the second and fourth equations of (1.32) by taking the curl (multiplying by the operator $\nabla\times$) of both equations, taking the partial derivative of the fourth equation with respect to t and using the result to eliminate $\nabla \times \partial\mathbf{B}/\partial t$ from the second. Upon using the vector identity $\nabla \times (\nabla \times \mathbf{E}) = \nabla (\nabla \cdot \mathbf{E}) - \nabla^2\mathbf{E} = -\nabla^2\mathbf{E}$, since $\nabla \cdot \mathbf{E} = 0$, we obtain

$$\bar{c}^2\nabla^2\mathbf{E} - \frac{\partial^2\mathbf{E}}{\partial t^2} = 0, \qquad \bar{c} = \frac{c}{\sqrt{\varepsilon\mu}}. \tag{1.33}$$

Similarly, we can eliminate $\mathbf{E}$ to obtain the generic equation

$$\bar{c}^2\nabla^2\mathbf{v} - \frac{\partial^2\mathbf{v}}{\partial t^2} = 0, \tag{1.34}$$

where $\mathbf{v}$ stands for $\mathbf{E}$ or $\mathbf{B}$. Both (1.29) and (1.31) satisfy (1.34) for $\acute{A}$ and $\acute{\Phi}$, respectively, in free space (nonconducting medium). Equation (1.34) is the three-dimensional wave equation for each of the three components of $\mathbf{E}$ or $\mathbf{B}$. The next chapter will present the mathematical properties of this pivotal equation in wave propagation in some detail. Suffice it to say at this point that we shall interpret (1.34) to represent the propagation of *planar waves*, which means that the wave front as a function of time consists of a continuous family of planar surfaces. We shall show below that electromagnetic waves are *transverse waves*, which means that both the $\mathbf{E}$ and the $\mathbf{B}$ fields oscillate perpendicular to the direction of the propagating wave.

One-Dimensional Wave Propagation. We first consider the special case where the electric and magnetic fields depend only on x and t so that $\mathbf{E} = \mathbf{E}(x, t)$

and $\mathbf{B} = \mathbf{B}(x, t)$. Let $\mathbf{E} = (E_x, E_y, E_z)$ and $\mathbf{B} = (B_x, B_y, B_z)$, where E_x is the x component of $\mathbf{E}$, etc. Maxwell's equations $\nabla \cdot \mathbf{E} = 0$ and $\nabla \cdot \mathbf{B} = 0$ imply $E_x = B_x = 0$ so that $\mathbf{E} = \mathbf{E}(0, E_y, E_z)$ and $\mathbf{B} = \mathbf{B}(0, B_y, B_z)$. This means that, for one-dimensional wave propagation in the x direction, the electric and magnetic fields $\mathbf{E}$ and $\mathbf{B}$ are perpendicular to the direction of wave propagation so that we have transverse electromagnetic waves. Again, if $\mathbf{v}$ stands for $\mathbf{E}$ or $\mathbf{B}$ we have $\nabla \times \mathbf{v} = -\mathbf{j} \, \partial v_z / \partial x + \mathbf{k} \, \partial v_y / \partial x$, where $(\mathbf{j}, \mathbf{k})$ are unit vectors in the (y, z) directions. This gives the following for the $\mathbf{j}$ and $\mathbf{k}$ components of the appropriate Maxwell equations (1.32):

$$\mathbf{j}: \quad -\frac{\partial E_z}{\partial x} + \left(\frac{1}{c}\right)\frac{\partial B_y}{\partial t} = 0, \qquad \frac{\partial B_z}{\partial x} + \left(\frac{\varepsilon\mu}{c}\right)\frac{\partial E_y}{\partial t} = 0,$$

$$\mathbf{k}: \quad \frac{\partial E_y}{\partial x} + \left(\frac{1}{c}\right)\frac{\partial B_z}{\partial t} = 0, \qquad \frac{\partial B_y}{\partial x} - \left(\frac{\varepsilon\mu}{c}\right)\frac{\partial E_z}{\partial t} = 0. \tag{1.35}$$

We now introduce the following more symmetric notation: Let $E = (0, v, w)$, $\mathbf{B} = (0, \bar{v}, \bar{w})$, where all the components are continuously differentiable functions of (x, t). Let a subscript designate partial differentiation with respect to the subscript.† For example, we represent $\partial E_z / \partial x$ by w_x, etc. We then rewrite (1.35) as

$$\mathbf{j}: \quad -w_x + \left(\frac{1}{c}\right)\bar{v}_t = 0, \qquad \bar{w}_x + \left(\frac{\varepsilon\mu}{c}\right)v_t = 0,$$

$$\mathbf{k}: \quad v_x + \left(\frac{1}{c}\right)\bar{w}_t = 0, \qquad \bar{v}_x - \left(\frac{\varepsilon\mu}{c}\right)w_t = 0. \tag{1.36}$$

We had shown above that the electromagnetic waves formed by $\mathbf{E}$ and $\mathbf{B}$ are transverse waves since their oscillations are in the plane of the wave front which is normal to the direction of wave propagation. We now prove that $\mathbf{E}$ and $\mathbf{B}$ are normal to each other (at least for the case of one-dimensional wave propagation). Let the scalar $u(x, t)$ stand for each of the components of $\mathbf{E}$ or $\mathbf{B}$. We expand the total time derivative of u as follows:

$$\frac{du}{dt} = \bar{c}u_x + u_t \qquad \text{where} \quad \frac{dx}{dt} = \bar{c} = \frac{c}{\sqrt{\varepsilon\mu}}. \tag{1.37}$$

In (1.37) we anticipate a result, namely, that each component of $\mathbf{E}$ or $\mathbf{B}$ is a traveling wave with a wave speed $\bar{c}$. We now manipulate (1.36), using (1.37), and obtain

$$\frac{d\bar{w}}{dt} = -(\sqrt{\varepsilon\mu})\frac{dv}{dt}, \qquad \frac{d\bar{v}}{dt} = (\sqrt{\varepsilon\mu})\frac{dw}{dt}. \tag{1.38}$$

Integrating (1.38) yields $\bar{w} = -(\sqrt{\varepsilon\mu})v$, $\bar{v} = (\sqrt{\varepsilon\mu})w$, neglecting the constants

---

† In conformity with the standard notation of PDEs, we henceforth (unless otehwise stated) use the subscript notation to represent partial differentiation. For example $\partial g(x, y, z, t)/\partial y = g_y$. See, for example, [11, II, Chap. 1]. Do not confuse this notation with E_z, for example, which represents the z component of $\mathbf{E}$.

of integration. This gives the result

$$v\bar{v} + w\bar{w} = 0. \tag{1.39}$$

Translating (1.39) into physical notation gives the result

$$\mathbf{E} \cdot \mathbf{B} = 0, \tag{1.40}$$

which tells us that $\mathbf{E}$ and $\mathbf{B}$ are normal to each other. This is the result we set out to prove.

We now prove that each component $u(x, t)$ satisfies the one-dimensional wave equation. By appropriate differentiations of the system (1.36) with respect to t and x and eliminations we deduce that each component of $\mathbf{E}$ and $\mathbf{B}$ satisfies the following one-dimensional wave equation:

$$\bar{c}^2 u_{xx} - u_{tt} = 0. \tag{1.41}$$

For example, by differentiating the first equation of (1.36) with respect to x, the fourth equation with respect to t, and eliminating B_y we easily see that w satisfies (1.41).

It will be proved in the next chapter that all solutions of (1.41) are of the form

$$u(x, t) = f(x - \bar{c}t) + g(x + \bar{c}t), \tag{1.42}$$

where f and g are arbitrary functions of their respective arguments. They are precisely determined if we prescribe the two initial conditions, i.e., if we set $u(x, 0) = F(x)$ and $u_t(x, 0) = G(x)$, where F and G are prescribed functions of x. In addition, we must prescribe two boundary conditions (one at each end if the one-dimensional medium is finite, or the conditions of boundedness if the medium is infinite). We need these conditions since (1.41) is second order in x and t. We shall also show in the next chapter that no solution that is not of the form given by (1.42) satisfies (1.41). In other words, (1.42) is the necessary and sufficient condition for (1.41).

The arguments $x - \bar{c}t$ and $x + \bar{c}t$ play a pivotal role in the theory of wave propagation in one dimension. If we set $\xi = x - ct$, $\eta = x + ct$, the new independent variables (ξ, η) are called *characteristic coordinates*. They are fundamental in the study of the wave equation vis-à-vis characteristic theory, which will be discussed in detail in the next chapter. Suppose $g(x + \bar{c}t) = 0$; then $u(x, t) = f(x - \bar{c}t) = f(\xi)$. If ξ is constant, then f is also constant. This means that $u(x, t)$ is constant if ξ is constant and $g(\eta) = 0$. This also means that the total derivative $du/dt = 0$. But du/dt is given by (1.37); this incidentally justifies the expansion given by (1.37). du/dt is called, in the language of characteristic theory, the *directional derivative*. Its meaning is that it represents the slope of a characteristic curve. But we leave the explanation to the next chapter. Suppose we choose a value of ξ such that f is a maximum. Then, if we increase t by an amount Δt, we must increase x by an amount Δx in order for ξ to be constant. We have $\xi = x - \bar{c}t = x + \Delta x - \bar{c}(t + \Delta t)$ which yields

$$\frac{\Delta x}{\Delta t} = \bar{c}.$$

This tells us that the maximum value of f travels in the direction of increasing x with a velocity $\bar{c}$. Choosing an arbitrary value of ξ, we see that each point on the wave form characterized by the function $f(x - \bar{c}t)$ travels in the direction of increasing x with the same velocity $\bar{c}$. This gives us the important result: The wave form represented by $f(x - \bar{c}t)$ is invariant with respect to a change in x and t, meaning that physically there is no dispersion or attenuation (dissipation). We then call the function $f(x - \bar{c}t)$ a *progressing wave* since the wave form characterized by $f(x - \bar{c}t)$ travels undeformed in the direction of increasing x. A similar analysis can be performed wherein we set $f(x - \bar{c}t) = 0$. We obtain $\Delta x / \Delta t = -\bar{c}$. The result is that $g(x + \bar{c}t)$ represents a *regressing wave* that travels undeformed in the direction of decreasing x. Thus we see that (1.42) tells us that the general solution of the one-dimensional wave equation is composed of progressing and regressing traveling waves. Note that this general solution of the wave equation says nothing about the frequencies of the wave forms. If we are interested in traveling waves that involve a frequency distribution, then we have the effect of dispersion and there exists a wave packet characterized by a group velocity, which will be described below.

Time Harmonic Waves. The simplest case of waves involving frequency distributions is to consider a wave form with a single frequency ω (in radians per sec.). Such wave forms are called *time harmonic waves* since they oscillate with a given frequency. They can be represented in complex form by

$$u(x, t) = A \exp[i(kx - \omega t)] + B \exp[-i(kx + \omega t)], \qquad (1.43)$$

where A and B are complex quantities representing the amplitude of the wave. (We suppress the bar over $\bar{c}$ for simplicity.) Note that the right-hand side of (1.43) is a special case of the general solution (1.42) of the wave equation for a time-harmonic wave form of a single frequency ω. The arguments $u(x, t)$ (which occur in the exponents of (1.43)) can be written as

$$kx \mp \omega t = k(x \mp ct) \qquad \text{where} \quad \omega = kc.$$

It is clear that we must take the real part of the right-hand side of (1.43) since $u(x, t)$ is real. However, it is more convenient to manipulate the complex quantities mathematically, and in a specific problem take the real part at the end of the calculations. Therefore, the solution for u in terms of real functions can be expressed as

$$u(x, t) = a \cos(kx - \omega t + \alpha) + b \sin(kx + \omega t + \beta), \qquad (1.44)$$

where a, b, α, and β are real, α and β being the phase angles. k is the *wave number*. In more than one dimension k generalized to a vector $\mathbf{k}$ is called the *wave vector*. The relationship between k, c, ω, and λ is

$$k = \frac{\omega}{c} = \frac{2\pi}{\lambda}, \qquad (1.45)$$

where λ is the wave length. Clearly, the dimension of k is L^{-1}. It is also clear that the first term on the right-hand side of (1.43) or (1.44) represents progressing waves and the second term, regressing waves. Note that the right-hand side of (1.44) can be given either in terms of sine or cosine functions; it is just a matter of properly defining the phase angles for specific problems. Also note that the treatment given above is for *monochromatic* or single frequency wave forms.

Group Velocity and Dispersion.† We saw that c is the velocity of wave propagation in an electromagnetic medium. We called c the wave speed; it is more correctly called the *phase velocity*. Up to now we have taken c to be constant, which means that we did not consider its dependence on frequency or wave length. We saw that any wave form maintains its shape for constant c. In general, however, in electromagnetic wave propagation, the dielectric constant depends on frequency or wave length so that c can be considered to be a differentiable function of ω. This means that wave forms travel with velocities that depend on frequency; those with higher frequency travel faster, for example. Therefore a *polychromatic* (more than one frequency) wave form does not maintain its shape, and becomes *dispersed* because each frequency component travels with a different phase velocity. However, there exists a *wave packet* that travels with a characteristic velocity called the *group velocity*. It is this phenomenon of dispersion that we shall now discuss.

We consider an "almost monochromatic" wave form, meaning that we study a sinusoidal wave that is the sum of two waves, one of frequency ω and wave number k, and the other of slightly different frequency ω' and wave number k'. To this end, we consider two particular solutions of the wave equation of the same amplitude but different frequencies and phase velocities defined by

$$u_1(x, t) = a\sin(k'x - \omega't), \qquad u_2(x, t) = a\sin(kx - \omega t).$$

The resultant wave form is given by $u(x, t) = u_1(x, t) + u_2(x, t)$. Using a trigonometric identity, we get

$$u(x, t) = 2a\cos[\tfrac{1}{2}[(k' - k)x - (\omega' - \omega)t]]\sin[\tfrac{1}{2}[(k' + k)x - (\omega' + \omega)t]].$$

Setting $\omega' - \omega = \Delta\omega$ and $k' - k = \Delta k$, where the changes in ω and k are very small, we get the following approximation for the above expression for $u(x, t)$:

$$u(x, t) = [2a\cos(x\Delta k - t\Delta\omega)]\sin(kx - \omega t). \tag{1.46}$$

The term in brackets is the amplitude of the sine wave. We see that the amplitude is a cosine wave. To obtain the velocity associated with this wave we set $\cos(x\Delta k - t\Delta\omega) = \text{const.}$ This means that $x\Delta k - t\Delta\omega = \text{const.}$ or

$$x\Delta k - t\Delta\omega = x'\Delta k - t'\Delta\omega,$$

$$\frac{\Delta x}{\Delta t} = \frac{\Delta\omega}{\Delta k} \qquad \text{where} \quad \Delta x = x' - x, \quad \Delta t = t' - t.$$

---

† This topic is treated in [14, pp. 206] from the point of view of wave propagation in thermoelastic media, but the same principle holds for electromagnetic media.

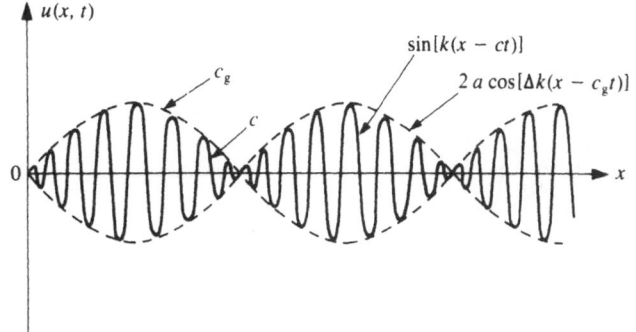

Fig. 1.1. Profile of $u(x, t)$ as a modulated sine wave.

In the limit we obtain

$$\frac{dx}{dt} = c_g = \frac{d\omega}{dk}, \tag{1.47}$$

where we call c_g the *group velocity*, since dx/dt is the velocity with which the wave associated with the amplitude of the sine wave travels. We can now rewrite (1.46) as

$$u(x, t) = 2a \cos[\Delta k(x - c_g t)] \sin[k(x - ct)]. \tag{1.48}$$

We interpret (1.48) as follows: $u(x, t)$ represents a progressing amplitude modulated wave traveling with a phase velocity c, as shown by the sine term which is modulated by the cosine term traveling with a velocity c_g. Figure 1.1 shows this amplitude modulated sine wave. The full-line amplitude modulated sine wave is a high-frequency wave (compared to the cosine wave) which travels with the phase velocity c. The broken-line curves are the envelopes of the sine curve; they represent the low-amplitude cosine wave which travels with the group velocity c_g.

Suppose we have a continuous distribution of progressing amplitude modulated waves (of the same amplitude) whose frequencies and wave numbers are within a narrow band defined by $\Delta\omega$ and Δk. Then we have a *wave packet*. This consists of an ensemble of amplitude modulated sine waves whose phase velocity c depends on ω (so that we have dispersion), and whose envelope travels at the group velocity c_g which also depends on ω. Figure 1.2 shows an example of a wave packet and its corresponding envelope.

Effect of Index of Refraction. Consider light waves (which Maxwell proved to be electromagnetic waves). Let c_0 be the velocity of light in a vacuum, and let $n(k)$ be the *index of refraction* in an optically transparent medium, expressed as a function of the wave number. The index of refraction is defined as the ratio of the velocity of light *in vacuo* to that in the medium, or

$$n(k) = \frac{c_0}{c}. \tag{1.49}$$

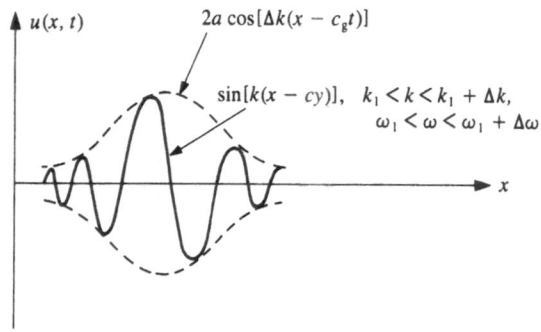

Fig. 1.2. Example of $u(x, t)$ as a wave packet within a narrow frequency range.

Since c_0 is the largest value of the phase velocity of light it is clear that $n > 1$ for any optically transparent medium other than a vacuum. Upon using (1.45), $\omega(k)$ then becomes

$$\omega(k) = \frac{c_0 k}{n(k)}. \tag{1.50}$$

Inserting (1.50) into (1.47) gives the following expression for the group velocity:

$$c_g = \frac{c_0}{n(\omega) + \omega(dn/d\omega)}. \tag{1.51}$$

In this equation it is more convenient to think of n as a function of ω rather than of k. For *normal dispersion* $dn/d\omega > 0$. Then $c_g < c$, so that the envelope of the wave packet travels with a smaller velocity than the phase velocity. For the case of *anomalous dispersion* $dn/d\omega$ becomes negative, and is so large in absolute value that $c_g > c$. However, a large value of $|dn/d\omega|$ is equivalent to a rapid variation of ω as a function of k, which is ruled out within the framework of our approximation of small changes in ω and k.

As a simple example we consider the explicit function $\omega(k)$ given by

$$\omega(k) = \omega_0 \left(1 + \frac{\lambda_0^2 k^2}{2} \right),$$

where ω_0 is a fixed frequency and λ_0 is a fixed wave length. Then

$$c_g = \omega_0 \lambda_0^2 k,$$

which tells us that the group velocity is proportional to the wave number or inversely proportional to the wave length.

Three-Dimensional Wave Propagation. We now address the more general problem of three-dimensional wave propagation in a nonconducting medium described by Maxwell's equations (1.32). Each component of $\mathbf{E}$ and $\mathbf{B}$ satisfies the three-dimensional wave equation (1.34). Restricting ourselves to harmonic progressing waves, the three-dimensional generalization of (1.43) becomes

$$E(x, t) = e_1 E_0 \exp[i(k \cdot x - \omega t)],$$
$$B(x, t) = e_2 B_0 \exp[i(k \cdot x - \omega t)],$$

(1.52)

where (e_1, e_2) are unit vectors in the (E, B) directions, (E_0, B_0) are the complex amplitudes of (E, B), and $k \cdot x = k_x x + k_y y + k_z z$, (k_x, k_y, k_z) being the (x, y, z) components of the wave number k which specifies the direction of wave propagation. Since $\nabla \cdot E = 0$ and $\nabla \cdot B = 0$, we have $k \cdot E = k \cdot B = 0$ so that E and B are both in the plane normal to k, and transverse waves are propagated in the general three-dimensional case.

We now sketch a proof that E and B are normal to each other. For simplicity, we consider the case where we can separate the time-dependent parts from the spatial-dependent parts of E and B so that $E = f(t)\bar{E}$ and $B = g(t)\bar{B}$, where f and g are arbitrary functions of t. Inserting these expressions into the curl equations of Maxwell's equations (1.32) it is easily seen that

$$\nabla \times E = \alpha B, \qquad \nabla \times B = \beta E$$

(where we have dropped the overbars for convenience). α and β are arbitrary scalar constants. From a vector identity we obtain

$$\nabla \cdot (E \times (\nabla \times B)) = (\nabla \times B) \cdot (\nabla \times E) - E \cdot \nabla \times (\nabla \times B).$$

Another vector identity yields

$$\nabla \times (\nabla \times B) = -\nabla^2 B = -\gamma^2 B,$$

where γ is a constant. The first equation arises from setting $\nabla \cdot B = 0$, the second from separating off the t-dependent variable in the vector wave equation for B. We obtain

$$\nabla \cdot (E \times E) = \alpha\beta E \cdot B + E \cdot \nabla^2 B,$$
$$0 = \alpha\beta E \cdot B + \gamma^2 E \cdot B,$$

which yields the condition $E \cdot B = 0$, meaning that E and B are normal to each other. We can therefore construct the set of orthogonal unit vectors (e_1, e_2, e_3) where $e_3 = k/k$. This orthogonal set corresponds to the directions of (E, B, k). This set of vectors forms a right-hand system in following sense: If we rotate the system about k then the direction of k is that of the motion of a right-handed screw so that E and B rotate as a rigid body in the clockwise direction, as seen in the direction of k. This is shown in Fig. 1.3.

Polarization. Since E and B are transverse waves, we have the phenomenon of polarization. In longitudinal wave propagation, such as sound waves, polarization does not exist since there is always symmetry around an axis in the direction of wave propagation. On the other hand, transverse waves are induced by a vibrating string (which oscillates in a plane normal to its equilibrium configuration). If the string were restricted to oscillate in only

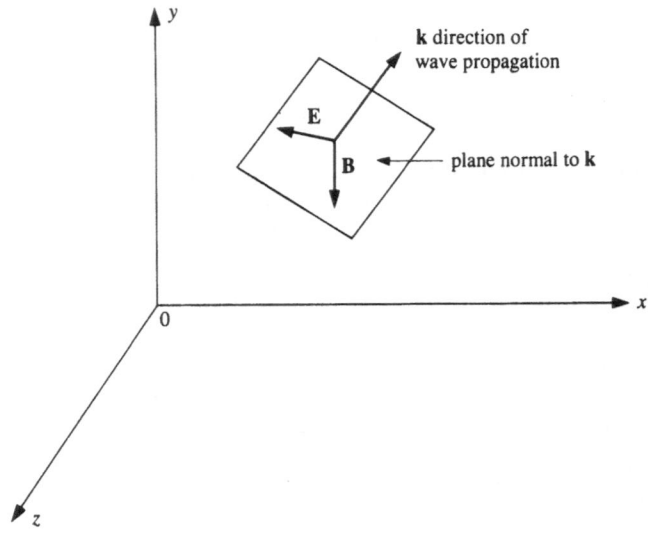

Fig. 1.3. Wave propagation vector **k** and the orthogonal vectors (**E, B**) in the normal plane.

one direction, then it would exhibit vertically polarized waves. For the case of light waves, we can exhibit polarization by sending a beam of light through a pair of "crossed" tourmaline crystals. The phenomenon of vertical polarization is shown in Fig. 1.4. In a nonpolarized light beam the transverse vibrations have no preferred direction in the plane normal to the direction of the propagating wave. This means that the different components of the light wave, that originated from the radiating source, have their vibrations randomly oriented in all possible directions in a plane normal to the direction of wave propagation. Then a pencil or beam of nonpolarized light approach-

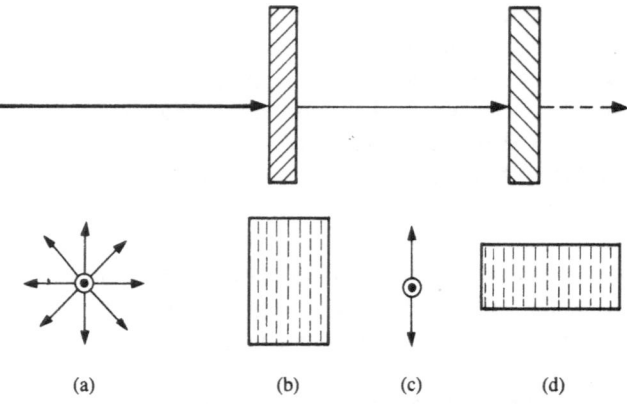

Fig. 1.4. Transmission of light waves through "crossed" tourmaline crystals.

ing the reader involves vibrations in all these directions, as shown in Fig. 1.4(a). The tourmaline slab shown in Fig. 1.4(b) is oriented in such a manner as to transmit only the components of light associated with oscillations in the vertical direction, and therefore absorb all light having oscillations in the horizontal direction, as indicated in Fig. 1.4(c). If a second tourmaline slab is oriented at right angles to the first, as shown in Fig. 1.4(d), it can only transmit light associated with the horizontal oscillations, which were shown to be blocked out by the first slab. Since no light reaches the observer, the transverse nature of light waves is demonstrated.

In 1935 Edwin Land developed a method of orienting large numbers of crystals of quinine iodosulphate into a material he called *Polariod*. This consists of a film of cellulose acetate in which are suspended a large number of these crystals which are given the same orientation by stretching the film in one direction during the manufacturing process. For optical use, the films are mounted between glass plates. Two such polaroid films can produce a more efficient polarizing effect.

We now consider two **E** and two **B** waves which we take as

$$\mathbf{E}_j = \mathbf{e}_j E_j \exp[i(\mathbf{k} \cdot \mathbf{x} - \omega t)],$$

$$\mathbf{B}_j = \sqrt{\mu \varepsilon} \mathbf{k} \times \frac{\mathbf{E}_j}{k}, \qquad j = 1, 2, \tag{1.53}$$

where the amplitudes E_1 and E_2 are complex numbers which involve the phase angles to allow for the possibility for a phase change between the waves. A general solution for a plane wave propagating in the direction $\mathbf{k}$ can be given in terms of the linearly independent solutions $\mathbf{E}_1$ and $\mathbf{E}_2$ to Maxwell's equations. (The $\mathbf{B}_j$'s can be treated similarly.) We therefore write a linear combination of the $\mathbf{E}_j$'s as

$$\mathbf{E}(\mathbf{x}, t) = (\mathbf{e}_1 E_1 + \mathbf{e}_2 E_2) \exp[i(\mathbf{k} \cdot \mathbf{x} - \omega t)]. \tag{1.54}$$

If E_1 and E_2 have the same phase, then $\mathbf{E}(\mathbf{x}, t)$ as given by (1.54) represents a *linearly polarized* wave. Its *polarization vector* **E** makes an angle $\alpha = \tan^{-1}(E_2/E_1)$ with respect to $\mathbf{e}_1$, and its magnitude is $E = \sqrt{E_1^2 + E_2^2}$. This is shown in Fig. 1.5. Now suppose that E_1 and E_2 have the same magnitude but differ in phase by 90°. The **E** wave given by (1.54) then becomes

$$\mathbf{E}(\mathbf{x}, t) = E_0(\mathbf{e}_1 \pm i\mathbf{e}_2) \exp[i(\mathbf{k} \cdot \mathbf{x} - \omega t)], \tag{1.55}$$

where E_0 is the common real amplitude. We choose an axis system where the direction of wave propagation $\mathbf{k}$ is in the z direction and $(\mathbf{e}_1, \mathbf{e}_2)$ are in the (x, y) directions. Then the components of the electric field are obtained by taking the real part of the right-hand side of (1.55) yielding

$$E_x(z, t) = E_0 \cos(kz - \omega t),$$

$$E_y(z, t) = \mp E_0 \sin(kz - \omega t). \tag{1.56}$$

At a fixed point in space the x and y components of **E**, as given by (1.56), yield

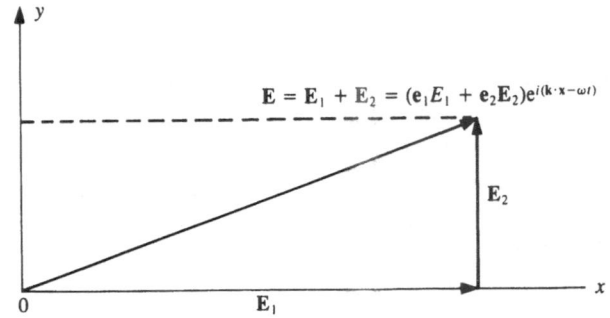

Fig. 1.5. Electric field of a linearly polarized wave resulting from E_1 and E_2 being in phase.

a constant amplitude, since $\sqrt{E_x^2 + E_y^2} = E_0$. **E** sweeps a circle with an angular velocity ω in the (x, y) plane, as shown in Fig. 1.6. For the upper sign $(\mathbf{e}_1 + \mathbf{e}_2)$ the rotation of **E** is counterclockwise when the observer is facing the progressing wave (in the direction $-\mathbf{k}$). Such a wave is said to exhibit *positive helicity*; it is sometimes called a *left circularly polarized* wave. For the lower sign $(\mathbf{e}_1 - \mathbf{e}_2)$ the rotation is clockwise when looking in the $-\mathbf{k}$ direction, and the wave has *negative helicity* or is *right circularly polarized*.

For the more general case of an *elliptically polarized wave* we generalize (1.55) to the following representation for **E**:

$$\mathbf{E}(\mathbf{x}, t) = (E_+ \mathbf{e}_+ + E_- \mathbf{e}_- \exp[i(\mathbf{k} \cdot \mathbf{x} - \omega t)], \qquad (1.57)$$

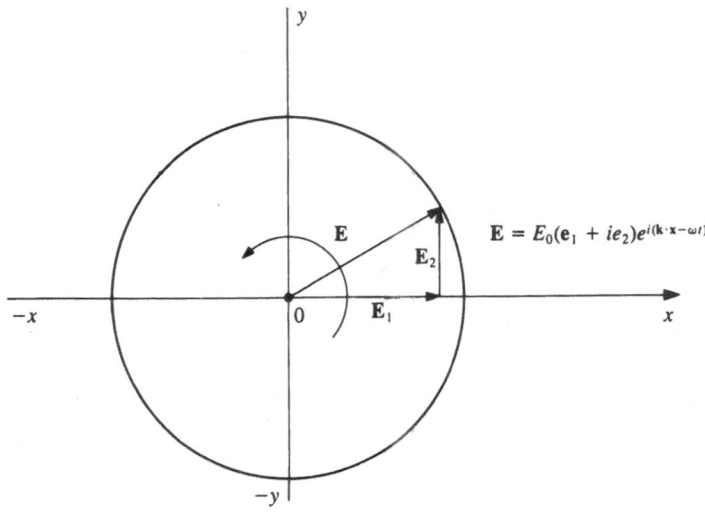

Fig. 1.6. Electric field of a circularly polarized wave resulting from E_1 and E_2 being 90° out of phase.

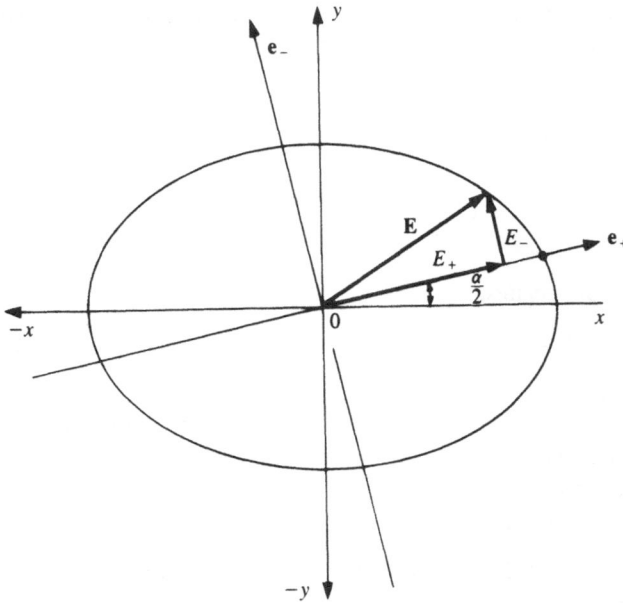

Fig. 1.7. Electric field for an elliptically polarized wave with phase difference α.

where (E_+, E_-) are complex amplitudes, and $(\mathbf{e}_+, \mathbf{e}_-)$ are complex orthogonal unit vectors defined by

$$\mathbf{e}_\pm = \frac{1}{\sqrt{2}}(\mathbf{e}_1 \pm i\mathbf{e}_2). \tag{1.58}$$

If E_+ and E_- have different amplitudes but the same phase, then (1.58) represents an elliptically polarized $\mathbf{E}$ wave with principle axes in the directions of $\mathbf{e}_1$ and $\mathbf{e}_2$. Letting $r = E_-/E_+$, we have

$$E_+\mathbf{e}_+ + E_-\mathbf{e}_- = \frac{E_+}{\sqrt{2}}[(1 + r)\mathbf{e}_1 + i(1 - r)\mathbf{e}_2].$$

We see that the ratio of semimajor to semiminor axis is $(1 + r)/(1 - r)$. For the special case of $r = \pm 1$ we recover a linearly polarized wave. If the amplitudes (E_+, E_-) have a phase difference so that $E_-/E_+ = re^{i\alpha}$, where α is the phase angle, then the ellipse traced out by $\mathbf{E}$ in the (x, y) plane has its axes rotated by an angle $\alpha/2$. Figure 1.7 shows the general case of elliptical polarization and the ellipses traced out by $\mathbf{E}$ and $\mathbf{B}$.

Reflection and Refraction of Electromagnetic Waves at a Plane Interface

We consider the reflection and refraction of plane electromagnetic waves at a plane interface between two media of different dielectric properties. The properties of the phenomena of reflection and refraction of electromagnetic waves can be divided into the following classes:

(1) Kinematic properties:

(a) Principle of reflection—angle of incidence equals angle of reflection.
(b) Principle of refraction, Snell's law: $\sin i/\sin r = n'/n$, where (i, r) is the (incidence, refraction), and (n, n') are the corresponding indices of refraction.

(2) Dynamic properties:

(a) Intensities of reflected and refracted waves relative to incident wave.
(b) Polarization effects and phase changes.

The kinematic properties follow from the geometric nature of the waves. On the other hand, the dynamic properties depend on the wave properties of the electromagnetic fields and the boundary conditions at the interface between the two different media.

The geometry of reflection and refraction of electromagnetic waves is shown in Fig. 1.8. The interface between the regions is represented by the (x, y) plane and is shown in the figure by the x axis. The normal to the interface is given by the unit normal $\mathbf{n}$ which is in the z direction. The upper half-plane is region (1) and the lower half-plane is region (2). An incident wave originating in region (1) is represented by the line AO which intersects the x axis at O from which

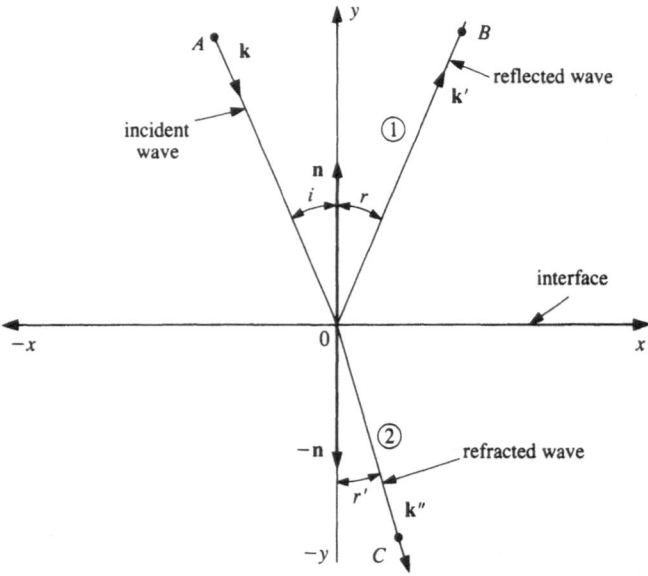

Fig. 1.8. Incident wave $\mathbf{k}$ hits interface and gives rise to a reflected wave $\mathbf{k}'$ and a refracted wave $\mathbf{k}''$.

a reflected wave OB travels back into region (1) and a refracted wave OC travels into region (2). Let $\mathbf{k}$, $\mathbf{k'}$, and $\mathbf{k''}$ be the wave vectors of the incident, relected, and refracted waves, respectively. Also, let i be the angle of incidence (angle between $\mathbf{k}$ and $\mathbf{n}$), r be the angle of reflection (angle between $\mathbf{k'}$ and $\mathbf{n}$), and r' be the angle of refraction (angle between $\mathbf{k''}$ and $-\mathbf{n}$). We assume that the reflected and refracted waves have the same frequency ω as the incident wave. Let c_1 be the phase velocity of the incident and reflected waves, and let c_2 be the phase velocity of the refracted wave. The wave numbers have the

$$|\mathbf{k}| = |\mathbf{k'}| = k = \frac{\omega}{c_1}, \qquad |\mathbf{k''}| = k'' = \frac{\omega}{c_2}. \tag{1.59}$$

These three $\mathbf{E}$ and $\mathbf{B}$ waves are progressing time harmonic waves. Using (1.52) they become:

Incident wave: $\mathbf{E} = \mathbf{e}E_0 \exp[i(\mathbf{k} \cdot \mathbf{x} - \omega t)], \qquad \mathbf{B} = \sqrt{\mu\varepsilon}\,\mathbf{k} \times \dfrac{\mathbf{E}}{k},$

Reflected wave: $\mathbf{E'} = \mathbf{e'}E_0' \exp[i(\mathbf{k'} \cdot \mathbf{x} - \omega t)], \qquad \mathbf{B'} = \sqrt{\mu\varepsilon}\,\mathbf{k'} \times \dfrac{\mathbf{E'}}{k'}, \tag{1.60}$

Refracted wave: $\mathbf{E''} = \mathbf{e''}E_0'' \exp[i(\mathbf{k''} \cdot \mathbf{x} - \omega t)], \qquad \mathbf{B''} = \sqrt{\mu_2\varepsilon_2}\,\mathbf{k''} \times \dfrac{\mathbf{E''}}{k''}.$

The boundary conditions at the planar interface $z = 0$ will be discussed in detail below. Whatever these boundary conditions are, they must be satisfied at all points on the plane at all times. This means that the phase factors for the incident, reflected, and refracted waves must all be equal, so that we set the exponents in the expressions in (1.60) equal to each other and obtain

$$(\mathbf{k} \cdot \mathbf{x})_{z=0} = (\mathbf{k'} \cdot \mathbf{x})_{z=0} = (\mathbf{k''} \cdot \mathbf{x})_{z=0}, \tag{1.61}$$

independent of the nature of the boundary conditions (as implied above). Writing (1.61) out gives

$$k_x x + k_y y = k_x' x + k_y' y = k_x'' x + k_y'' y,$$

which clearly means that all three wave vectors $\mathbf{k}$, $\mathbf{k'}$, and $\mathbf{k''}$ must lie in the (x, y) plane (all three waves are coplanar). From Fig. 1.8 we get $k \sin i = k' \sin r$. Since $k' = k$ we obtain the law of reflection, i.e., the angle of incidence i equals the angle of reflection r.

It is instructive to give a simple geometric proof of this simple law of reflection of an electromagnetic wave (which will, of course, hold for any planar wave reflection such as a stress wave). Figure 1.9 shows an incident wave AO reflected at O (on the interface $z = 0$) into a reflected wave OB. We perform the following construction: Drop perpendicular lines from A and B to the x axis and extend them. Let A' be the image of A and B' be the image of B so that $AC = CA'$ and $BC = CB'$. Draw the straight lines AOB' and BOA'. (It is easily seen that they intersect at O.) From the figure we see that the angle

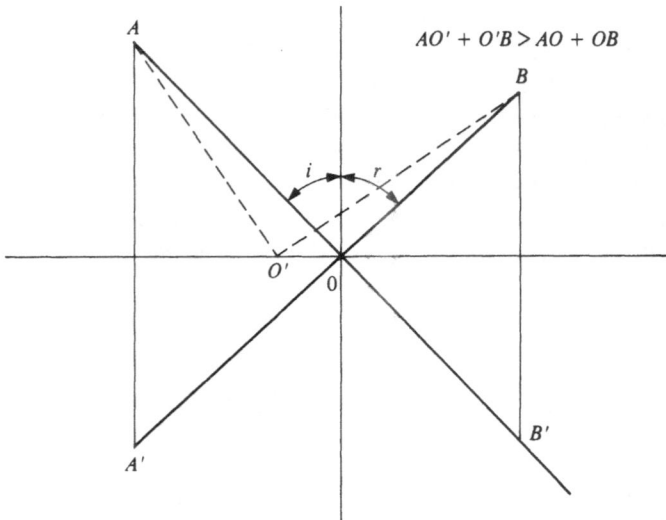

Fig. 1.9. Two-dimensional reflection.

of incidence i equals the angle of reflection r. Suppose an electromagnetic wave has its source at A and sends out an incident beam which is reflected at O into a reflected beam which is observed at B. The x axis acts as a mirror. The construction in the figure tells us that, of all possible paths $AO' + O'B$ which stem from A and is reflected by the mirror to B, the actual path is line $AO + OB$. Since $OB = OB'$, an easily seen from the construction, and AOB' is a straight line, it follows that the actual path $AO + OB$ is the shortest path taken by the beam emanating from the fixed point A and observed at the fixed point B. This *minimum principle*, involving the reflection of light or elastic waves in a plane, goes back to antiquity. It was discovered by Heron (Hero of Alexandria, first century A.D.), and can be formulated purely geometrically. Given a line—the x axis—and two points A and B on the same side of this line. For what point O on the line is $AO + OB$ the shortest path? As shown, it is the path such that the line acts like a mirror. Incidentally, this reflection principle can be applied to the game of billiards by constructing multiply reflecting elastic waves, as long as we assume perfectly elastic collisions; we neglect English or other rotational motion.

We now give a geometric proof of Snell's law which is formulated as

$$n_1 \sin i = n_2 \sin r', \tag{1.62}$$

using the notation given in Fig. 1.8, where n_i is the refractive index in region (1) for $i = 1, 2$. Figure 1.10 shows two incident waves AO and BC making an incident angle i with the y axis. Let OD be the common wave fronts of these incident waves so that OD is normal to AO and BC. In a time t the incident wave front OD travels a distance $DC = c_1 t$ so that the refracted wave front after the time t is EC, which is normal to the refracted wave OE which makes

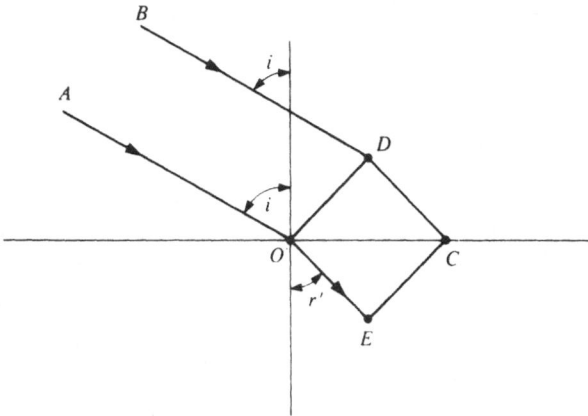

Fig. 1.10. Two-dimensional refraction, Snell's law.

the angle of refraction r' with the negative y axis. Since c_2 is the velocity of the refracted wave in region (2) we have $OE = c_2 t$. From the figure we obtain

$$\sin i = \frac{DC}{OC}, \qquad \sin r' = \frac{OE}{OC},$$

from which we obtain Snell's law (1.62) upon using the definition of the indices of refraction.

The boundary conditions at the interface involve the following continuity conditions: the normal component of **D** and **B** and the tangential component **E** and **H** must be continuous across the surface $z = 0$. These conditions can be proved from Gauss's law for charges in dielectric media. In terms of the electric and magnetic fields given by (1.60) the boundary conditions become

$$(eE_0 + e'E_0') \cdot \mathbf{n} = (e''E_0'') \cdot \mathbf{n} \qquad \text{at} \quad z = 0,$$

$$\left(\sqrt{\mu_1 \varepsilon_1} \frac{\mathbf{k}}{k} \times eE_0 + \frac{\mathbf{k}'}{k'} \times e'E_0' \right) \cdot \mathbf{n} = \left(\sqrt{\mu_2 \varepsilon_2} \frac{\mathbf{k}''}{k''} \times e''E_0'' \right) \cdot \mathbf{n},$$

$$(eE_0 + e'E_0') \times \mathbf{n} = (e''E_0'') \times \mathbf{n}, \tag{1.63}$$

$$\left(\sqrt{\mu_1 \varepsilon_1} \frac{\mathbf{k}}{k} \times eE_0 + \frac{\mathbf{k}'}{k'} \times e'E_0' \right) \times \mathbf{n} = \left(\sqrt{\mu_2 \varepsilon_2} \frac{\mathbf{k}''}{k''} \times e''E_0'' \right) \times \mathbf{n}.$$

The first equation represents the continuity of the tangential component of **E** across $z = 0$, the second represents the continuity of the tangential component of **B**, the third represents the continuity of the normal component of **E**, and the fourth represents the continuity of the normal component of **B**.

We now apply these boundary conditions (1.63) to two cases.

Case 1. **E** is normal to the *plane of incidence* which is defined as the plane containing the unit normal **n** and (**k**, **k**′, **k**″). (Note that the three wave vectors are coplanar.) Referring to Fig. 1.8, for the three waves, **E** is directed normal to the plane of incidence toward the reader, and **B** is in the plane of incidence normal to the waves such that **E** × **B** is directed in the direction of the appropriate wave vectors. Then the Poynting vector, which represents the flow of energy, is directed in the same direction.

Case 2. **E** is in the plane of incidence normal to the appropriate wave vector, and **B** is normal to both **E** and the wave vector in the direction such that **E** × **B** is in the direction of wave propagation. We now discuss these cases in detail.

Case 1. Since **E**, for each of the three waves, is parallel to the planar interface, the first boundary condition in (1.63) yields nothing. Using Snell's law we see that the second equation duplicates the third. (In fact, this is another way to prove Snell's law.) The third and fourth equations of (1.63) yield

$$E_0 + E_0' - E_0'' = 0$$

$$\sqrt{\frac{\varepsilon_1}{\mu_1}}(E_0 - E_0')\cos i - \sqrt{\frac{\varepsilon_2}{\mu_2}}E_0''\cos r' = 0. \tag{1.64}$$

The ratio of the amplitude of the reflected to the incident wave E_0'/E_0 is called the *reflection coefficient*. The ratio of the amplitude of the refracted to the incident wave E_0''/E_0 is called the *refraction coefficient*. Using (1.64) these coefficients become

$$\frac{E_0'}{E_0} = \frac{1 - \mu_1\tan i/\mu_2\tan r'}{1 + \mu_1\tan i/\mu_2\tan r'}$$

$$\rightarrow \frac{\sin(r' - i)}{\sin(i + r)} \quad \text{for} \quad \mu_2 \rightarrow \mu_1,$$

$$\frac{E_0''}{E_0} = \frac{2}{1 + \mu_1\tan i/\mu_2\tan r'} \tag{1.65}$$

$$\rightarrow \frac{2\cos i\sin r'}{\sin(i + r')}.$$

The approximations on the right in each case occur for the case $\mu_1 = \mu_2$, which is valid for optical frequencies. Use was made of Snell's law in the form $\sqrt{\varepsilon_2/\varepsilon_1} = n_2/n_1 = \sin i/\sin r'$.

Case 2. **E** is in the plane of incidence, so that the continuity of the tangential components of **E** and **H** yield

$$(E_0 - E_0')\cos i - E_0''\cos r' = 0,$$

$$\frac{\varepsilon_1}{\varepsilon_1(E_0 + E_0')} - \frac{\varepsilon_2}{\varepsilon_2 E_0''} = 0. \tag{1.66}$$

The reflection coefficient and the refraction coefficient therefore become

$$\frac{E_0'}{E_0} = \frac{(\mu_1/\mu_2)\sin 2i - \sin 2r'}{\sin 2r' + (\mu_1/\mu_2)\sin 2i}$$

$$\rightarrow \frac{\tan(i - r')}{\tan(i + r')},$$

$$\frac{E_0''}{E_0} = \frac{2\sqrt{\mu_1\varepsilon_1/\mu_2\varepsilon_2}\,\sin 2i}{\sin 2r' + (\mu_1/\mu_2)\sin 2i}$$

$$\rightarrow \frac{2\cos i \sin r'}{\sin(i + r')\cos(i - r')}.$$

(1.67)

Again the approximations apply for the case $\mu_2 \rightarrow \mu_1$.

For the case of normal incidence ($i = 0$) both Case 1 and Case 2 reduce to the following expressions for the reflection and refraction coefficients:

$$\frac{E_0'}{E_0} = \frac{n_2 - n_1}{n_2 + n_1},$$

$$\frac{E_0''}{E_0} = \frac{2n_1}{n_2 + n_1}.$$

(1.68)

Brewster's Angle. For simplicity, we consider the special case $\mu_2 = \mu_1$. For Case 2 the expression for the reflection coefficient in (1.67) vanishes when $i + r' = \pi/2$. Therefore, for this case there is no reflected wave when $i = i_B = \pi/2 - r'$. This angle of incidence i_B is called *Brewster's angle*. Using Snell's law we obtain

$$i_B = \tan^{-1}\left(\frac{n_2}{n_1}\right).$$

(1.69)

The practical implication of Brewster's angle is: If a plane wave of mixed polarization (having nonzero parallel and transverse components of the polarization vector **E**) is incident on the interface at Brewster's angle, then the reflected radiation is completely plane polarized with the polarization vector **E** normal to the plane of incidence. Clearly this property can be used to produce beams of plane-polarized light, although it is not as efficient as other methods which use the anisotropic properties of dielectric media.

We now discuss the phenomenon of *total internal reflection* which implies that the incident and reflected waves are in a medium of larger index of refraction than the refracted wave, or $n_1 > n_2$. Snell's law tells us that if this inequality holds then $r' > i$, which means that $r' = \pi/2$ when $i = i_0$, the angle of total reflection. We thus have

$$i_0 = \sin^{-1}\left(\frac{n_2}{n_1}\right).$$

(1.70)

For incident waves at an angle $i = i_0$ the refracted wave is propagated along the interface surface. Since there can be n$\cdot$ energy flow across the surface i_0 must also be the angle of total reflection. For $i > i_0$, r' must be a complex angle since $\sin i > 1$. The refracted wave $\cdot$ not only propagated along the surface but it is attenuated exponentially in $\cdot$ region (2). This is seen by writing

$$\exp(i\mathbf{k}'' \cdot \mathbf{x}) = \exp[ik''(x \sin r' + z \cos r')]$$

$$= \exp\left[-k'' \sqrt{\left(\frac{\sin i}{\sin r'} - 1\right)} z\right] \exp\left[ik''\left(\frac{\sin i}{\sin r'}\right)x\right], \qquad i > i_0.$$

The fact that there is no energy flow across the surface can be verified by calculating the time-average normal component of Poynting's vector in region (2) near the surface and recognizing that $\cos r'$ is purely imaginary.

1.5. Plane Waves in a Conducting Medium

We now investigate the propagation of plane electromagnetic waves in a conducting medium. Let the medium be characterized by a dielectric constant ε, permeability μ, and conductivity σ. We use all the terms in Maxwell's equations (1.32), replace $\mathbf{B}$ by $\mu\mathbf{H}$, and supplement them by Ohm's law in the form

$$\mathbf{J} = \sigma\mathbf{E}. \tag{1.71}$$

Maxwell's equations become

$$\nabla \cdot \varepsilon\mathbf{E} = 0, \qquad \nabla \times \mathbf{E} + \frac{\mu}{c}\frac{\partial \mathbf{H}}{\partial t} = 0,$$

$$\nabla \cdot \mu\mathbf{H} = 0, \qquad \nabla \times \mathbf{H} - \frac{\varepsilon}{c}\frac{\partial \mathbf{E}}{\partial t} - \left(\frac{4\pi\sigma}{c}\right)\mathbf{E} = 0. \tag{1.72}$$

Note that the system (1.72) reduces to (1.32) for the nonconducting medium upon setting $\sigma = 0$ and replacing $\mathbf{H}$ by $\mu^{-1}\mathbf{B}$. Recall that for the nonconducting medium we were able to separate $\mathbf{E}$ and $\mathbf{B}$ and thereby obtain three-dimensional wave equations for $\mathbf{E}$ and $\mathbf{B}$. Using the same technique of operating on the second and fourth equations of (1.72) by the curl, taking the appropriate partial derivative with respect to t, and eliminating $\mathbf{H}$ or $\mathbf{E}$ we obtain the following generic equation:

$$\nabla^2\mathbf{V} - \frac{\mu\varepsilon}{c^2}\frac{\partial^2 \mathbf{V}}{\partial t^2} - \frac{4\pi\mu\sigma}{c^2}\frac{\partial \mathbf{V}}{\partial t} = 0, \tag{1.73}$$

where $\mathbf{V}$ stands for $\mathbf{E}$ or $\mathbf{H}$. Equation (1.73) is the *damped wave equation* in three dimensions. Clearly, for a nonconducting medium, $\sigma = 0$, so that (1.73) reduces to (1.34).

For a nonconducting medium we saw that we could only have transverse $\mathbf{E}$ and $\mathbf{H}$ or $\mathbf{B}$ waves. There is no *a priori* reason why we cannot have longitudinal electromagnetic waves as well as transverse waves. To see what happens we make use of a fact in vector analysis which says that any vector can be decomposed into a *solenoidal* and an *irrotational* vector. A solenoidal vector is defined as one whose divergence is zero. An irrotational vector is one whose curl is zero so that it is the gradient of a potential. We note that, for both a nonconducting and a conducting medium, we have div $\mathbf{V} = 0$ so that $\mathbf{E}$ and $\mathbf{H}$ are solenoidal, and therefore transverse waves are propagated. However, the fourth equation of (1.72) tells us that there exists a longitudinal part of $\mathbf{E}$ which satisfies the equation

$$\left(\frac{\partial}{\partial t} + \frac{4\pi\sigma}{\varepsilon}\right)\mathbf{E}_L = 0, \tag{1.74}$$

where $\mathbf{E}_L$ is the longitudinal part of $\mathbf{E}$. Since div $\mathbf{E} = 0$, $\mathbf{E}$ is uniform in space but varies in time as

$$\mathbf{E}_L = E_0 \exp\left(\frac{-4\pi\sigma t}{\varepsilon}\right), \tag{1.75}$$

where E_0 is the amplitude of $\mathbf{E}_L$. This means that no static longitudinal fields can exist in a conducting medium in the absence of an applied current density σ. For good conductors, such as copper, $\sigma \sim 10^{17}$ sec^{-1}, so that these longitudinal effects are damped out in a very short time.

We now discuss transverse electromagnetic fields in a conducting medium. Again, assuming these fields to be time-harmonic we refer to the generic damped wave equation (1.73). Time-harmonic solutions are of the form

$$\mathbf{V} = \mathbf{v}V_0 \exp[i(\mathbf{k} \cdot \mathbf{x} - \omega t)], \tag{1.76}$$

where $\mathbf{v}$ is a unit vector representing the direction of $\mathbf{E}$ or $\mathbf{H}$ and V_0 is the amplitude. Inserting (1.76) into (1.73) yields the following equation relating the wave propagation vector or wave number $\mathbf{k}$ to the frequency, the current density, and the material parameters:

$$k^2 = \left(\frac{\mu\varepsilon\omega^2}{c^2}\right)\left[1 + i\left(\frac{4\pi\sigma}{\omega\varepsilon}\right)\right]. \tag{1.77}$$

In this analysis we assume that the frequency ω, the current density σ, and the material parameters μ and ε are real. Then (1.77) tells us that the wave number k is complex for a conducting medium.

For a nonconducting medium, $\sigma = 0$, so that (1.77) reduces to

$$k = \frac{\sqrt{\mu\varepsilon}\omega}{c}, \qquad \sigma = 0, \tag{1.78}$$

which is (1.45), where c is replaced by $c/\sqrt{\mu\varepsilon}$ (c is the velocity of light in a vacuum).

For a conducting medium let $k = \alpha + i\beta$, where α is the real and β is the imaginary part of k. Then $k^2 = \alpha^2 - \beta^2 + 2i\alpha\beta$. Inserting this expression into (1.77) and equating real and imaginary parts yields

$$k = \alpha + i\beta, \tag{1.79}$$

where

$$\alpha = \left(\frac{\sqrt{\mu\varepsilon}\,\omega}{2c}\right)\left[\sqrt{1 + \left(\frac{4\pi\sigma}{\omega\varepsilon}\right)^2} + 1\right]^{1/2},$$

$$\beta = \left(\frac{\sqrt{\mu\varepsilon}\,\omega}{2c}\right)\left[\sqrt{1 + \left(\frac{4\pi\sigma}{\omega\varepsilon}\right)^2} - 1\right]^{1/2}.$$

(1.80)

One-Dimensional Case.† For a simplicity, we consider the one-dimensional case where the damped **E** and **H** waves are propagated in the x direction. Inserting $k = \alpha + i\beta$ into (1.76), $V(x, t)$ becomes

$$V(x, t) = V_0 \exp(-\beta x) \exp[i(\alpha x - \omega t)].$$

(1.81)

It is clear from (1.81) that $\text{Im}[k] = \beta$ represents the damping or attenuation of the wave with respect to x, and $\beta > 0$ for positive damping for $x > 0$. Also $\text{Re}[k] = \alpha$ represents the real part of the wave number. The imaginary part of the exponential in (1.81) can be written as $\alpha x - \omega t = \alpha(x - \bar{c}t)$, where the phase velocity is $\bar{c}$, and

$$\bar{c} = \frac{\omega}{\alpha},$$

(1.82)

Setting $\sigma = 0$ in (1.80) for α shows us that (1.82) reduces to (1.78) for the nonconducting case.

Referring to (1.81) we now consider two approximations: Case 1—a poor conductor. Case 2—a good conductor.

Case 1. For a poor conductor the conductivity σ is very small compared to the relevant quantities, so that we have the inequality $4\pi\sigma/\omega\varepsilon \ll 1$. Expanding α and β in powers of $(4\pi\sigma/\omega\varepsilon)$ and taking the first approximation, we get the following approximation for k:

$$k \sim \frac{\sqrt{\mu\varepsilon}\,\omega}{c} + i\left(\frac{2\pi}{c}\right)\sqrt{\frac{\mu}{\varepsilon}}\,\sigma.$$

(1.83)

From (1.83) we observe that β is independent of ω so that the attenuation of the **E** and **H** waves are independent of frequency for a poor conductor. The expression for α in (1.83) is the same as for the nonconducting case (1.78). This case also tells us that $\alpha/\beta \gg 1$.

Case 2. For a good conductor the conductivity is very large in the sense of the inequality $4\pi\sigma/\omega\varepsilon \gg 1$. Expanding α and β in powers of $(\omega\varepsilon/4\pi\sigma)$ in (1.80), the first approximation for k becomes

$$k = \alpha + i\beta \sim (1 + i)\left(\frac{\sqrt{2\pi\mu\sigma\omega}}{c}\right).$$

(1.84)

This tells us that the attenuation factor and the real part of the wave number

---

† The phenomenon of dissipation.

vary as the square root of the frequency. We can write (1.81) as

$$V(x, t) = V_0 \exp\left[-\left(\frac{\sqrt{2\pi\mu\sigma\omega}}{c}\right)x\right]\exp\left[\left(\frac{\sqrt{2\pi\mu\sigma\omega}}{c}\right)i(x - \bar{c}t)\right], \quad (1.85)$$

where the phase velocity $\bar{c}$ is

$$\bar{c} = \left(\frac{\omega}{\sqrt{2\pi\mu\sigma\omega}}\right)c. \qquad (1.86)$$

We now relate the phases and amplitudes of **E** and **H**. To this end, we write the time-harmonic solutions for **E** and **H** as

$$\mathbf{E} = \mathbf{E}_0 \exp(-\beta\mathbf{n}\cdot\mathbf{x})\exp[i(\alpha\mathbf{n}\cdot\mathbf{x} - \omega t)],$$
$$\mathbf{H} = \mathbf{H}_0 \exp(-\beta\mathbf{n}\cdot\mathbf{x})\exp[i(\alpha\mathbf{n}\cdot\mathbf{x} - \omega t)], \qquad (1.87)$$

where **n** is a unit vector in the direction of **k** (the direction of wave propagation).

Using (1.87) the relationship between **E** and **H** can be obtained by rewriting the second equation of (1.72) as

$$i k \mathbf{n} \times \mathbf{E} - i\left(\frac{\mu\omega}{c}\right)\mathbf{H} = 0.$$

This gives

$$\mathbf{H}_0 = \left(\frac{c}{\mu\omega}\right)(\alpha + i\beta)\mathbf{n} \times \mathbf{E}_0. \qquad (1.88)$$

This shows that there is a phase difference between **E** and **H**. Using (1.80) the magnitude and phase ϕ of the complex wave number become

$$|k| = \sqrt{\alpha^2 + \beta^2} = \sqrt{\mu\varepsilon}\left(\frac{\omega}{c}\right)\left[1 + \left(\frac{4\pi\sigma}{\omega\varepsilon}\right)^2\right]^{1/4},$$

$$\tan\phi = \frac{\beta}{\alpha}, \qquad \tan 2\phi = \frac{4\pi\sigma}{\omega\varepsilon}. \qquad (1.89)$$

Note the obvious fact that for a nonconductor $\sigma = 0$, so that $\beta = 0$ and **E** and **H** are in phase. Equation (1.89) tells us that **H** lags **E** by the phase angle ϕ given by (1.89). The amplitude of **H** relative to **E** is

$$\frac{|\mathbf{H}|}{|\mathbf{E}|} = \sqrt{\frac{\varepsilon}{\mu}}\left[1 + \left(\frac{4\pi\sigma}{\omega\varepsilon}\right)^2\right]^{1/2} \qquad (1.90)$$

From (1.90) we see that in a very good conductor the magnetic field is very large compared to the electric field and lags in phase by approximately 45°.

As shown β is the damping factor for the **E** and the **H** waves with respect to x. We define the *depth of penetration* δ by

$$\delta = \frac{1}{\beta}. \qquad (1.91)$$

For a good conductor high-frequency waves are attenuated very fast. This fact tells us that in high-frequency circuits current flows only at the surface of the conductor (the *skin effect*).

CHAPTER 2

Hyperbolic Partial Differential Equations in Two Independent Variables

Introduction

It was shown in Chapter 1 that **E** and **H** both satisfy vector wave equations for a nonconducting medium. This obviously means that each component of the electric and magnetic fields satisfy a scalar wave equation. It was also shown that **E** and **H** satisfy the damped wave equations for a conducting medium, where the damping terms depend on the first time derivative of **E** and **H** and is proportional to the conductivity. These equations reduce to the corresponding wave equations for a nonconducting medium.

In this chapter we shall investigate one-dimensional wave propagation from a mathematical point of view, in the generalized setting of second-order partial differential equations (PDEs) of the hyperbolic type (to be defined below) in two independent variables, which constitute the field equations of wave propagation. We shall start with the one-dimensional wave equation and first derive its general solution. Characteristic coordinates will be introduced which will be essential to the method of characteristics to be developed later. We discuss *D'Alembert's method*† of solving the so-called *Cauchy initial value problem*. This Cauchy problem is the pivotal problem in wave propagation phenomena. Since D'Alembert's method and other analytical approaches, such as Laplace and Fourier transforms, apply only to the linear Cauchy problem they are thereby somewhat restrictive in their applications. In order to attack nonlinear problems we must resort to more powerful mathematical tools such as the method of characteristics, which will be treated in detail in this chapter.

In discussing the theory of quasilinear PDEs, we start with a single first-order PDE, generalize to a system of first-order PDEs, and then show the correspondence between a first-order system and a second-order PDE. The reader will see that the method of characteristics is a natural way of treating PDEs, since it directly uses the process of integrating a function of several variables (in this case x and t) and makes use of the geometric properties of the solutions. The importance of these geometric properties will be stressed, and illustrative examples will be given. We start with the linear wave equation.

---

† See, for example, [7].

2.1. General Solution of the Wave Equation

For simplicity we consider a single scalar function $u = u(x, t)$ where u may stand for any component of $\mathbf{E}$ or $\mathbf{H}$. The one-dimensional linear wave equation for $u(x, t)$ is

$$c^2 u_{xx} - u_{tt} = 0, \tag{2.1}$$

where c is the constant wave speed or phase velocity in the appropriate electromagnetic medium. Note that we use the subscript notation to represent partial derivatives. Since we seek a general solution of (2.1) for $u(x, t)$ the range of x is given by $-\infty < x < \infty$, which means that we do not consider boundary conditions, other than that u must be bounded at infinity. In addition, for the time being, we shall not be interested in initial conditions since we just want the general solution.

We observe that (2.1) involves a second-order linear operator L with constant coefficients, namely

$$L = c^2 \frac{\partial^2}{\partial x^2} - \frac{\partial^2}{\partial t^2} = L_1 L_2 = \left(c \frac{\partial}{\partial x} - \frac{\partial}{\partial t} \right)\left(c \frac{\partial}{\partial x} + \frac{\partial}{\partial t} \right).$$

This means that (2.1) can be reformulated as

$$L(u) = L_1 L_2(u) = (cu_x - u_t)(cu_x + u_t) = 0. \tag{2.2}$$

We first consider what happens if

$$cu_x + u_t = 0. \tag{2.3}$$

This is a first-order equation with constant coefficients. Geometrically speaking, the solution of this equation is a family of *integral surfaces* in (u, x, t) space. On each surface u is a particular constant. Another way of looking at the problem of finding the general solution of (2.3) is to consider the (x, t) plane. A solution of (2.3) is characterized by a single family of curves in this plane such that u is a constant along each curve (a different constant for each curve). We must first find this family of curves and then, for a given initial or boundary value problem, find the values of u for each curve. If u is constant on a particular curve then $du/dt = 0$ on that curve. Expanding this total derivative gives

$$\frac{du}{dt} = \left(\frac{dx}{dt} \right) u_x + u_t = 0, \tag{2.4}$$

u must satisfy (2.3). If u is constant along a particular curve then it must also satisfy (2.4). Correlating (2.3) and (2.4) yields

$$\frac{dx}{dt} = c. \tag{2.5}$$

Equation (2.5) is a simple first-order ordinary differential equation (ODE); it is called the *characteristic equation* corresponding to the given PDE (2.3). The solution of (2.5) is a one-parameter family of straight lines $x - ct = \alpha$, wher α is the constant of integration and is thus the parameter that defines the curve. In general, the solution of the characteristic equation is called the family of *characteristic curves*.† For our first-order equation we obtain a single family of characteristic curves. It is therefore natural to introduce a variable ξ defined by $\xi = x - ct$. Therefore the general solution of (2.3) is $u(x, t) = F(x - ct) = F(\xi)$.

We now set the other factor of (2.2) equal to zero and obtain

$$cu_x - u_t = 0. \tag{2.6}$$

By a similar analysis it is easy to show that the characteristic equation is $dx/dt = -c$ whose solution is the one parameter of characteristic curves $x + ct = \beta$, where β is the parameter defining each curve. Setting $x + ct = \eta$, the general solution of (2.6) becomes $u(x, t) = G(x + ct) = G(\eta)$. (Note: To shorten the notation u is used for the solution of (2.2), (2.3), and (2.6). There should be no confusion.)

The general solution of (2.2) is the sum of the above two solutions, so that we obtain

$$u(x, t) = F(\xi) + G(\eta), \tag{2.7}$$

where

$$x - ct = \xi, \qquad x + ct = \eta. \tag{2.8}$$

(ξ, η) are called the *characteristic coordinates* of the one-dimensional wave equation, and (2.8) represents the transformation from the (x, t) plane to the *characteristic plane* (ξ, η). As long as the Jacobian of the mapping $(x, t) \rightleftarrows (\xi, \eta)$ is not zero there exists a unique inverse transformation.

We now use (2.8) to transform the wave equation given by (2.1) to characteristic coordinates. By the chain rule, using (2.8), we obtain

$$u_x = u_\xi + u_\eta, \qquad u_{xx} = \left(\frac{\partial}{\partial \xi} + \frac{\partial}{\partial \eta}\right)^2 = u_{\xi\xi} + 2u_{\xi\eta} + u_{\eta\eta},$$

$$u_t = u_\xi \xi_t + u_\eta, \qquad u_{tt} = c^2(u_{\xi\xi} - 2u_{\xi\eta} + u_{\eta\eta}).$$

Substituting these expressions for u_{xx} and u_{tt} into (2.1) yields

$$u_{\xi\eta} = 0. \tag{2.9}$$

Equation (2.9) is the one-dimensional wave equation in characteristic coordinates. It is clearly easier to determine the general solution of (2.9) than that of (2.1). In fact, (2.9) can be written as $(\partial/\eta)u_\xi = 0$ so that u_ξ is independent of η. However it depends on ξ in an arbitrary way, which means $u_\xi = f(\xi)$, where f is an arbitrary function of ξ. Integrating this expression and letting $\int f(\xi)\, d\xi = F(\xi)$ we obtain

$$u(\xi, \eta) = F(\xi) + G(\eta), \tag{2.10}$$

where F and G are arbitrary functions of their respective arguments. They can be uniquely determined from a specific problem which specifies two boundary conditions and two initial conditions. Equation (2.10) is the general solu-

---

† See, for example, [11, Vol. II], [2], and [7].

tion of the one-dimensional wave equation in characteristic coordinates. Using (2.8) we obtain the general solution of this wave equation in physical coordinates

$$u(x, t) = F(x - ct) + G(x + ct). \tag{2.11}$$

It is easy to check that (2.11) does indeed satisfy (2.1). We therefore obtain the statement made in Chapter 1 that all solutions of the wave equation are of the form given by (2.11), and no functions $u(x, t)$ not of that form are solutions of (2.1).

The germ of characteristic theory for the one-dimensional wave equation lies in the above treatment, which involves the characteristic equation whose solution gives *two families* of curves which are straight lines of slopes $\pm c$. Further insight is obtained by referring to (1.42) and the accompanying discussion in Chapter 1.

2.2. D'Alembert's Solution of the Cauchy Initial Value Problem†

We now define the Cauchy initial value problem for the one-dimensional wave equation. Find $u(x, t)$, which satisfies the following initial value problem:

$$c^2 u_{xx} - u_{tt} = 0, \qquad -\infty < x < \infty,$$
$$u(x, 0) = f(x), \qquad u_t(x, 0) = g(x), \tag{2.12}$$

where the initial conditions are defined by the prescribed functions $f(x)$ and $g(x)$. Since the range of x is infinite, this is a pure initial value problem.

The first question to ask is: How do we know there is only one solution to (2.12)? In general, for linear PDEs, the answer is "yes". But it is instructive to prove uniqueness directly from the problem. Suppose that u and v are two different solutions of (2.12). Then by the linearity of the wave equation, $w = v - u$ is also a solution of the wave equation. But w satisfies the initial conditions $w(x, 0) = 0$ and $w_t(x, 0) = 0$. We know that all solutions are of the form given by (2.11). To satisfy the initial conditions we must have $F(x) + G(x) = 0$ and $-F'(x) + G'(x) = 0$. Integrating the second equation gives $-F(x) + G(x) = 0$, which shows that $F = G = 0$ for all values of the arguments so that $w(x, t) = 0$, which is a contradiction. This completes the proof.

We now proceed with D'Alembert's solution of (2.12). The general solutions for u and u_t are

$$u(x, t) = F(x, t) + G(x, t),$$

$$u_t(x, t) = -cF'(x, t) + cG'(x, t), \qquad F' = \frac{dF}{d\xi}, \qquad G' = \frac{dG}{d\eta}.$$

---

† The Cauchy initial value problem is of fundamental importance in the solution of wave propagation problems in electromagnetic media.

Using the initial conditions in (2.12) we get

$$F(x) + G(x) = f(x),$$
$$-cF'(x) + cG'(x) = g(x).$$

Integrating the second equation from 0 to x gives

$$-F(x) + G(x) = \frac{1}{c} \int_0^x g(z) \, dz - F(0) + G(0).$$

This gives us two algebraic equations for $F(x)$ and $G(x)$. Solving them yields

$$F(x) = \frac{f(x)}{2} - \frac{1}{2c} \int_0^x g(z) \, dz + F(0) - G(0),$$

$$G(x) = \frac{f(x)}{2} + \frac{1}{2c} \int_0^x g(z) \, dz - F(x) + G(0).$$

Note that x is generic variable so that the argument of F can be replaced by $x - ct$ and that of G by $x + ct$. This allows us to get off the x axis. The expressions for F and G then become

$$F(x - ct) = \frac{f(x - ct)}{2} - \frac{1}{2c} \int_0^{x-ct} g(z) \, dz + F(0) - G(0),$$

$$G(x + ct) = \frac{f(x + ct)}{2} + \frac{1}{2c} \int_0^{x+ct} g(z) \, dz - F(0) + G(0).$$

The solution given by (2.11) then becomes

$$u(x, t) = \tfrac{1}{2}[f(x - ct) + f(x + ct)] + \frac{1}{2c} \int_{x-ct}^{x+ct} g(z) \, dz. \tag{2.13}$$

Equation (2.13) is *D'Alembert's solution* to the Cauchy initial value problem for the one-dimensional wave equation with constant phase velocity.

We now give a geometric interpretation of (2.13) in terms of traveling waves. We first observe that the linearity of the wave equation tells us that the solution of the Cauchy problem for nonhomogeous initial conditions is equal to the sum of two solutions:

(1) $f(x)$ not zero and $g(x)$ identically zero.
(2) $f(x)$ identically zero and $g(x)$ not zero.

We shall discuss these two cases separately.

Case 1. $u_t(x, 0) = 0$ for all x. Then (2.13) reduces to

$$u(x, t) = \tfrac{1}{2}[f(x - ct) + f(x + ct)]. \tag{2.14}$$

Equation (2.14) can be interpreted geometrically as follows: At $t = 0$ the wave form is composed of two equal waves $u(x, 0) = f(x)/2 + f(x)/2$. As t increases from zero, one-half of the wave, namely, $f(x - ct)/2$, travels in the direction of increasing positive x (it is a progressing wave). The other half, $f(x + ct)/2$, is a regressing wave and therefore travels in the direction of decreasing x. As an example we take $f(x)$ to be a square wave pulse of amplitude $\bar{u}$ and pulse width $2a$ which is given by

$$f(x) = \begin{cases} \bar{u}, & |x| < |a|, \\ 0, & |x| > |a|. \end{cases}$$

Figure 2.1(a) shows this pulse at $t = 0$. Figure 2.1(b) shows the same pulse at t, split into two symmetric waves of amplitude $\bar{u}/2$ (a progressing and regressing wave) and the remnant of the original wave on the interval $(-a + ct_1, a - ct_1)$. The progressing wave occupies the interval $(a - ct_1, a + ct_1)$ and the

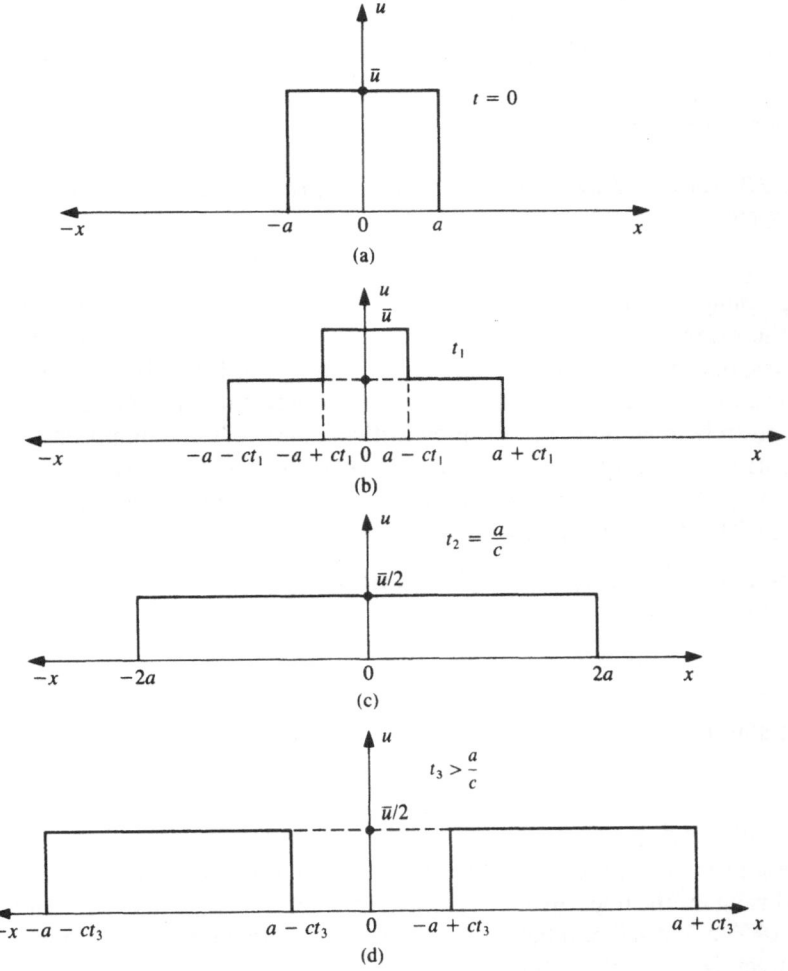

Fig. 2.1. (a) Initial wave form; (b) wave form at t_1; (c) wave form at t_2; (d) wave form at t_3.

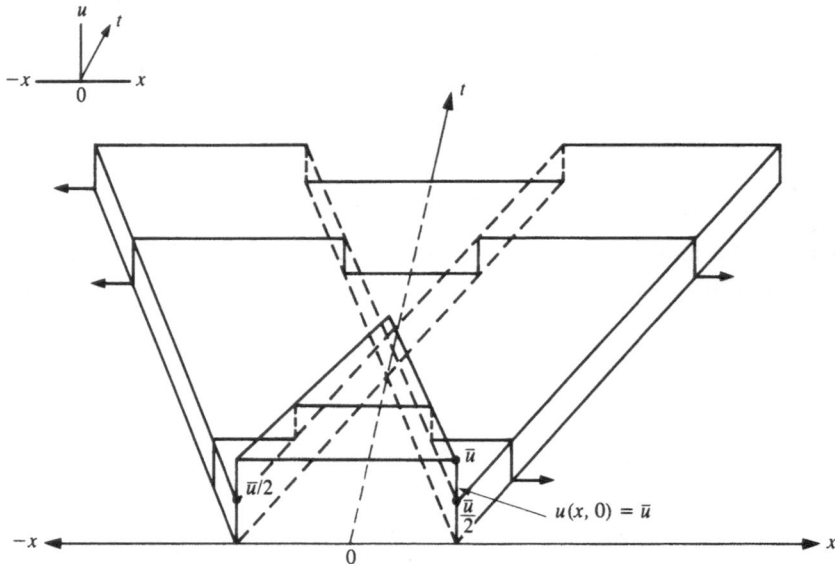

Fig. 2.2. Isometric drawing of surface $u(x, t) = \frac{1}{2}[f(x - ct) + f(x + ct)]$ for a square wave pulse.

regressing wave the interval $(-a - ct_1, -a + ct_1)$. At $t_2 = a/c$ the remnant of the original wave is entirely eaten up by the progressing and regressing waves, but the waves do not split. This is shown in Fig. 2.1(c). Figure 2.1(d) shows the situation for $t_3 > t_2$ where the progressing and regressing waves are entirely separated; they progress with equal speeds and amplitudes $\bar{u}/2$. It is instructive to give an isometric drawing of the surface formed by the solution given by the square wave pulse for $f(x)$. This is shown in Fig. 2.2 in (u, x, t) space. The different surfaces are clearly shown.

Case 2. $u(x, 0) = 0$, $u_t(x, 0) = g(x)$, where we take $g(x)$ to be the square wave pulse given by

$$g(x) = \begin{cases} \bar{v}, & |x| < |a|, \\ 0, & |x| > |a|. \end{cases}$$

The solution for this case is

$$u(x, t) = \frac{1}{2c} \int_{x-ct}^{x+ct} g(z) \, dz. \tag{2.15}$$

The solution for this case can be expressed geometrically as shown in Fig. 2.3. In this figure the lines through $\pm a$ are drawn of slopes $\pm 1/c$ in the half-plane $t > 0$. The line AB parallel to the x axis is also drawn. For field points lying between D and F the value of $u(x, t)$ is proportional to the interval RS. This value will increase as t increases to a maximum, when P is at the point K when $u(x, t) = (1/2c)\int_{x-ct}^{x+ct} g(z) \, dz$. For points on AC or on GB $u(x, t) = 0$. In regions

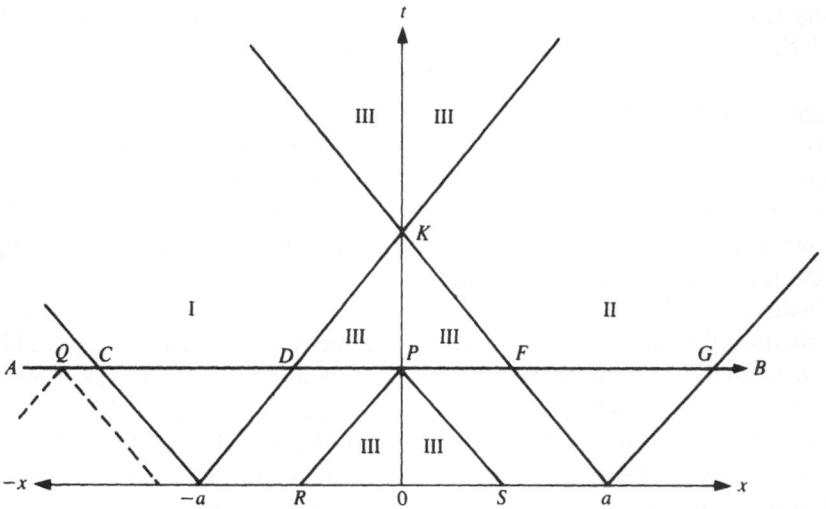

Fig. 2.3. Regions in (x, t) plane for initial condition

$$u_t = \begin{cases} 0, & |x| > |a|, \\ 1, & -a < x < a. \end{cases}$$

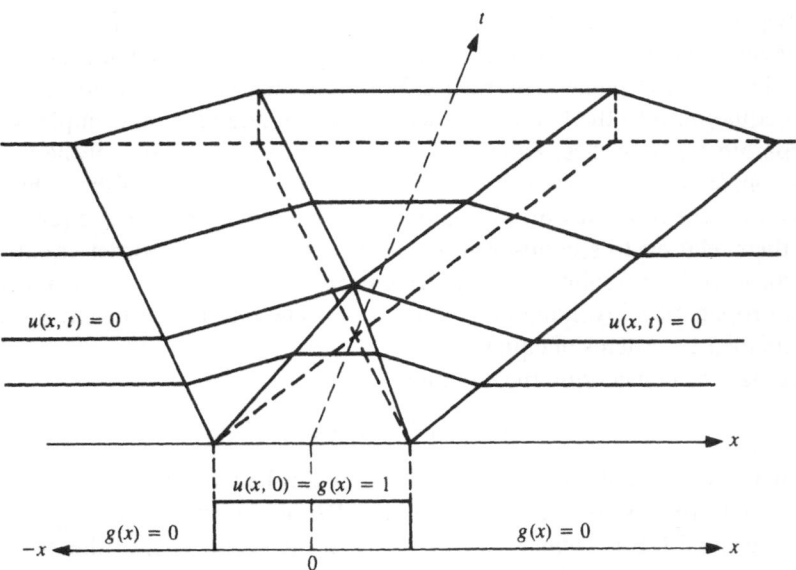

Fig. 2.4. Isometric drawing for the surface $u(x, t) = \dfrac{1}{2c} \displaystyle\int_{x-ct}^{x+ct} g(z)\, dz.$

I and II $u(x, t)$ can be determined from the fact that u varies linearly on CD and FG, while $u = u_K$ for all points in region III. We now use Fig. 2.3 to construct an isometric drawing for the surface $u = u(x, t)$ in (u, x, t) space. This is described in Fig. 2.4.

We have given a geometric interpretation of these two cases where one initial condition is identically zero and the other is given by a square wave pulse. The solution for the case where the two nonhomogeneous initial conditions are combined is obtained by summing the individual solutions according to the law of superposition of solutions.

In discussing D'Alembert's solution we have not described a more powerful geometric approach which is given by the method of characteristics. This interpretation (vis-à-vis characteristic theory) will have to be reserved for a later part of this chapter, after we formulate the proper background in characteristic theory. It will be seen (in the appropriate place) that applying characteristic theory to D'Alembert's solution will add great insight to the nature of solutions of the wave equation in general, and give us a better understanding of wave propagation phenomena.

2.3. Method of Characteristics for a Single First-Order Equation

We begin our study of characteristic theory, in relation to the PDEs of wave propagation in an electromagnetic medium, by considering a single first-order PDE in the (x, y) plane. The physical motivation lies in our factoring the wave equation into the two first-order PDEs (2.3) and (2.6) of a simple type. A separate study of these equations gives insight into solutions of the wave equation. Moreover, a study of a single first-order equation furnishes a background for the more advanced treatment of systems of first-order equations and their relation to second-order equations. Also, in a first-order equation we can introduce nonlinear effects which are important in electromagnetic wave propagation involving large amplitudes. These nonlinearities can then be extended to systems of equations.

We start be considering the equation

$$au_x + bu_y = cu + d, \tag{2.16}$$

where the coefficients a, b, c, and d are prescribed functions of (x, y, u). Equation (2.16) is said to be a *quasilinear PDE* of the first order. In general, an nth-order PDE is defined as quasilinear if the coefficients of the highest derivative terms (nth-order derivatives) are prescribed functions of the coordinates and, at most, all the derivatives through the $(n - 1)$st. For example, $yu_{xx}u_{xxx} + (u_x)^2 u_{yy} = 0$ is quasilinear, but $(u_{xxx})^2 + u_{yyy} + u_{xy} + u_x = 0$ is nonlinear but not quasilinear. Implicit in the definition of quasilinearity is the fact that the highest derivative terms are displayed linearly. It will be shown later that characteristic equations that will be developed depend only on the *principal part* of the PDE, which is defined as being the highest derivative terms. If the principal part depends at most on the coordinates then the PDE is said to be *semilinear*. It will be shown below that the characteristic equations do not depend on the solution, as they do for the quasilinear case that is not semilinear.

Coming back to (2.16) we note that the left-hand side is a linear combination of the derivatives u_x and u_y for any fixed values of (x, y, u). (Indeed, this arises from the definition of quasilinearity.) Therefore, the expression $au_x + bu_y$ (which is the principal part of (2.16)) can be written as

$$au_x + bu_y = (a, b) \cdot (u_x, u_y),$$

which is interpreted as the scalar product of the vectors $\mathbf{r} = (a, b)$ and **grad** $u = (u_x, u_y)$. It is clear that the expression $\mathbf{r} \cdot \mathbf{grad}$ u represents the derivative of u in the direction given by $\mathbf{r}$. This will be made clear by the following simple analysis: We first expand du/dx, obtaining

$$\frac{du}{dx} = \left(\frac{dy}{dx}\right)u_y + u_x. \tag{2.17}$$

We now divide (2.16) by $a(x, y, u)$ (assuming that $a \neq 0$ in the domain x, y of the solution) and coordinate (2.16) with (2.17), obtaining

$$\frac{du}{dx} = \left(\frac{dy}{dx}\right)u_y + u_x = \left(\frac{b}{a}\right)u_y + u_x. \tag{2.18}$$

In order for (2.18) to be valid in the domain of the solution we must obtain the important result

$$\frac{dy}{dx} = \frac{b(x, y, u)}{a(x, y, u)}. \tag{2.19}$$

Equation (2.19) is an ODE whose solution can be put in the symmetric form $f(x, y, u, \alpha) = 0$ where α is a constant of integration. du/dx is called the *directional derivative*. Equation (2.18), coupled with (2.19), makes it clear why this terminology is used, since du/dx is the total derivative of u with respect to x *in the direction given by the slope dy/dx of a one-parameter family of curves* obtained by integrating (2.19). Dividing (2.16) by $a(x, y, u)$ and relating the result to (2.17) yields another important result

$$\frac{du}{dx} = \left(\frac{c}{a}\right)u + \frac{d}{a}. \tag{2.20}$$

Equations (2.19) and (2.20) are the two pivotal equations of characteristic theory; they are called the *characteristic equations*. They are the ODEs associated with the original PDE (2.16). If we can solve these characteristic equations then we have solved the original equation (2.16) for $u(xy)$. *Therefore, the problem of solving a single first-order PDE is converted to that of solving a pair of ODEs, the characteristic equations.*

Clearly for the linear case, since $a = a(x, y)$, $b = b(x, y)$, we know the right-hand side of (2.19) as a function of x and y. Therefore (2.19) is an ODE (albeit nonlinear) whose solution is $y = y(x) + \alpha$, where α is the integration constant. Hence, by solving (2.19) for the linear case, we obtain a one-parameter family of *characteristic curves*, and that family is independent of u. Clearly this family of characteristic curves does not depend on (2.20). Thus the characteristic equations (2.19) and (2.20) are *uncoupled for the linear case*. Once (2.19) is solved for the family of characteristic curves, then (2.20) can be solved to obtain $u(xy)$, given an initial value curve (which, incidentally, must never have the direction of any characteristic curve) and given u along that curve. This will be discussed in detail below.

For the quasilinear case the solution of the characteristic equation (2.19), as mentioned above, depends on the solution $u(x, y)$. In other words, (2.19) must be used in conjunction with (2.20). We must solve a pair of simultaneous differential equations. This means that the characteristic equations (2.19) and (2.20) are *coupled for the quasilinear case*.

Geometric Considerations. We first consider the case $c = 0$ in (2.16). Following standard notation in the field of PDEs we set

$$p = u_x, \qquad q = u_y. \tag{2.21}$$

Then (2.16) becomes

$$ap + bq - d = 0. \tag{2.22}$$

We now give a geometric interpretation of (2.22) in (u, x, y) space. Suppose a solution to (2.22) exists in some domain in this space. The solution then generates a one-parameter family of *integral surfaces* in this domain. An integral surface is defined by the property that every point on this surface is a solution to (2.22). Now consider any field point $P: (u, xy)$ on an integral surface. Equation (2.22) can be put in the vector form

$$(a, b, d) \cdot (p, q, -1) = 0, \tag{2.23}$$

which tells us that the scalar product of two vectors is zero. Since P is on the integral surface (a, b, c, d, p, q) are known functions of (u, x, y) at P. As seen from (2.23) the vector $(p, q, -1)$ is normal to the direction given by the vector (a, b, d) at P. This clearly means that (a, b, d) defines the tangent plane of an element of the integral surface at P. Now suppose we have a family of integral surfaces going through the field point P. Then, in the neighborhood of P all these integral surfaces are tangent planes through P, so that they form an *axial pencil of planes* through a straight line, since p and q appear linearly in the quasilinear equation (2.22). This line is called the *Monge† axis* and the family of planes is called the *Monge pencil*. The directions of the Monge axes everywhere on the integral surface form a *direction field* in our domain. The integral curves of this direction field are determined by integrating the characteristic

---

† See [11, Vol. II].

equations (2.19) and (2.20). We thus have the situation that every surface generated by a one-parameter family of characteristic curves (obtained by integrating the characteristic equations) is an integral surface of the PDE. Conversely, every integral surface $u = u(x, y)$ is generated by a one-parameter family of characteristic curves. Clearly, the parameter is the constant of integration.

This last statement is easily proved. On every integral surface $u = u(x, y)$ which is a solution of (2.22), a one-parameter family of curves is obtained by integrating the characteristic equations. Along such a curve (2.20) becomes

$$\frac{du}{dx} = \frac{d(u, x, y)}{a(u, x, y)},$$

upon using the definition of the directional derivative du/dx. Therefore our one-parameter family of curves satisfies the characteristic equations; hence it is a one-parameter family of characteristic curves.

A uniqueness theorem in ODEs tells us that, under general conditions, the characteristic equations generate *unique solutions*, given an initial value curve and a defined set of initial values of (u, x, y) along that curve. From this uniqueness theorem we deduce that every characteristic curve that has one point in common with an integral surface lies entirely on that surface. Moreover, every integral surface is generated by a one-parameter family of characteristic curves for our first-order PDE. We will see below when we study second-order PDEs that two distinct families of characteristic curves exist for the equations that propagate waves.

The *Cauchy problem* for a single first-order equation is: Find the solution $u(x, y)$ of the initial value problem

$$au_x + bu_y = cu + d,$$

$$u = f(x) \quad \text{on } \Gamma : y = g(x). \tag{2.24}$$

This means we desire the solution to the first-order quasilinear PDE (2.16) subject to the prescribed initial condition for u on the initial value curve Γ given by $y = g(x)$. The characteristic equations (2.19) and (2.20) will be used to solve completely this initial value problem. However, for a more symmetric discussion we shall reformulate this Cauchy initial value problem by introducing parametric coordinates.

Parametric Coordinates. Let C_α be a characteristic curve in the (x, y) plane corresponding to the parameter α. The equation of this one-parameter family of characteristic curves, obtained by solving (2.19) and (2.20), is put in the form $y = y(x, \alpha)$. We now introduce two parameters (t, s), where t generates a particular characteristic curve C_α (for a particular s) and s generates the initial value curve (for $t = 0$). This notation has the advantage of symmetrizing the formulation in the sense that x and y would play equal roles. This means that the solution of the characteristic equations can be put in the form

$$x = x(s, t), \qquad y = y(s, t), \qquad u = u(x(s, t), y(s, t)). \tag{2.25}$$

The first two equations of (2.25) represent a mapping from the (x, y) to the (s, t) plane. The inverse transformation

$$s = s(x, y), \qquad t = t(x, y), \qquad u = u(s(x, y), t(x, y)), \qquad (2.26)$$

allows us to map from the (s, t) to the (x, y) plane. Such an inverse exists and is unique if and only if the determinant of the Jacobian of the transformation $\det[J(x_s, x_t/y_s, y_t)] \neq 0$, where

$$\det[J(x_s, x_t, y_s, y_t)] = x_s y_t - x_t y_s.$$

The geometry of the family of C_α characteristics can be expressed by replacing the parameter α by s so that we deal with a C_s family. Figure 2.5 shows such a family where we have chosen the parameter s to have its origin at the origin of the (x, y) coordinate system. As s increases (for simplicity we take $s > 0$) we generate the initial value curve Γ along which we take $t = 0$. The family of C_s characteristics cross Γ. Each characteristic is defined by the value of s which it picks up at its intersection with Γ (where $t = 0$). As t varies for a particular s, we generate a unique characteristic having a particular s as its defining parameter. This is shown in Fig. 2.5, where three characteristics denoted by C_1, C_2, and C_3 cross the initial value curve at s_1, s_2, and s_3, respectively. In general, we can extend the characteristic family C_s for $t < 0$; but if t is designated as time then we do not know the past history of u, and we are therefore only interested in the domain for $t > 0$. Part of the formulation of the Cauchy initial value problem is the condition that u is prescribed along the initial value curve. Therefore the initial condition is that u is a prescribed function of the parameter s for $t = 0$ along Γ.

The *crux of the characteristic method* in this parametric formulation as

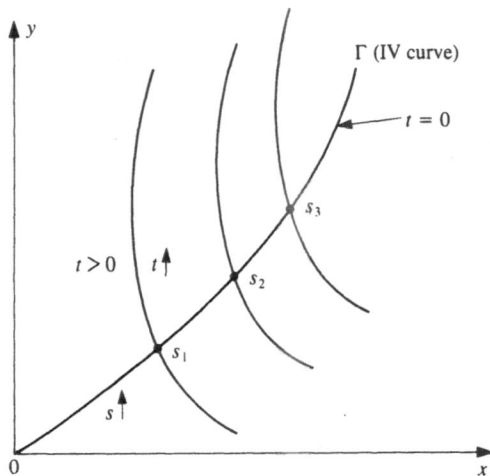

Fig. 2.5. C_s characteristic curves crossing initial value curve Γ in two dimensions.

applied to the Cauchy problem is: Given an initial value curve Γ and given u along Γ, the solution at any field point off Γ can be obtained by integrating the characteristic equations in parametric form along the unique characteristic that comes from a given point on Γ and goes through the field point. It is clear that no two characteristics can intersect; for at the point of intersection there would be two different solutions, which violates the uniqueness theorem for ODEs. Our immediate task now is to reformulate the characteristic equations in parametric form.

To obtain the characteristic equations in parametric form we first expand $du(s, t)/dt$. We get

$$\frac{du(s, t)}{dt} = \left(\frac{dx}{dt}\right) u_x + \left(\frac{dy}{dt}\right) u_y = au_x + bu_y = cu + d$$

upon using (2.24). We immediately obtain the characteristic equations in the (s, t) plane, which are

$$\frac{dx}{dt} = a(x(s, t), y(s, t), u(s, t)), \qquad \frac{dy}{dt} = b(x(s, t)y(s, t)u(s, t)), \quad (2.27)$$

$$\frac{du}{dt} = cu + d. \tag{2.28}$$

Equations (2.27) and (2.28) are the counterparts of (2.19) and (2.20), respectively, in parametric form. Integrating the system (2.27) and (2.28) yields the solution in the form given by (2.25). For the quasilinear case (2.27) is coupled with (2.28) which, upon integration, yields $u = u(x(s, t), y(s, t))$. It is clear that s plays the role of the integration constant which defines a particular C characteristic. (Note: We call this a "C" characteristic to distinguish it from another type of characteristic which is associated with the hodograph or velocity plane, occurring in our subsequent study of two dependent variables.) It is clear that our two formulations of the characteristic equations are related, since $(dy/dt)/(dx/dt) = dy/dx$ and $du/dt = (dx/dt)u_x + (dy/dt)u_y$. We now give some examples to illustrate the method.

EXAMPLE 1. Our first example will involve a linear PDE. We set $a = x, b = y$, $c = $ const., $d = 0$ in (2.16), and define the initial condition that $u = \sin x$ on $\Gamma: y = 1$. The formulation of this Cauchy problem in parametric form becomes: Solve the following problem for $u = u(x, y)$:

$$xu_x + yu_y = cu,$$

$$\text{IC:} \quad u = \sin s \quad \text{on } \Gamma: (t = 0), \qquad x = s, \quad y = 1.$$

The characteristic equations (2.27) and (2.28) become

$$\frac{dx}{dt} = x, \qquad \frac{dy}{dt} = y, \qquad \frac{du}{dt} = cu.$$

Integrating these characteristic equations and using the initial condition yields

$$x = se^t, \qquad y = e^t, \qquad u = (\sin s)e^{ct},$$

which is a particular case of (2.25). Thus we have x, y, and u as functions of s and t. Note that, since this is an example of a linear equation, u depends explicitly on s and t only. To get $u = u(x, y)$ we invert this transformation and obtain the solution

$$u(x, y) = \sin\left(\frac{x}{y}\right)y^c,$$

from

$$s = \frac{x}{y}, \qquad e^{ct} = y^c.$$

It is of interest to see how this problem is formulated in the (s, t) plane. By the chain rule, we have

$$u_x = u_s s_x + u_t t_x, \qquad u_y = u_s s_y + u_t t_y.$$

The above transformation gives $s_x = y^{-1} = e^{-t}$, $s_y = -x/y^2 = -se^{-t}$, $t_x = 0$, $t_y = e^{-t}$. This gives $u_t = cu$. Integrating and using the initial condition gives $u = (\sin s)e^{ct}$.

It is instructive for the reader to solve this problem directly in the (x, y) plane by using (2.19) and (2.20).

EXAMPLE 2. We now give an example of a Cauchy problem involving a quasilinear PDE. Find $u(x, y)$ that satisfies the following initial value problem:

$$uu_x + u_y = 1,$$

$$\text{IC:}\quad u(s, 0) = Us \quad \text{on } \Gamma\colon x(s, 0) = s, \qquad y(s, 0) = 0,$$

where U is a given constant. The characteristic equations in parametric form become

$$\frac{dx}{dt} = u, \qquad \frac{dy}{dt} = 1, \qquad \frac{du}{dt} = 1.$$

The third equation gives

$$u(s, t) = t + Us.$$

This is the solution in parametric form. It clearly satisfies the initial condition $u(s, 0) = Us$. To determine $s = s(x, y)$ $t = t(x, y)$, all we need do is integrate the first two characteristic equations and use the above solution for $u(s, t)$. We get

$$x = \frac{t^2}{2} + Ust + s, \qquad y = t.$$

Using these results we obtain the solution for $u(x, y)$, which is

$$u(x, y) = y + \left[\frac{U}{2(Uy + 1)}\right](2x - y^2).$$

Since the solution for u is simpler in the (s, t) plane we transform the Cauchy problem to (s, t) coordinates (as we did in Example 1). Using the above transformations, we finally obtain

$$(u - t - Us)u_s + (Ut + 1)u_t = Ut + 1, \qquad \text{IC:} \quad u(s, 0) = Us.$$

The characteristic equations become

$$\frac{ds}{u - t - Us} = \frac{dt}{Ut + t} = \frac{du}{Ut + t},$$

The solution $u - t - Us = 0$ yields $ds/dt = 0$. It follows that the characteristics in the (s, t) plane consist of a family of straight lines $s = \text{const.}$ along which $u = t + Us$. The integral surface in (u, s, t) space is therefore a plane which depends on the given value of U.

A special case of this example is the homogeneous equation

$$uu_x + u_y = 0.$$

It is easily seen that the general solution of this quasilinear PDE is

$$u = f(-x + uy),$$

where f is an arbitrary function of the argument $-x + uy = \xi$. A physical interpretation in mechanics can be given for this PDE. If y is time and u is the velocity of a particle as a function of one-dimensional space and time, then the equation represents the fact that the particle acceleration is zero. The u_y term is the part of the acceleration at a fixed point in space as time varies. The uu_x term is called the *convective* term and represents the spatial variation of the particle velocity at a fixed time. If we set $\xi = \text{const.}$ then $dx/dy = u$, since $du/dy = (dx/dy)u_x + u_y = 0$. This means the particle moves with a velocity u along the curve $uy - x = \text{const.}$ in the (x, y) plane in such a manner that its acceleration is zero. Suppose we have the initial condition that $u = \sin x$ on the initial curve $\Gamma: y = 0$. Then $\sin x = f(-x)$, so that $\sin(uy - x) = f(x - uy)$, and the solution is $u(x, y) = -\sin(uy - x)$.

Mention was made above that the Jacobian of the mapping $(x, y) \rightleftarrows (\xi, \eta)$ must have a nonzero determinant. Now we ask the question: What would happen if this determinant is zero everywhere on an initial value curve? The answer leads to an important theorem which is:

If $\det[J] = x_s y_t - x_t y_x = 0$ *everywhere along an initial value curve* Γ *then the Cauchy initial value problem cannot be solved unless* Γ *is a characteristic curve, in which case there are infinitely many solutions.*

The proof of this theorem is: If the Cauchy initial value problem has a solution and $\det[J] = 0$ on an initial value curve Γ, then Γ is a characteristic curve. This is seen by the fact that $x_s y_t - x_t y_s = x_s b - y_s a = 0$, upon using (2.27). This means $y_s/x_s = dy/dx = b/a$, so that Γ is a characteristic curve. But if Γ is a characteristic curve there is no unique solution to the Cauchy problem (since $\det[J] = 0$), and hence there must be infinitely many integral surfaces passing through Γ.

2.4. Method of Characteristics for a First-Order System

We now extend the method of characteristics to a system of first-order quasilinear PDEs. The motivation for studying such a system is contained in the following observations: The field equations of electromagnetic phenomena (Maxwell equations) arise from the three conservation laws: mass or continuity, momentum, and energy, as discussed in Chapter 1. They are generally expressed as sets of second-order PDEs and are usually quasilinear when nonlinear phenomena are considered. If we take a single scalar function $u(x, y)$, which may represent a component of $\mathbf{E}$ or $\mathbf{H}$, for example, then the generic equation in two dimensions that we shall consider is

$$au_{xx} + 2bu_{xy} + cu_{yy} + d = 0, \tag{2.29}$$

where the coefficients $a, \ldots, d$ are known functions of (x, y, u, u_x, u_y), so that (2.29) is a second-order quasilinear PDE. Note that all the second-order derivative terms on the left-hand side constitute the principal part of the equation. Therefore the term d does not play any role as far as characteristic theory is concerned.

We now convert (2.29) to a set of first-order quasilinear PDEs by introducing the two dependent variables (p, q) defined by (2.21) $(u_x = p, u_y = q)$. Then (2.29) becomes

$$L_1 = ap_x + bp_y + bq_x + cq_y + d = 0. \tag{2.30}$$

Since the derivatives are assumed to be continuous we also have the equation

$$L_2 = p_y - q_x = 0. \tag{2.31}$$

Equations (2.30) and (2.31) are the two quasilinear first-order PDEs equivalent to the single second-order equation (2.29). In this notation we see that L_1 and L_2 are forms that are linear in (p_x, p_y, q_x, q_y) where the coefficients (a, b, c) are given functions of (x, y, u, p, q) for the quasilinear case, and (x, y) for the linear case.

We now introduce the *directional derivatives*† dp/dx and dq/dx. We attempt to find conditions for which these derivatives are *in the same direction*—the

---

† See, for example, [11, Vol. II].

characteristic direction which is tangent to the characteristic curves in the (x, y) plane. It will be shown that the characteristic directions (there turns out to be two unique directions for the case of wave propagation) are the roots of a quadratic equation. This quadratic is the condition that dp/dx and dq/dx are in the same direction. The directions are the slopes of the appropriate characteristic curves. It will also be shown that it is the character of these roots that allows us to classify our PDE system into three types.

To carry out this program of finding these characteristic directions, we seek the proper combination of the forms L_1 and L_2 (since p_x, p_y, q_x, q_y are embedded in these expressions). To this end, we introduce the parameter λ which defines the linear combination $L = L_1 + \lambda L_2$. For the quasilinear case λ depends on (x, y, u, p, q) at each field point. This means that, in general, there is a different linear combination L at each field point. L is obtained from (2.30) and (2.31). We get

$$L = ap_x + (b - \lambda)q_x + (b + \lambda)p_y + cq_y + d = 0. \tag{2.32}$$

Upon expanding dp/dx and dq/dx, it is easily shown that the condition, that these two directional derivatives be in the same direction (dy/dx), is given by the following rule:

$$\frac{\text{coef.}(p_y)}{\text{coef.}(p_x)} = \frac{\text{coef.}(q_y)}{\text{coef.}(q_x)}. \tag{2.33}$$

The set of ratios given by (2.33) is the key to the method of characteristics applied to our system. Applying this rule to (2.32) yields

$$\frac{b + \lambda}{a} = \frac{c}{b - \lambda} = \frac{dy}{dx} = z, \tag{2.34}$$

where we have defined the slope of the characteristic curve by $dy/dx = z$. From the system (2.34) we obtain the following quadratic equation for the slope z:

$$az^2 - 2bz + c = 0. \tag{2.35}$$

Equation (2.35) is called the *characteristic equation* for the system of two first-order PDEs for p and q. It is a pivotal result. Note that it depends only on the coefficients (a, b, c) which make up the principal part of the PDE. The roots z of (2.35) are the characteristic ODEs associated with the first-order system (2.30) and (2.31) which is equivalent to the single second-order equation (2.29). Clearly these roots depend on the discriminant $D = b^2 - ac$. The nature of D allows us to categorize our system (2.30) and (2.31) into three separate types according to the three possibilities for D:

(1) $D > 0$; real and unequal roots, system is hyperbolic.
(2) $D = 0$; real and equal roots, system is parabolic.
(3) $D < 0$; complex conjugate roots, system is elliptic.

The *hyperbolic* system is the only set of quasilinear equations that allow for wave propagation. Consider the special case where the generic equation (2.29)

Table 2.1. Classification of PDEs of type (2.30), (2.31).

Discriminant	Type of PDE	Nature of roots	Example
$D = b^2 - ac$			
> 0	hyperbolic	real unequal	wave equation $b = 0, ac < 0$
$= 0$	parabolic	real equal	Fourier's equation $b = c = 0, a \neq 0$
< 0	elliptic	complex conjugates	Laplace equation $b = 0, ac > 0$

is the wave equation $au_{xx} + cu_{yy} = 0$ where $ac < 0$ and $b = 0$. $D > 0$ so that the wave equation is hyperbolic and wave phenomena is exhibited. An example of type (2) is the unsteady heat equation or Fourier's equation obtained from (2.29) by setting $b = c = 0$. $D = 0$ so that diffusion phenomena is governed by a parabolic equation since Fourier's equation represents a diffusion process. For type (3), we may consider Laplace's equation $au_{xx} + cu_{yy} = 0$, so that $b = 0, ac > 0$, and $D < 0$, which means Laplace's equation is parabolic. Therefore any potential such as that representing a gravitational field is governed by an elliptic equation. Table 2.1 summarizes the classification of the system (2.30) and (2.31) according to the nature of the roots.

From the nature of the roots z of the characteristic equation we see that:

(1) The hyperbolic equation has two distinct families of characteristics.
(2) The parabolic has one familiy of characteristics.
(3) The elliptic equation has no real characteristics.

For the hyperbolic case, if we had an nth-order hyperbolic PDE there would be n distinct families of characteristics. We saw that in dealing with a single first-order equation, there was a single family of characteristics.

2.5. Second-Order Quasilinear Partial Differential Equation

We return to our second-order quasilinear PDE (2.29) and develop the method of characteristics directly for this equation (without transforming it to an equivalent first-order system). The basic idea for this approach is contained in our treating the Cauchy initial value problem for a second-order equation. We recall that, for the corresponding Cauchy problem for a single first-order equation, the initial data is that u is prescribed along the initial value curve. For the second-order equation, which we treat here, the Cauchy problem is defined by the fact that the initial data for (u, p, q) are prescribed along the initial value curve, since the PDE is second order in x and y. From p and q we can calculate the derivative of u normal to the initial value curve Γ. We define a *strip* as a neighborhood of Γ. The idea is that characteristic theory allows us to solve (2.29) for $u(x, y)$ in this strip by using the initial data on Γ. Having obtained the solution in this strip, we can then calculate as many of the higher-order partial derivatives as we need for a Taylor series expansion to determine the solution extended outside the strip. Therefore the initial value curve must be one that allows us to calculate all the derivatives we need. This is an important statement; for, as we shall see below, there are curves which do not let us calculate the higher-order derivatives—these turn out to be the characteristic curves. We extend the notation to include second-order derivatives (standard in PDEs).

$$u_x = p, \qquad u_y = q, \qquad u_{xx} = r, \qquad u_{xy} = s, \qquad u_{yy} = t.$$

Figure 2.6 shows the situation for an initial value curve, the initial data and the surrounding strip.

Our first task is to derive algebraic equations that will allow us to solve for r, s, t on Γ, and then use these values to calculate the higher-order derivatives on Γ. Equation (2.29) becomes

$$ar + 2bs + ct + d = 0. \tag{2.36}$$

We need two additional equations since we have three unknowns. They are obtained by expanding dp/dx and dq/dx. We get

$$r + zs = \frac{dp}{dx} = p', \qquad z \equiv \frac{dy}{dx},$$

$$s + zt = \frac{dq}{dx} = q'. \tag{2.37}$$

Since z, p', and q' are known on Γ (2.36) and (2.37) are the three equations to

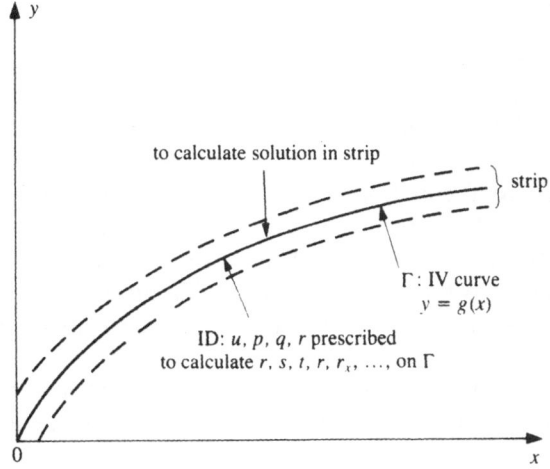

Fig. 2.6. Initial value curve Γ showing initial data and solution strip in two dimensions.

determine r, s, t on Γ. The solution is

$$r = -\frac{\Delta_1}{\Delta_4}, \qquad s = \frac{\Delta_2}{\Delta_4}, \qquad t = -\frac{\Delta_3}{\Delta_4}, \tag{2.38}$$

where the determinants are

$$\Delta_1 = \begin{vmatrix} 2b & c & d \\ z & 0 & -\acute{p} \\ 1 & z & -\acute{q} \end{vmatrix}, \qquad \Delta_2 = \begin{vmatrix} a & c & d \\ 1 & 0 & -\acute{p} \\ 0 & z & -\acute{q} \end{vmatrix},$$

$$\Delta_3 = \begin{vmatrix} a & 2b & d \\ 1 & z & -\acute{p} \\ 0 & 1 & -\acute{q} \end{vmatrix}, \qquad \Delta_4 = \begin{vmatrix} a & 2b & c \\ 1 & z & 0 \\ 0 & 1 & z \end{vmatrix}. \tag{2.39}$$

As long as $\Delta_4 \neq 0$, r, s, t are finite and can be uniquely determined from (2.38). Now an interesting question arises. What happens when $\Delta_4 = 0$? For this case we cannot solve (2.38) for r, s, t on Γ. This means that we cannot obtain the solution in the strip surrounding Γ since we cannot expand in a Taylor series using the derivatives on Γ. Therefore such a curve, for which $\Delta_4 = 0$, is not an admissible initial value curve. Setting $\Delta_4 = 0$ yields the equation $az^2 - 2bz + c = 0$ which is the same as (2.35). Since the roots z of (2.35) give the characteristic directions, we obtain the important result: The initial value curve Γ shall never be in a characteristic direction z, since the curves for which the initial data do not allow us to obtain solutions are the characteristic curves. We see that this approach vis-à-vis the second-order PDE gives new insight into the meaning of characteristic curves, as those curves along which we cannot solve the Cauchy initial value problem.

Suppose that in addition to $\Delta_4 = 0$ we have $\Delta_2 = 0$, which means

$$az\acute{p} + c\acute{q} + zd = 0. \tag{2.40}$$

It follows that $\Delta_1 = \Delta_3 = 0$ (which yield no new results) so that r, s, t are indeterminate. Equation (2.40) can also be obtained by multiplying the first equation of (2.37) by az, the second by c, adding the result, and using (2.35) and (2.36). Equation (2.40) is an important result. It is the characteristic (ordinary differential) equation relating the directional derivatives $\acute{p}$ and $\acute{q}$. This equation and (2.35) give us the following situation: If (2.36) is hyperbolic then there are two distinct families of characteristics in the (x, y) plane whose directions are given by the roots of (2.35). In addition, along each characteristic there is a relationship between the directional derivatives $\acute{p}$ and $\acute{q}$ given by (2.40).

We now develop a finite difference method† using the characteristic equations (2.35) and (2.40) to solve (2.36) by constructing a *characteristic net* in the

---

† See, for example, [13, Chap. 3].

(x, y) plane. The net is composed of two families of characteristics, the C_+ family corresponding to the positive slope z_+, and the C_- family corresponding to the negative z_- slope. In general, the characteristics are curvilinear except for the special case where (2.36) is linear with constant coefficients. In general, (2.36) is quasilinear so that the line segments (from neighboring field points) forming the net must be determined from the solution at two neighboring field points at the previous time step. This will be shown in the graphical description for the construction of the solution field. But first we show how to construct the solution at a typical field point in the strip surrounding Γ. We start with prescribed initial data (u, p, q) along a given initial value curve, by considering these data at a finite number of points on (for simplicity, a uniform distribution of data points). Let Γ be defined by $y = g(x)$. Figure 2.7 shows a C_+ segment coming from point A on Γ and a C_- from B on the same side of the strip. Their intersection defines the field point P in the strip. The z_+ slope at A and the z_- slope at B are calculated from the coordinates of these points which are (x_A, y_A), (x_B, y_B) and the initial data which are (u_A, p_A, q_A), (u_B, p_B, q_B). First, we must calculate the coefficients (a, b, c) at A and B which are known functions of (x, y, u, p, q) at these points. Next we use (2.35) to calculate z_{+A} and z_{-B} as functions of these coefficients at A and B. Using the definition of z we get

$$y_P - y_A = z_{+A}(x_P - x_A),$$
$$y_P - y_B = z_B(x_P - x_B). \qquad (2.41)$$

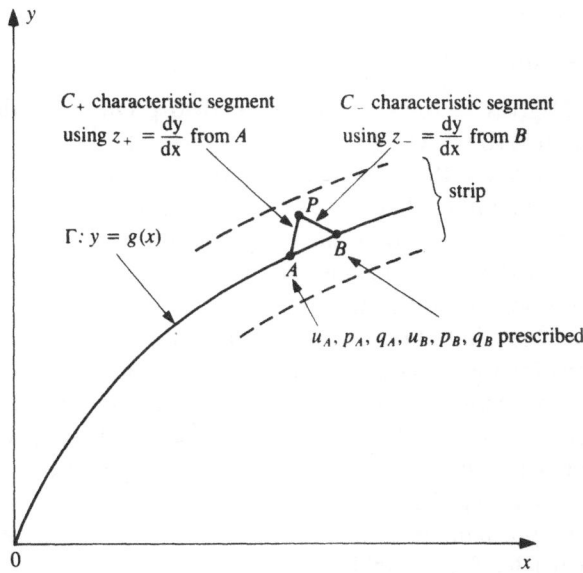

Fig. 2.7. Portion of strip showing how to calculate u_P in strip from initial data at A and B.

Equation (2.41) is used to solve for the coordinates (x_P, y_P) of the field point P. Again we see that the coordinates of P are obtained from the intersection of the C_+ segment from A and the C_- segment from B. (Obviously, these characteristic segments emanate on the same side of Γ in the strip.) We observe that if any of the initial data for u, p, or q are changed at A or B then the position of P is changed, which clearly means that the coordinates of the field points in the strip depend on these initial conditions. For the linear case, the situation is much more simple in that the positions of the field points in the strip only depend on the curve $y = g(x)$ (are independent of the initial data). We next solve for p and q at P by using (2.40) in differential form, by approximating dp and dq by finite differences. We get

$$a_A z_{+A}(p_P - p_A) + c_A(q_P - q_A) + d_A(x_P - x_A) = 0,$$
$$a_B z_{-B}(p_P - p_B) + c_B(q_P - q_B) + d_B(x_P - x_B) = 0. \tag{2.42}$$

Since the coordinates of P have already been calculated we use (2.42) to calculate (p_P, q_P) in terms of the data at A and B.

Next we compute the solution u_P by integrating

$$du = p\, dx + q\, dy.$$

One method of calculating u_P is to first integrate this differential expression along the C_+ segment AP using the average values of p_A, p_P, q_A, q_P. We then integrate along the C_- segment BP using the average of p_B, p_P, q_B, q_P. We then get two equations for u_P, one using AP and the other using BP. Solving them for u_P yields

$$u_P = \tfrac{1}{2}(u_A + u_B) + \tfrac{1}{4}(p_A + p_P)(x_P - x_A) + \tfrac{1}{4}(p_B + p_P)(x_P - x_A)$$
$$+ \tfrac{1}{4}(q_A + q_P)(y_P - y_A) + \tfrac{1}{4}(q_B + q_P)(y_P - y_B). \tag{2.43}$$

Equation (2.43) gives us the solution at the field point P for the first finite difference step in y. (If we let y play the role of time then P is at the first time step.) We then perform the same calculations for all the field points corresponding to the first finite difference step in y using all our selected initial data points on Γ. If y represents time then the solution at these field points are sufficient to define the strip (which is one-sided) surrounding Γ. It is now easy to build up solutions by using the results in the strip. We do this by using these results as "new initial conditions" and continue the procedure to construct solutions for the second time step, etc. It is clear that this numerical method is an approximate one whose accuracy depends on the number of initial data points chosen; the more dense these points are, the better the approximation. Also, the degree of approximation depends on the method of integration to obtain u_P. The curvilinear families of characteristics is thereby replaced by an approximate characteristic net made up of characteristic line segments constructed from neighboring points. For more details on this finite difference method including stability of solutions, the reader is referred to [13].

2.6. Domain of Dependence and Range of Influence

In discussing the geometric interpretation of D'Alembert's solution of the one-dimensional wave equation we gave two examples:

(1) $g(x) = 0$ and $f(x)$ a square wave pulse.
(2) $f(x) = 0$ and $g(x)$ a square wave pulse.

For both these cases Fig. 2.1(a), (b) showed that there are regions in the (x, t) plane in which the field points are not affected by the initial data so that these regions are "dead" in the sense that the solution is zero there. We shall investigate this property of the solution of the wave equation in detail in this section.

The first observation derived from D'Alembert's solution is that the value of the solution at a field point P depends only upon the values of the initial data on the segment of the initial curve Γ (which in this case is the x axis or $t = 0$), obtained by drawing the C_+ and C_- backward in time from P to where they intersect the x axis. This segment is the interval along the x axis whose left end point is the intersection of the C_+ characteristic and whose right end point is the intersection of the C_- characteristic. This interval is called the *domain of dependence* of the field point P. It is clear that no points outside this domain of dependence affects the solution at P, for no characteristics from any of these points can touch P, and the characteristics from the initial data curve are the only curves which carry initial data to the field point. Figure 2.8(a) shows the domain of dependence, which is the interval $[x - ct, x + ct]$ arising from the field point P whose coordinates are (x, t). Figure 2.8(b) shows the triangular region bounded by the two characteristics from Q on Γ and the domain of dependence. This region is called the *domain of determinacy*.

Another observation arising from the properties of D'Alembert's solution is as follows: Consider an interval $[x_1, x_2]$ on which we change the initial data. Draw a C_- characteristic line from x_1, and a C_+ characteristic line from x_2 (both lines going into the region $t > 0$). Then the region in the (x, t) plane bounded by the interval $[x_1, x_2]$ and these characteristics is called the *range of influence* of this interval. This is shown in Fig. 2.8(b). From this definition of the range of influence we deduce the following: None of the field points in the range of influence of this interval is affected by changing the initial data on the portion of the x axis outside $[x_1, x_2]$. Suppose we shrink the interval $[x_1, x_2]$ to the point x_0 at which we have nonzero initial data on u or u_t or both. We also suppose that the initial data is identically zero everywhere else on the x axis. Then the range of influence of the point x_0 is the open triangular region shown in Fig. 2.8(c). This gives us the important fact that a disturbance in the initial data (at x_0) can only affect the field points in the range of influence bounded by the C_+ and C_- characteristics stemming from x_0. These above-mentioned properties of the one-dimensional wave equation, vis-à-vis the domain of dependence and range of influence, are important characteristics

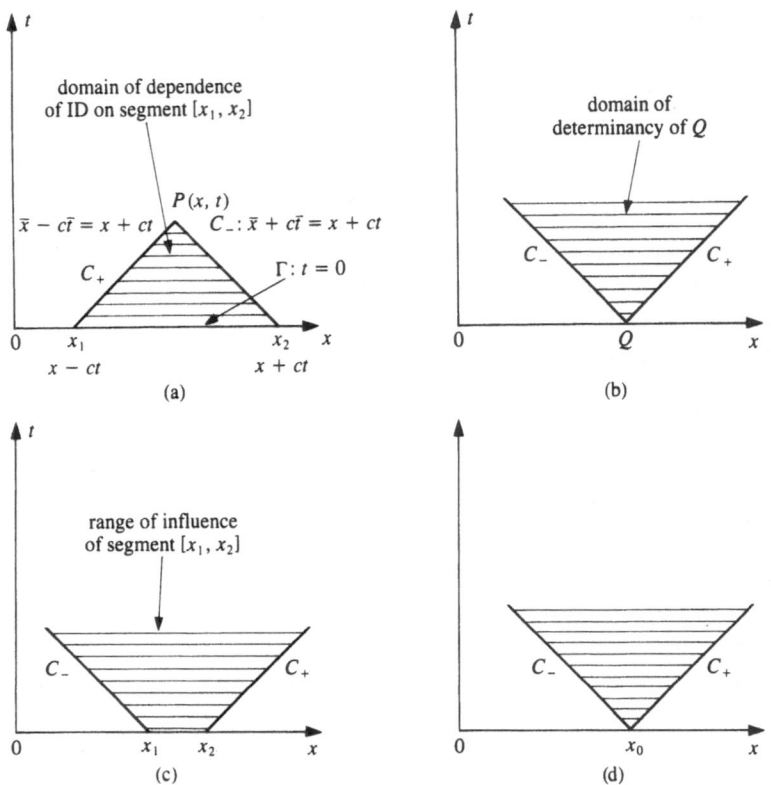

Fig. 2.8. (a) Domain of dependence; (b) domain of determinacy; (c) range of influence of segment $[x_1, x_2]$; (d) range of influence of point x_0.

of the hyperbolic PDE. The fact that changes in the initial data can only affect a certain region (the range of influence) and not all of space is a crucial one in distinguishing the hyperbolic set of equations from the elliptic and parabolic ones. Thus, these concepts of domain of dependence and range of influence are pivotal ones in the sense of giving the hyperbolic equation wave properties. With this background we shall show in a subsequent section that discontinuities can only be propagated along characteristic curves. These discontinuities have wave-like properties.

2.7. Some Basic Mathematical and Physical Principles

At this point it is instructive to discuss some basic principles concerning the relationship between mathematical models, physical reality, and the three types of PDEs. In [11, Vol. II] it is pointed out that a mathematical problem that corresponds to physical reality should satisfy the following requirements:

(1) The solution must exist. This is an obvious requirement and expresses a point in logic that ideally there be no contradictory properties demanded of the solution. It also tells us that the mathematical model describes the physics of the problem in a realistic, although simplified, manner.

(2) The solution should be unique. The mathematical model should lead to a unique or unambiguous mathematical solution to the physical situation. However, there are cases where there are multiple solutions. For example, in the case of multiple eigenvalues, whole families of solutions of the eigenvalue problem exist. But, in general, a mathematical description of a given physical situation should uniquely describe that situation.

(3) The solution should depend continuously on the given data. This is a condition that the solution be *stable*. This is a particularly important requirement. It is necessary if the mathematical model is to describe adequately the physics of the problem. For in any physical situation the data are not prescribed exactly. There are always errors in measurements so that the data vary between certain bounds. A mathematical model should be posed in such a way that the solution should depend continuously on the variations of the data. This is the requirement of stability. It is not only essential for a proper mathematical formulation of a physical problem, but is also essential in dealing with approximation or numerical methods. It is clear why we say that this is a "stability requirement". For, if the solution to a given problem varies wildly for a small variation in the given data, we say the solution is unstable and we have no confidence in it.

This continuous dependence of the solution on given data does not rule out another type of situation: A mathematical description of a given physical situation may show jumps in some of the variables across certain surfaces. For example, in studying shock wave phenomena we describe a shock wave as a surface across which there is a discontinuity in pressure, entropy, etc. But this is an idealization. A closer examination shows that a shock wave has a certain structure to it wherein the pressure gradients, etc., across the front are finite although large. However, in most cases the idealization of finite jumps across surfaces or curves is a good enough approximation to physical reality.

Any problem that satisfies these three requirements is said to be a *properly posed* problem. It is therefore important to understand what a non-properly posed or "ill-posed" problem is. To this end, we define the Cauchy initial value problem as one consisting of a second-order PDE in the (x, y) plane with two initial conditions. We used as our PDE a hyperbolic equation such as the wave equation and learned that there exists a unique solution in a range of influence which depends on the domain of dependence. Now suppose our PDE is elliptic such as the equation $u_{xx} + u_{yy} = 0$, which is Laplace's equation in two dimensions. We further prescribe the initial data $u(x, 0) = f(x)$, $u_y(x, 0) = g(x)$ on the initial value curve Γ: $y = 0$. Since u satisfies Laplace's equation it is a potential function. According to a principle of potential theory any solution in the upper half-plane $y > 0$ can be continued into the lower half-plane by reflection and would therefore be automatically analytic on the x axis. But $g(x)$ is an arbitrary continuously differentiable function of x and need not be analytic. Analyticity means that there is a relation between $f(x)$ and $g(x)$; but for the Cauchy problem $f(x)$ and $g(x)$ are independent. Therefore the Cauchy problem for Laplace's equation is an ill-posed problem.

As an example of an ill-posed problem let u_n be the solution of $\nabla^2 u_n = 0$ (where ∇^2 is the two-dimensional Laplacian) subject to the initial conditions

$$u_n(x, 0) = f_n(x) = \left(\frac{1}{n^2}\right) \sin nx, \qquad y = 0, \quad n = 1, 2, \ldots,$$

$$u_{n,y}(x, 0) = g_n(x) = 0.$$

Let $u(x, y)$ be the solution to this initial value problem for the limiting case case $n \to \infty$. It is clear that the initial conditions for this case are $u(x, 0) = f(x) = 0$, $u_y(x, 0) = g(x) = 0$, so that the solution is $u(x, y) = 0$. But, from the initial conditions for $f_n(x)$ and $g_n(x)$, we have $u_n(x, y) = (1/n^2) \cosh ny \sin nx$. This solution for u_n does not yield $u_\infty(x, y) = 0$. This is a clear example of an ill-posed problem, since the solution does not depend continuously on the initial data in the sense that the limit function $u_\infty(x, y)$ is not equal to $u(x, y)$.

As an example of a well-posed problem we consider $u_n(x, y)$ as a solution to the one-dimensional wave equation $u_{n,xx} - u_{n,yy} = 0$ (where y is the dimension of time) with the same initial conditions, namely, $u_n(x, 0) = (1/n^2) \sin nx$, $u_{n,y}(x, 0) = 0$. The solution is

$$u_n(x, t) = \left(\frac{1}{n^2}\right) \cos ny \sin nx.$$

We see that the limit solution $u_n(x, t) \to 0$ as $n \to \infty$, which is the solution of $u_{xx} - u_{yy} = 0$ for $u(x, 0) = u_y(x, 0) = 0$. In general, the Cauchy initial value problem for the hyperbolic PDE satisfies all three of the above requirements.

These two examples clearly indicate that the elliptic PDE presents an ill-posed problem and the hyperbolic equation presents a well-posed problem with respect to the Cauchy initial value problem. For the parabolic equation the Cauchy initial value problem is a well-posed problem but only one initial condition is needed. For the elliptic equation the boundary value problem satisfies our requirement of uniqueness. The domain of dependence of the solution is the whole boundary, which means that the value of the solution for the elliptic equation in any closed domain depends on the boundary values on every point of the boundary surrounding the domain.

It is again worthwhile to point out the differences in the characteristics for these three types of PDE: For the hyperbolic case there are two distinct families of characteristic curves. For the elliptic case there are no real characteristics. For the parabolic case there is one family of characteristic curves of infinite slope, i.e., $dx/dt = \infty$, so that disturbances are propagated along characteristics with infinite speed. This tells us that a diffusion process, which is associated with the parabolic equation, propagates information instantaneously throughout all space even though the intensity is exponentially damped as we go away from the source.

The basic principle in PDEs is that boundary value problems are associated with elliptic equations while initial value problems, mixed problems, and problems with radiation effects at the boundaries are associated with hyperbolic and parabolic equatons.

The conditions of existence, uniqueness, and continuous dependence of the solution on given data (stability) dominate classical physics. In fact, Laplace made a statement to the effect that he could calculate the future course of the universe if he were given the initial coordinates and velocities and forces on all the particles in the universe! This *deterministic view* meant that he could write down the equations of motion for each particle and then use the initial conditions to calculate the motion of each particle for all time by solving this system of equations. Since the equation of motion for each particle is a second-order vector differential equation, its solution yields the particle trajectory, given its initial position and velocity. With the advent of quantum mechanics this completely deterministic view of Laplace was proved false. Heisenberg taught us, in his famous *uncertainty principle*, that it is impossible to know with complete accuracy the simultaneous position and momentum of a particle. In fact, the product of the error in position and momentum cannot be less than Planck's constant. A far-reaching consequence of the uncertainty principle is that any measurement of a natural phenomenon in the laboratory has an effect on the measuring instrument, so that the experimenter cannot measure any piece of data with infinite accuracy. A simple example is that a photon of light affects the sight of the observer in such a way as to preclude the infinitely accurate measurement of a beam of light, since we cannot measure how the photon effects our retina. In this sense quantum mechanics is an experimental science rooted in our experience with measuring natural phenomena and deducing realistic consequences. Also, at the core of quantum mechanics is the use of probability theory from which we derive, for example, the path of an electron around a proton as a probability distribution of paths which is obtained from wave functions arising from the solution of Schrödinger's equation. The eigenvalues associated with these wave functions are observable energy levels which are quantized.

In addition to quantum mechanics, nonlinear phenomena arising in classical mathematical physics, which are best solved by powerful numerical methods, have shown that ill-posed problems that reflect natural phenomena can sometimes be attacked and solved. We showed that Laplace's equation leads to an ill-posed problem when considered as an initial value problem. Yet G.I. Taylor showed that an important question of stability arises when considering such an ill-posed problem. He studied a system of two incompressible fluids of different densities separated by an interface which moves toward the lighter fluid. Since the fluids are incompressible the flow is governed by a potential which means Laplace's equation must be satisfied. Taylor showed that the potential turns out to be a solution of an ill-posed initial value problem for Laplace's equation. One of the key problems in numerical analysis is to devise approximate or finite difference methods of constructing solutions to initial value problems, where the data are only approximately known and we want to make use of improved updated data. These problems arise particularly in meteorology where masses of inaccurate data must be assessed.

In general, *future research in electromagnetic phenomena as a venerated branch of mathematical physics will be concerned with the application of non-linear methods.*† These methods must involve characteristic theory and modern approaches to numerical methods.

2.8. Propagation of Discontinuities

It was mentioned at the end of Section 2.6 that discontinuities in initial data can only be propagated along characteristic curves. We now take up this important aspect of characteristic theory.

The PDEs of the elliptic and parabolic type have analytic solutions, even when the boundary or initial conditions are discontinuous. However, hyperbolic equations are different in the sense that discontinuities in the initial data are propagated along characteristics into the solution domain. A more exact statement is that only *weak* (small amplitude) discontinuities are propagated along characteristics. *Strong* (large amplitude) discontinuities correspond to strong shock waves; they are propagated along different curves which tend to the characteristic curves as the shock waves become infinitely weak in amplitude; these are sound waves. (The amplitude of a shock wave is defined as the ratio of the difference in pressure across the wave divided by the ambient pressure.)

We now give a geometric proof that any discontinuity in the initial data can only be propagated along a characteristic curve. Consider Fig. 2.9 which shows the first quadrant in the (x, t) plane. Let A and B be two points on the initial value curve which we take as the x axis. We draw a C_+ characteristic from A and a C_- characteristic from B and let them intersect at P as shown in the figure. More exactly, for the quasilinear case, the C_+ curve AP and the C_- curve BP are built up numerically as characteristic line segments from neighboring points at previous time steps, because the characteristic curves depend on the solution. In any case we shall call the curves AP and BP *frontier characteristics*. Every point in the curvilinear triangle APB is a field point affected by the initial data (u, p, q) in the interval $[A, B]$ which is the domain of dependence for the range of influence (the triangle APB). Now let there be a discontinuity in one or more of the initial data at points A and B. Let C be a point on the x axis to the left of A and let E be a point to the right of B.

---

† See, for example [1].

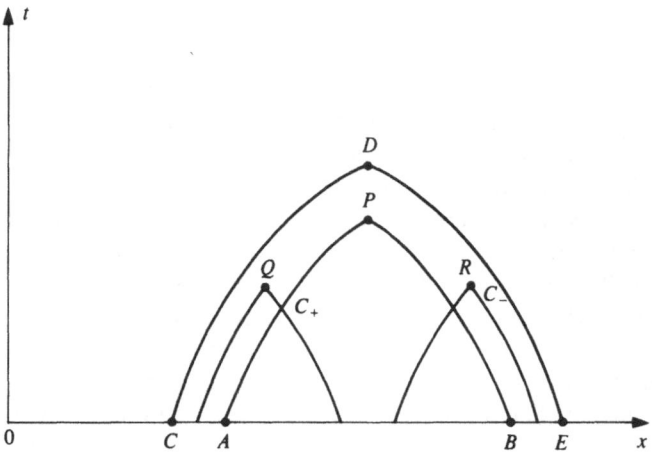

Fig. 2.9. Propagation of discontinuities.

Draw the frontier characteristics CD and ED. By the uniqueness of the solution these characteristics cannot penetrate the triangular region APB. Let Q be an interior point of the strip $ACDP$ and let R be an interior point of $BEDP$. Clearly the initial data on CA and BE are different from those on AB because of the discontinuities at A and B. Fixing points A and B, we let $C \to A$, $E \to B$, and $D \to P$. Then Q tends to a point on the frontier characteristic AP and R tends to a point on BP. This demonstrates that a discontinuity in the data at A propagates along a C_+ characteristic and a discontinuity at B propagates along a C_- characteristic. Clearly a C_- characteristic stems from A and a C_+ characteristic stems from B, along which discontinuities in the initial data propagate.

The above argument clearly shows that a characteristic curve can separate two different solutions. The solution u to (2.36), together with the first-order derivatives p and q, can be continuous across a characteristic. But we shall now show that the second-order derivatives (r, s, t) are *discontinuous* across a characteristic. The system (2.36) and (2.37) consists of three equations for (r, s, t). We showed that we can solve for these three second-order derivatives along any curve that is not a characteristic. If the curve is a characteristic then (r, s, t) are not independent. The system (2.36) and (2.37) can then be manipulated to eliminate r and t, for instance, yielding:

$$s[az^2 - 2bz + c] - [ap'z + cq' + dz] = 0. \qquad (2.44)$$

Once a value of s is assigned then r and t can be calculated from any pair of the system (2.36) and (2.37). We assume $s \neq 0$. Then the brackets on the left-hand side of (2.44) vanish, thus yielding the characteristic differential equations (2.35) and (2.40). This is seemingly another way of deriving these characteristic equations, but a little reflection will show that this is equivalent to setting the determinant $\Delta_4 = 0$. Clearly, the same analysis applies to r or

t, i.e., we can eliminate s and t or s and r to allow for arbitrary values of r or t, respectively.

Suppose we attempt to use a characteristic as an initial value curve. To obtain a solution for u in the neighborhood of one side of the characteristic, we expand in a Taylor series in terms of the derivatives $p, \ldots, t$ and higher-order derivatives on the characteristic. We showed above that the second-order derivatives are not independent of the characteristic, which means that one of the terms r, s, or t is arbitrary and can thus be prescribed. If we now do a similar Taylor series expansion on the other side of the characteristic, with respect to the same data, we are then free to choose another value of r, s, or t. This tells us that we can obtain two different solutions for u on either side of the characteristic, meaning that there is a jump discontinuity of u on the characteristic. This violates the uniqueness condition for the solution of (2.36). This analysis again illustrates the basic principle that a characteristic curve cannot be used as an initial value curve (we cannot prescribe (u, p, q) on a characteristic to obtain solutions off the characteristic.

The following example will illustrate what happens if we prescribe initial data on a characteristic. Suppose we desire a solution of $u_{xx} - u_{yy} = 0$ which satisfies the initial data $u = 1$, $p = 2$, $q = -2$ along the initial value curve $C: y - x = 0$. Characteristic theory tells us that C is a characteristic curve. We know that all solutions of the one-dimensional wave equation must be of the form $u(x, y) = f(x - t) + g(x + t)$. The following two solutions satisfy the initial data on $y - x = 0$:

$$u_1(x, y) = \cos(y - x) - 2\sin(y - x), \qquad u_2 = 1 - 2(y - x).$$

Since we cannot have two different solutions at a given field point we interpret these solutions as follows: $u_1(x, y)$ is the solution on one side of the characteristic, and $u_2(x, y)$ is the solution on the other side for the initial data at the same point on the curve $y - x = 0$. Both u, p, and q are continuous across C. We have $r = s = t = 0$ from $u_2(x, y)$. However, $r = -1$, $s = 1$, $t = -1$ on C when calculated from $u_1(x, y)$. The conclusion is that we have a discontinuity in the second derivatives if we use initial data on a characteristic curve. This means we cannot expand in a Taylor series to obtain solutions off that curve. It is obvious that $r - s = 0$ so that the wave equation is satisfied.

We now show analytically that any curve in the (x, y) plane which allows for a jump in the second derivatives of $u(x, y)$ must be a characteristic curve. Let C be a curve represented by the equation $\phi(x, y) = 0$. The C separates the region $\phi > 0$ from the region $\phi < 0$. As usual, $u(x, y)$ is a solution of the quasilinear PDE (2.36). Let $[f]$ denote the jump in the function f across C in the direction of increasing ϕ. We assume that u, p, and q are continuous across C but that $[r]$, $[s]$, and $[t]$ are finite and nonzero jumps across C. Consider any point P on C. Across C at P we have $dp = [r] + [s]z = 0$ (where $z = dy/dx$) and $dq = [s] + [t]z = 0$. We also have $\phi_x + \phi_y z = 0$ on C. This immediately gives

$$[r] = k\phi_x^2, \qquad [s] = k\phi_x\phi_y, \qquad [t] = k\phi_y^2, \qquad z = -\frac{\phi_x}{\phi_y}, \qquad (2.45)$$

where the constant of proportionality k is easily seen to be given by $k = [u_{\phi\phi}]$.

We now apply the PDE (2.36) to the points P_1 and P_2 on either side of P on C. Subtracting the resulting two equations from each other and letting $P_1 \to P$ and $P_2 \to P$ gives

$$a[r] + 2b[s] + c[t] = 0, \qquad (2.46)$$

since the term d is continuous across C and therefore vanishes. Using (2.45) and (2.46) we obtain

$$a\phi_x^2 + 2b\phi_x\phi_y + c\phi_v^2 = Q(\phi_x, \phi_y) = 0, \qquad (2.47)$$

which tells us that the quadratic expression $Q(\phi_x, \phi_y)$ as given by the left-hand side of (2.47) vanishes. Since $\phi_x/\phi_y = -z$, (2.47) becomes

$$Q(\phi_x, \phi_y) = Q(z) = az^2 - 2bz + c = 0,$$

which is (2.35), the characteristic equation corresponding to (2.36). The conclusion is that the curve $C: \phi(x, y) = 0$, across which r, s, and t have jump discontinuities, is a characteristic curve, which means that jump discontinuities in the second derivatives of u can only occur across characteristics.

The reader will note that nothing was said about jumps in u, p, or q across a characteristic curve (although we discussed jumps in these quantities across an initial value curve when they were considered as prescribed data). Discontinuities in u and the first derivatives are excluded across C because of the meaning of the PDE for which u is a solution. However, solutions of the wave equation that admit of discontinuities in the first derivatives involve a more generalized conception of what we mean by "solution". Indeed, in the next section we shall extend this concept to include the so-called "weak solutions" which allow for such discontinuities. This will be discussed in the context of a more generalized setting for the conservation laws.

2.9. Weak Solutions and the Conservation Laws†

We start by developing a generalized treatment of the three conservation laws of physics in connection with weak solutions which, we recall, are those that allow for discontinuities in the first derivatives. In Chapter 1 a discussion of these conservation laws was given in connection with electromagnetic theory. These fundamental laws of physics are the very heart of mathematical physics and give the foundations of all aspects of physics from electromagnetic theory

---

† See, for example, [11, Vol. II].

to mechanics to fluid dynamics, elasticity, thermal effects, etc. All the field equations of physics may be derived from these laws. They are important enough to repeat here.

(1) *Conservation of mass* which leads to the continuity equation.
(2) *Conservation of momentum* which leads to Newton's equations of motion. By momentum we mean linear momentum. The conservation of angular momentum applies to rotating systems and leads to the equation of motion for such systems which equates the moment of inertia times the angular acceleration to the external torque.
(3) *Conservation of energy* which leads to the energy balance equations and invokes the second law of thermodynamics which predicts a jump in entropy across a shock wave.

To get a feel for the meaning of the conservation laws in a generalized setting we interpret u as a physical quantity which may represent mass, momentum, energy, or any other quantity with which we associate a flux or flow of that quantity across a closed surface bounding a given volume of material. It is clear from vector analysis that a flux of u involves the divergence of u out of the volume (across the bounding surface). This is described mathematically by

$$(d/dt) \int_{x_1}^{x_2} u \, dx = f(u(x_2, t), x_2, t) - f(u(x_1, t), x_1, t), \tag{2.48}$$

where f is a prescribed function of u, x, and t. Equation (2.48) expresses the fact that the total physical quantity represented by u contained in the closed interval $[x_1, x_2]$ changes at a rate equal to the flux f of u through the end points of the interval. Therefore (2.48) is a *generalized conservation law* which applies to the conservation of mass, momentum, and energy for the one-dimensional case where the "volume" is the interval $[x_1, x_2]$ and the bounding "surface" consists of the end points x_1, x_2. There is no difficulty in extending (2.48) to three dimensions (we merely replace the integral by a volume integral and use Gauss's divergence theorem to convert the volume integral of the divergence of u to the surface integral of the normal component of u across the surface).

We first consider the case where $u(x, t)$ is a differentiable solution of the general conservation law (2.48). This means that there are no interior points of discontinuity in u in the interval $[x_1, x_2]$. Then, by using the mean value theorem, we can write (2.48) as

$$\int_{x_1}^{x_2} u_t(x, t) \, dx = u_t(\bar{x}, t)(x_2 - x_1) = f(u(x_2, t)x_2, t) - f(u(x_1, t), x_1, t),$$

where $\bar{x}$ is some interior point. Dividing this equation by $x_2 - x_1$ and passing to the limit as $x_2 \to x$ and $x_1 \to x$ (where $x_1 \leq x \leq x_2$), we obtain

$$u_t = f_u u_x + f_x = \left(\frac{\partial}{\partial x}\right) f(u, x, t). \tag{2.49}$$

Equation (2.49) is a quasilinear PDE for $u(x, t)$ where f is a given function of u, x, and t.

Physical Interpretation of the Conservation Laws

At this stage we digress somewhat from our discussion of the generalized conservation laws to describe briefly the conservation laws in a physical setting. For a continuous electromagnetic medium (or any other continuous medium) there are two ways of describing the behavior of a particle with respect to the conservation laws:†

(1) The Lagrangian representation.
(2) The Eulerian representation.

In general, in order to describe wave propagation in a conducting medium we need to relate the equilibrium state of a particle to its dynamic state where it is acted on by forces and moments. In order to characterize this relation between the equilibrium and disturbed states we introduce two coordinate systems:

(1) A spatial or laboratory fixed coordinate system that describes the passage of the particles with time at any point in space.
(2) The particle fixed coordinate system that identifies each particle with respect to a fixed reference frame.

For simplicity we make the origins of the two coordinate systems coincide.

In the *Lagrange representation* the independent variables are the particle coordinates, called the *Lagrange coordinates* and time. The dependent variables such as particle velocity, pressure, etc., are functions of the Lagrange coodinates. The spatial coordinates are called the *Eulerian coordinates*. If a problem is solved in this representation, we then obtain the Eulerian as a function of the Lagrangian coordinates and time. The proper mapping function between these two coordinate systems then gives the particle trajectories. This representation is best used in problems of elasticity where the boundaries are essentially stationary and there is relatively small scale particle motion, as opposed to the situation that occurs, for example, in magnetohydrodynamics where the motion of the charged particles are superimposed on the large scale motion of the fluid.

In the *Eulerian representation* the independent coordinates are the Eulerian coordinates and time so that the dependent variables are to be determined as functions of the Eulerian coordinates and time from the field equations which arise from the conservation laws. These conservation laws are expressed in the Eulerian system. In Chapter 1 they were described for an electromagnetic medium in this system. If we linearize the conservation laws, which we can do

---

† See, for example, [14] and [15].

for small amplitude particle motions, then it can easily be shown that the Lagrangian and Eulerian representations are identical. The difference between the two representations only shows up in the nonlinear terms of the conservation laws. In [15] these conservation laws are discussed in some detail from the point of view of wave propagation in a nonconducting medium.

We are now in a position to apply (2.48) to the three conservation laws for the one-dimensional case. Recall that the generic function u represents mass, momentum, or energy so that we have the following correspondence:

$$\begin{array}{lll} \textit{mass:} & u \sim \rho(x, t), & f = 0, \\ \textit{momentum:} & u \sim \rho u, & f = p, \\ \textit{energy:} & u \sim \rho(e + \tfrac{1}{2}u^2), & f = pu, \end{array}$$

where ρ is the fluid density, u is the particle velocity, p is the pressure, and e is the specific internal energy (per unit mass). Then (2.48) becomes

Conservation of Mass (Continuity Equation)

$$\frac{d}{dt} \int_{x_1}^{x_2} \rho(x, t)\, dx = 0. \tag{2.50}$$

$\rho\, dx$ is the element of mass occupying the interval $(x, x + dx)$ at time t where $x_1 < x < x_2$. In this form the integration is with respect to x over the interval (x_1, x_2) for a fixed t. Clearly, x_1, x_2 depend on t.

Conservation of Momentum (Equation of Motion)

$$\frac{d}{dt} \int_{x_1}^{x_2} \rho u\, dx = p_0 - p_1, \tag{2.51}$$

where p_0, p_1 is the pressure at x_1, x_2, respectively.

Conservation of Energy

$$\frac{d}{dt} \int_{x_1}^{x_2} \rho(e + \tfrac{1}{2}u^2)\, dx = p_0 u_0 - p_1 u_1, \tag{2.52}$$

where the integrand is the total energy which is the sum of the internal and kinetic energy, and $p_i u_i$ is the power input due to the pressure at x_i, $i = 1, 2$. Therefore (2.52) tells us that the rate of change of the total energy is equal to the difference of power at the end points.

Generalized Conservation Laws†

We now go back to the generalized conservation law given by (2.48). We first give an example of how discontinuities can arise in hyperbolic systems, by examining the Cauchy initial value problem applied to (2.49) for the case where $f = f(u)$. (Recall that u is a generic variable and not necessarily particle

---

† Cf. Ref. [11], Vol. II.

velocity.) Letting $f_u = -a(u)$, we get the following Cauchy problem:

$$u_t + au_x = 0, \qquad -\infty < x < \infty, \quad t > 0,$$

$$u(x, 0) = f(x) \quad \text{on the initial value curve } \Gamma: t = 0.$$

(2.53)

Let $z = dt/dx$ be the slope of the characteristic curve in the (x, t) plane. The characteristic equation yields

$$z = \frac{1}{a(u)}. \qquad (2.54)$$

We have a one-parameter family of characteristics which depends on the solution. However, (2.53) tells us that

$$u_t + au_x = u_t + \left(\frac{1}{z}\right)u_x = du\, dt = 0$$

on the characteristic whose slope is given by (2.54). Clearly this means that u is a constant along a characteristic, so that from (2.54) the characteristics become a single family of straight lines. We now prescribe the initial data $f(x)$ that leads to a *singular point* at a field point, meaning that the solution is not unique at that point. Let $f(x_1) = f_1$, $f(x_2) = f_2$, $a(f_1) = a_1$, $a(f_2) = f_2$. Suppose f_1 and f_2 are given such that $0 < a_2 < a_1$. Constructing characteristics from x_1 and x_2 into the region $t > 0$, we see that they will intersect at some field point $P_I: (x_I, t_I)$. This is shown in Fig. 2.10. The coordinates of P_I are

$$x_I = \frac{a_1 x_2 - x_2 a_1}{a_1 - a_2}, \qquad t_I = \frac{x_2 - x_1}{a_1 - a_2}.$$

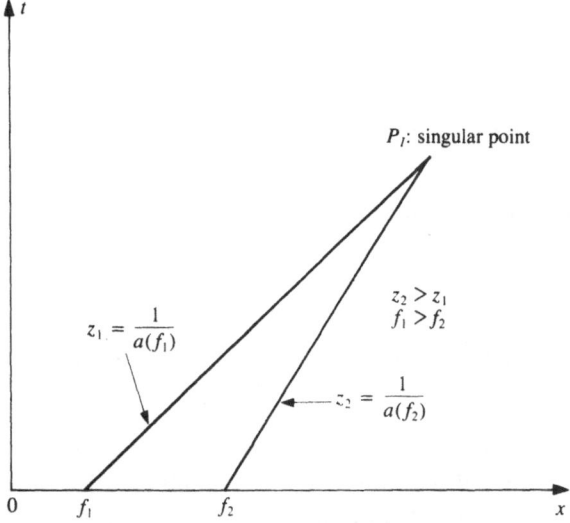

Fig. 2.10. Showing a solution that leads to a singular point at P_I.

Since the characteristics from x_1 and x_2 carry different values of u, namely f_1 and f_2, we see that P_I is a singular point, meaning that the solution is not unique there. This means that a continuous solution cannot be given for $t > t_I$. This example illustrates the fact that continuous solutions of quasilinear PDEs in general do not exist "in the large" (globally). The operative word is "continuous", for if we relax the condition of continuity by allowing for discontinuous or *weak solutions*, we then admit a wider class of solutions to the quasilinear system. For example, in the study of shock wave phenomena it is seen that shock waves are curves (or surfaces in higher dimensions) across which discontinuities in pressure, entropy, etc., occur. These are called the "jump conditions" across the shock wave.

We now extend the generalized conservation law given by (2.48) to weak solutions. We rewrite (2.48) in notation compatible with (2.50), (2.51), and (2.52).

$$\left(\frac{d}{dt}\right)I = f_2 - f_1, \quad \text{where } I = \int_{x_1}^{x_2} F(x, t)\, dx, \tag{2.55}$$

where the integrand $F(x, t)$ stands for mass, momentum, or energy. We assume that there exists an interior point $x = \bar{x}$ in the interval (x_1, x_2) where the integrand F is discontinuous. Specifically, at a fixed t let there be a finite jump discontinuity in $F(x, t)$ at $\bar{x}$. Let F_+ be the value of F as it approaches $\bar{x}$ from the right or left, respectively. Then $F_+ - F_- = [F] \neq 0$ at $\bar{x}$. We now calculate dI/dt, using the formula for differentiating an integral with respect to the parameter t.

$$\begin{aligned}
\frac{dI}{dt} &= \left(\frac{d}{dt}\right)\left[\int_{x_1}^{\bar{x}_-} + \int_{\bar{x}_+}^{x_2}\right]F(x, t)\, dx \\
&= \left(\frac{d\bar{x}}{dt}\right)_- F(\bar{x}_-, t) - \left(\frac{dx}{dt}\right)_1 F(x_1, t) + \left(\frac{dx}{dt}\right)_2 F(x_2, t) \\
&\quad - \left(\frac{d\bar{x}}{dt}\right)_+ F(\bar{x}_+, t) + \int_{x_1}^{x_2} \frac{\partial F(x, t)}{\partial t}\, dx.
\end{aligned} \tag{2.56}$$

$\partial F/\partial t$ is assumed to be continuous across $\bar{x}$ so that

$$\frac{\partial F_+}{\partial t} - \frac{\partial F_-}{\partial t} = 0.$$

Taking the limit as $x_2 \to x_1$ (for $x_1 \leq \bar{x} \leq x_2$) in (2.56) we obtain

$$[f] = -U[F], \tag{2.56'}$$

where $U = dx/dt$ is the velocity of propagation of the discontinuity at x and $[f]$ is the finite jump discontinuity of f at $\bar{x}$. This is called the *jump condition* for the generalized conservation law (2.48). Discontinuities of this type are important in obtaining the jump conditions that occur across shock waves.

Going back to the equation $u_t + au_x = 0$ where $a = a(u)$, we shall apply the jump condition (2.56) to the case of small discontinuities. We obtain the approximation $[f]/[u] = df/du$, so that

$$-\frac{df}{du} = a(u). \tag{2.57}$$

Equation (2.57) tells us that small disturbances are propagated along a characteristic curve since $a(u) = 1/z$, so that $1/U$ is the slope $(z = dt/dx)$ of the characteristic.

2.10. Normal Forms for Second-Order Partial Differential Equations

A *normal form* or *canonical form* of a PDE is its simplest or standard form. We learned previously that second-order PDEs in two variables, or their equivalent first-order systems, can be classified into three types according to the nature of the associated characteristic equation, which is a quadratic for the slopes of the characteristics. This classification is related to these normal forms which gives us a deeper insight into these types.

We shall start with a linear second-order PDE with constant coefficients. We turn to (2.29) and write the principal part as the operator $L(u)$ where

$$L(u) = au_{xx} + 2bu_{xy} + cu_{yy}. \tag{2.58}$$

Then (2.29) can be written as

$$L(u) + d = 0, \tag{2.59}$$

where d is a given function of u, p, and q and is not necessarily linear in these variables. We need only determine the normal form for $L(u)$ in (2.59).

In determining the normal form we introduce a new set of independent variables (ξ, η) defined by the transformations

$$\phi(x, y) = \xi, \qquad \psi(x, y) = \eta. \tag{2.60}$$

The transformations $\phi(x, y)$ and $\psi(x, y)$ are to be determined in such a way as to put $L(u)$ in normal form.

Application to the Wave Equation. In order to explain the meaning of (2.60) we first consider the one-dimensional wave equation with unit phase velocity $L(u) = u_{xx} - u_{yy} = 0$ which is a special case of (2.59) for $a = -c = 1$, $b = d = 0$ (y plays the role of time). The characteristic equation (2.35) becomes $z^2 = 1$ whose roots are $z = \pm 1$. The characteristics are $x - t = \xi$, $x + t = \eta$ (where ξ, η are the constants of integration). This gives $\phi(x, y) = x - y$, $\psi(x, y) = x + y$, so that (2.60) becomes

$$\phi(x, y) = \xi, \qquad \psi(x, y) = \eta. \tag{2.61}$$

When we studied the general solution of the wave equation in Section 2.1 we introduced (ξ, η) as the *characteristic coordinates* defined by (2.8), which are those given by (2.61) for the phase velocity $c = 1$. Using the chain rule in applying (2.61) to $u_{xx} - u_{yy}$ (as shown in Section 2.1), we have

$$u_{\xi\eta} = 0, \tag{2.62}$$

which is the same as (2.9). Equation (2.62) is the *normal form* of the one-dimensional wave equation.

We recall that, in general, the transformations (2.60) have the following properties for the hyperbolic case: The first equation of (2.60) is a one-parameter (ξ) family of solutions of the characteristic equation (2.35), given by the root $dy/dx = z_+$. The second equation is a one-parameter (η) family of solutions given by $dy/dx = z_-$. This means that if ξ is assigned various fixed values and η is varied continuously, a family of C_+ characteristics $\phi(x, y) = \xi$ is generated, one curve for each value of ξ. Alternatively, if η is given various values and ξ is varied, a family of C_- characteristics $\psi(x, y) = \eta$ is generated. The two-parameter family of $C_\pm$ characteristics span the solution domain in (x, y) space. The transformations given by (2.60) allow us to map these families of characteristics $C_\pm$ from (x, y) space into (ξ, η) space. The form of the characteristics are much simpler in the transformed space, forming a rectangular grid. In the (ξ, η), or *characteristic coordinate system*, ξ is the abscissa and η the ordinate, so that the C_+ curves map into vertical lines, and the C_- curves map into horizontal lines. Therefore, for any real transformation given by the functions $\phi(x, y)$ and $\psi(x, y)$, which have a nonzero Jacobian so that there is a unique inverse transformation, $C_\pm$ curves in the (x, y) plane are transformed into a rectangular grid in the *characteristic plane* (ξ, η). For example, for the wave equation with phase velocity c, the $C_\pm$ characteristics in the (x, y) plane, where y is the ordinate and x is the abscissa, have slopes $dy/dx = \pm 1/c$. These characteristics are transformed into a rectangular grid in the (ξ, η) plane.

We therefore see why the transformation (2.60), that maps (2.59) into a PDE in the characteristic coordinates (ξ, η), is called a mapping into the "normal form". For this normal form of the PDE is the one where the two families of characteristics form a rectangular grid in (ξ, η) space.

We now transform (2.59) from (x, y) to (ξ, η) space by using the mapping given by (2.60). It is only necessary to transform the principal part given by (2.58) from the (x, y) to the characteristic coordinates (ξ, η). It is not necessary to transform the terms in d in (2.59), since they do not involve the principle part and therefore do not contribute to the characteristic equations. Using the transformation (2.60) and expanding $u_x, \ldots, u_{yy}$ by the chain rule, the second derivative term transform into:

$$u_{xx} = u_{\xi\xi}(\phi_x)^2 + 2u_{\xi\eta}\phi_x\psi_x + u_{\eta\eta}(\psi_x)^2 + \cdots,$$

$$u_{xy} = u_{\xi\xi}\phi_x\phi_y + u_{\xi\eta}(\phi_x\psi_y + \phi_y\psi_x) + u_{\eta\eta}\psi_x\psi_y + \cdots,$$

$$u_{yy} = u_{\xi\xi}(\phi_y)^2 + 2u_{\xi\eta}\phi_y\psi_y + u_{\eta\eta}(\psi_y)^2 + \cdots,$$

where the dots mean terms in which no second-order derivatives of u appear. Then the differential operator $L(u)$ given by (2.58) takes the form

$$\Lambda(u) = Au_{\xi\xi} + 2Bu_{\xi\eta} + Cu_{\eta\eta}, \tag{2.63}$$

where

$$\begin{aligned}
A &= a(\phi_x)^2 + 2b\phi_x\phi_y + c(\phi_y)^2 = Q(\phi_x, \phi_y),\\
B &= a\phi_x\psi_x + b(\phi_x\psi_y + \phi_y\psi_x) + c\phi_y\psi_y,\\
C &= a(\psi_x)^2 + 2b\psi_x\psi_y + c(\psi_y)^2 = Q(\psi_x, \psi_y).
\end{aligned} \tag{2.64}$$

An important observation about the form of A and C is that A is the same quadratic form Q in (ϕ_x, ϕ_y) as C is in (ψ_x, ψ_y). The significance of this statement will be seen below. B is a mixed quadratic form in $(\phi_x, \phi_y, \psi_x, \psi_y)$. The normal form of the quasilinear equation (2.59) becomes

$$Au_{\xi\xi} + 2Bu_{\xi\eta} + Cu_{\eta\eta} + D = 0, \tag{2.65}$$

where D involves terms (not necessarily linear) up to u_ξ and u_η. It is clear that the characteristic equation associated with (2.65) is

$$Q(Z) = AZ^2 - 2BZ + C = 0, \qquad Z = \frac{d\eta}{d\xi}, \tag{2.66}$$

where $Q(Z)$ stands for the quadratic expression on the left-hand side. We recall that the characteristic equation for (2.59) is

$$Q(z) = az^2 - 2bz + c = 0.$$

Thus we see that $Q(Z)$ has the same form, with respect to the coefficients A, B, C, that $Q(z)$ has with respect to a, b, c. In this sense, we say that the characteristic equation is *invariant* with respect to the transformation (2.60).

As mentioned, for the hyperbolic case the transformations given by (2.60) are real, since we have real characteristics. We have $d\phi = 0$ on a C_+ characteristic (on which ξ is constant). Similarly, $d\psi = 0$ on a C_- (where η is constant). This gives

$$\phi_x + \left(\frac{dy}{dx}\right)\phi_y = 0 \quad \text{or} \quad \frac{dy}{dx} = z_+ = -\frac{\phi_x}{\phi_y}, \qquad \xi = \text{const.: } C_+,$$

$$\psi_x + \left(\frac{dy}{dx}\right)\psi_y = 0 \quad \text{or} \quad \frac{dy}{dx} = z_- = -\frac{\psi_x}{\psi_y}, \qquad \eta = \text{const.: } C_-. \tag{2.67}$$

For the hyperbolic case the roots z_+ of the characteristic equation (2.35) are real and unequal. The system (2.67) is of the form $dy/dx = -f_x/f_y$, where $f = \phi$ for $dy/dx = z_+$ and $f = \psi$ for $dy/dx = z_-$. Inserting this expression for z into (2.35) yields

$$a(f_x)^2 + 2bf_xf_y + c(f_y)^2 = Q(f_x, f_y) = 0. \tag{2.68}$$

Comparing the quadratic function Q on the left-hand side of (2.68) with that for A and C in (2.64), we note that they are the same quadratic functions of their respective arguments. This gives us the important result

$$A = Q(\phi_x, \phi_y) = 0, \qquad B = Q(\psi_x, \psi_y) = 0, \qquad (2.69)$$

for the hyperbolic case. Then (2.63) becomes

$$\Lambda(u) = 2Bu_{\xi\eta}, \qquad (2.70)$$

which is the normal form for the principal part of the second-order equation (2.59) for the hyperbolic case. The normal form for the hyperbolic quasilinear equation (2.59) becomes

$$2Bu_{\xi\eta} + D = 0. \qquad (2.71)$$

For the homogeneous wave equation, (2.71) reduces to (2.62).

It is of interest to consider the other two types of second-order PDEs.

For the *elliptic case* we have $b^2 - ac < 0$ and $B^2 - AC < 0$, so that the roots z of the characteristic equation for (2.59) and also the roots Z of the characteristic equation for (2.65) are complex conjugates. To prove this statement we state the result of an algebraic calculation (which we do not reproduce here) which yields the following relation between (a, b, c) and (A, B, C):

$$(AC - B^2) = (ac - b^2)(\phi_x\psi_y - \phi_y\psi_x)^2. \qquad (2.72)$$

This shows that the discriminants $ac - b^2$ and $AC - B^2$ have the same sign, which means that the elliptic, hyperbolic, or parabolic character of the roots z are the same as those for Z. It is sufficient to set

$$A = C, \qquad B = 0,$$

for the elliptic case. This gives the normal form for the elliptic second-order equation as

$$\Lambda(u) + \cdots = u_{\xi\xi} + u_{\eta\eta} + \cdots = 0.$$

For the *parabolic case* it is sufficient to get

$$B = C = 0.$$

The normal form for the second-order equation then becomes

$$\Lambda(u) + \cdots = u_{\xi\eta} + \cdots = 0.$$

To complete this description we can give another normal form for the *hyperbolic* equation by setting

$$A = -C, \qquad B = 0.$$

This gives

$$\Lambda(u) = u_{\xi\xi} - u_{\eta\eta} + \cdots = 0.$$

EXAMPLES. As our first example we consider a second-order linear equation with variable coefficients, namely

$$xyu_{xx} + (x^2 - y^2)u_{xy} - xyu_{yy} = 0.$$

The characteristic equation is

$$xyz^2 - (x^2 - y^2)z - xy = 0, \qquad z = \frac{dy}{dx}.$$

The roots are

$$z_{1,2} = \frac{x}{y}, \; -\frac{y}{x}.$$

By solving these ODEs we obtain the transformations

$$\phi(x, y) = x^2 - y^2 = \xi, \qquad \psi(x, y) = xy = \eta.$$

Since the roots are real and unequal the PDE is hyperbolic. Incidentally, this two-parameter family of characteristic curves form orthogonal trajectories in the (x, y) plane. This property is due to the transformations $\phi(x, y)$ and $\psi(x, y)$ and is clearly not an intrinsic property of characteristic curves in general. We now calculate A, B, and C from (2.64) and obtain $A = C = 0$, $B = -4\eta\xi$, so that the normal form for our PDE becomes

$$\Lambda(u) + d = -4\eta\xi u_{\xi\eta} + d = 0.$$

If $d = 0$ then the normal form is $u_{\xi\eta} = 0$, which is the same as for the one-dimensional wave equation with constant phase velocity.

The next *example* is an example of a "mixed type". It is of different type in different domains, being elliptic for $x > 0$, parabolic for $x = 0$, and hyperbolic for $x < 0$. The equation is

$$u_{xx} + xu_{yy} = 0.$$

This is known as "Tricomi's" equation; it is of interest in fluid mechanics, especially in the transitional or transonic zone which separates supersonic from subsonic flow. In this case $a = 1$, $b = 0$, $c = x$. The characteristic equation is

$$z^2 + x = 0.$$

The roots are

$$z_+ = \begin{cases} \pm\sqrt{-x}, & \text{for} \quad x < 0, \\ \pm i\sqrt{-x}, & \text{for} \quad x > 0, \\ 0, & \text{for} \quad x = 0. \end{cases}$$

The character of these roots tells us that Tricomi's equation is hyperbolic for negative x, elliptic for positive x, and parabolic for $x = 0$, which separates these two domains. Fluid mechanics tells us that the flow is supersonic for $x < 0$, transonic for $x = 0$, and subsonic for $x > 0$.

For the half-plane $x < 0$ we obtain the transformation into characteristic coordinates by integrating $dy/dx = \pm\sqrt{-x}$. We obtain

$$\phi(x, y) = \tfrac{3}{2}y + (-x)^{3/2} = \xi,$$

$$\psi(x, y) = \tfrac{3}{2}y - (-x)^{3/2} = \eta.$$

Inserting this transformation into (2.64) we transform Tricomi's equation into the normal form

$$u_{\xi\eta} - (\tfrac{1}{6}(\xi - \eta))(u_\xi - u_\eta) = 0 \qquad \text{for} \quad \xi > \eta.$$

The characteristic curves are the cubical parabolas

$$y = \mp\tfrac{2}{3}\sqrt{(-x)^3} + c \qquad \text{for} \quad x < 0,$$

where the constant c specifies the curve. It is easily seen that the branches with a downward direction yield the curves $\phi = $ const. and those directed upward yield the curves $\psi = $ const.

For $x > 0$ the equation is elliptic so that there are no real characteristics. (ξ, η) are complex conjugates given by

$$\xi = \tfrac{3}{2}y - i\sqrt{x^3},$$
$$\eta = \tfrac{3}{2}y + i\sqrt{x^3}.$$

Setting

$$\alpha = \frac{\xi + \eta}{2} = \tfrac{3}{2}y,$$

$$\beta = \frac{\xi - \eta}{2i} = -\sqrt{x^3},$$

we obtain Tricomi's equation in the normal form

$$u_{\alpha\alpha} + u_{\beta\beta} + \left(\frac{1}{3\beta}\right)u_\beta = 0.$$

2.11. Riemann's Method

In this section we investigate a very important method (by Bernhard Riemann, the great German mathematician) of solving hyperbolic PDEs in two dimensions. The material we shall present was essentially given in [15, Chap. 3]; however, for the convenience of the reader, we summarize these ideas here. The modern theory of PDEs was initiated by Riemann's representation of the solution of Cauchy's problem. He developed his theory in the appendix to his classic paper on the fluid dynamics of compressible gases. The title, in English, is: "Propagation of Aerial Waves of Finite Amplitude".[†] Riemann's paper does not give an existence proof; but assuming the existence of a solution to the hyperbolic system that arises in compressible flow problems, he offers an elegant explicit integral representation of the solution in a form analogous to the solutions of boundary value problems for elliptic equations by the use of

[†] "Uber die Fortpflanzung ebener Luftwellen von endlicher Schwingungsweite", *Abhandl. Konigl. Ges. Wiss., Gottingen*, Vol. 8 (1860).

Green's function. In order to get an appreciation of Riemann's method, especially his *Riemann function* we need to review some mathematical concepts concerning the divergence theorem, Green's identity, adjoint operators. etc.

Divergence Theorem and Green's Identity

We first review the divergence theorem and therefrom develop Green's identity. For the sake of simplicity it is only necessary to consider the two-dimensional case. Roughly speaking, the main idea of this important theorem is that it transforms the volume integral of the divergence of a vector in a given region to the integral over the bounding surface of the normal component of that vector. Specifically, let $\mathbf{V}(x, y)$ be a vector, let $\mathbf{i}, \mathbf{j}$ be unit vectors in the (x, y) directions, and let P, Q be the (x, y) components of $\mathbf{V}$. Then

$$\mathbf{V} = \mathbf{i}P(x, y) + \mathbf{j}Q(x, y).$$

Let $\mathscr{R}$ be a simply connected region in the (x, y) plane bounded by a closed curve C, and let $\mathbf{n}$ be the unit outward normal along C. Then the divergence theorem, which was developed by Gauss, can be formulated as

$$\iint_{\mathscr{R}} \nabla \cdot \mathbf{V} \, dA = \oint_C \mathbf{V} \cdot \mathbf{n} \, ds, \quad \nabla = \mathbf{i}\frac{\partial}{\partial x} + \mathbf{j}\frac{\partial}{\partial y}, \quad \mathbf{n} = \left(\frac{dy}{ds}\right)\mathbf{i} - \left(\frac{dx}{ds}\right)\mathbf{j},$$

$$(2.73)$$

where dA is the element of area, ds is the element of arc length of C, and dy/ds, $-dx/ds$, are the direction cosines of $\mathbf{n}$. (2.73) becomes

$$\iint_{\mathscr{R}} (P_x + Q_y) \, dx \, dy = \oint_C (P \, dy - Q \, dx). \quad (2.74)$$

This is another form of the divergence theorem.

We now develop Green's identity. To this end, we consider the scalar functions $u(x, y)$, $v(xy)$ with continuous second derivatives. The two-dimensional gradient operator is $\mathbf{grad} = \nabla = \mathbf{i}\,\partial/\partial x + \mathbf{j}\,\partial/\partial y$. Choose a vector $\mathbf{V}$ such that $\mathbf{V} = u\nabla v + v\nabla u$. Using the vector identity

$$\nabla \cdot u\nabla v = u\nabla^2 v + (\nabla u) \cdot (\nabla v),$$

in (2.73) we obtain

$$\iint_{\mathscr{R}} [u\nabla^2 v + (\nabla u) \cdot (\nabla v)] \, dA = \oint_C \mathbf{n} \cdot u\nabla v \, ds, \quad (2.75)$$

which is sometimes called the *first form of Green's identity*. To obtain a more symmetric form we interchange u and v in (2.75) and subtract the result from (2.75). We express the result as

$$\iint_{\mathscr{R}} [v\nabla^2 u - u\nabla^2 v] \, dA = \oint_C \left[v\frac{\partial u}{\partial n} - u\frac{\partial v}{\partial n} \right] ds, \quad (2.76)$$

where the left-hand side involves integration over the region and the right-hand side involves integration over the closed curve C bounding $\mathcal{R}$. Comparing (2.76) with the divergence theorem (2.73), we want to find a vector $\mathbf{V}$ such that the integrand of the left-hand side of (2.76) equals div($\mathbf{V}$) or $\nabla \cdot \mathbf{V}$. This is easily obtained by writing out this integrand in extended form, obtaining the following identity:

$$v(u_{xx} + u_{yy}) - u(v_{xx} + v_{yy}) = (vu_x - uv_x)_x + (vu_y - uv_y)_y = P_x + Q_y. \quad (2.77)$$

We have thus found a vector $\mathbf{V} = \mathbf{i}P + \mathbf{j}Q$, where

$$P = vu_x - uv_x, \qquad Q = vu_y - uv_y.$$

This yields

$$v\nabla^2 u - u\nabla^2 v = \nabla \cdot \mathbf{V} = P_x + Q_y, \quad (2.78)$$

and (2.76) can thus be identified with the divergence theorem given in the form given by (2.74).

In general, for any second-order linear operator L, we may write the second form of Green's identity as

$$\iint_{\mathcal{R}} [vL(u) - uL(v)]\, dA = \oint_C \left[v\frac{\partial u}{\partial n} - u\frac{\partial v}{\partial n} \right] ds. \quad (2.79)$$

For $L(u) = u_{xx} + u_{yy}$, $L(v) = v_{xx} + v_{yy}$, (2.79) reduces to (2.76).

Adjoint operators

In the above case of the Laplacian operator, $L(u) = L(v)$. However, the Laplacian is not the only second-order operator. For a general second-order linear differential operator it is not necessarily true that $L(v) = L(u)$ in order for the integrand $vL(u) - uL(v)$ in the left-hand side of (2.79) to be of the form $\nabla \cdot \mathbf{V}$. Then $L(v)$ is replaced by a more general second-order differential operator $L^*(v)$ such that the integrand is of the divergence form. $L^*(v)$ is called the *adjoint operator*. It is the operator on v corresponding to $L(u)$ such that $vL(u) - uL^*(v)$ is of the form $\nabla \cdot \mathbf{V}$. If $L(u)$ is such that $L^*(v) = L(u)$, then L is called a *self-adjoint operator*. If $L(u)$ is such that this equality does not hold, then L is a *non-self-adjoint operator*. As indicated above, $L^*(v)$ has the property that the following equation is satisfied:

$$\iint_{\mathcal{R}} [vL(u) - uL^*(v)]\, dA = \oint_C \left[v\frac{\partial u}{\partial n} - u\frac{\partial v}{\partial n} \right] ds. \quad (2.80)$$

This means that (2.80) is actually the divergence integral given by (2.73) or (2.74).

The important point is this: If we can construct $L^*(v)$ that depends on $L(u)$, then we can integrate along the boundary C surrounding $\mathcal{R}$. The function $v(x, y)$ which has the property that $L^*(v)$ is the adjoint of $L(u)$ is called *Riemann's function*. Its purpose is to allow us to determine the solution $u(x, y)$

to the hyperbolic quasilinear equation. Indeed, this is the crux of Riemann's method, i.e., to construct the Riemann function for a specific problem that has the required properties in order to solve for $u(x, y)$. Later on we show how this is done.

We now give an example of a non-self-adjoint operator. It is the one-dimensional unsteady Fourier heat transfer operator L operating on u given by

$$L(u) = u_{xx} - u_y,$$

where x is the spatial coordinate and y represents time. We assume that $L^*(v)$ takes the form

$$L^*(v) = v_{xx} + (\alpha v)_x + (\beta v)_y,$$

where α and β are functions of x and y to be determined. We want to find the components $P(x, y)$ and $Q(x, y)$ of $\mathbf{V}$ such that $vL(u) - uL^*(v) = P_x + Q_y$, or

$$v(u_{xx} - u_y) - u[v_{xx} + (\alpha v)_x + (\beta v)_y] = P_x + Q_y,$$

which gives

$$P = vu_x - uv_x, \qquad Q = -uv, \qquad \alpha = 0, \qquad \beta = 1,$$

yielding

$$L^*(v) = v_{xx} + v_y.$$

Physically, the adjoint expression $L^*(v)$ for the heat transfer operator has the rather curious interpretation: If $L(u) = u_{xx} - u_y = 0$ is the one-dimensional unsteady heat conduction equation for the temperature $u(x, y)$, then the *adjoint problem* is the solution of the equation $L^*(v) = v_{xx} + v_y = 0$, where the "temperature" $v(x, y)$ is the solution of the heat conduction equation with time "running backwards". The author has investigated and successfully used this method in the more general study of unsteady thermal stresses, where $v(x, y)$ plays the role of a Green's function.

The general linear second-order hyperbolic equation in normal form (2.71) can be rewritten as

$$L(u) = u_{xy} + au_x + bu_y + cu = f, \tag{2.81}$$

where a, b, c, f are known functions of the characteristic coordinates x, y. For the case $a = b = c = 0$, (2.81) reduces to the nonhomogeneous wave equation and it is easily seen that $L(u)$ is self-adjoint. For the general case of (2.81), $L(u)$ is not self-adjoint and we must construct the adjoint $L^*(v)$. To do so, we assume the following expression:

$$L^*(v) = v_{xy} + (\alpha v)_x + (\beta v)_y + cv. \tag{2.82}$$

Our task is now to determine P and Q such that

$$vL(u) - uL^*(v) = v(u_{xy} + au_x + bu_y + cu) - u[v_{xy} + (\alpha v)_x + (\beta v)_y + cv]$$

$$= (-uv_y + auv)_x + (vu_x + buv)_y = P_x + Q_y. \tag{2.83}$$

This gives

$$P = -uv_y + auv, \qquad Q = vu_x + buv, \qquad \alpha = -a, \qquad \beta = -b,$$

so that

$$L^*(v) = -(av)_x - (bv)_y + cv. \tag{2.84}$$

We now consider the first-order linear operator given by

$$L(u) = au_x + bu_y. \tag{2.85}$$

It is easily seen that the adjoint is

$$L^*(v) = -(av)_x - (bv)_y. \tag{2.86}$$

Riemann's Method of Integration

We are now in a position to pursue the main task of this section, i.e., to investigate Riemann's method of finding solutions to the Cauchy problem by making use of the Riemann function. The Cauchy problem is formulated as follows: Find the solution to the linear hyperbolic equation in characteristic coordinates given by (2.81) with the initial data (u, p, q) prescribed on the initial value curve Γ. The geometric interpretation is given by Fig. 2.11. The initial value curve Γ: $\eta = \xi$ is shown in the characteristic or (ξ, η) plane. We consider the points A and B on Γ. From A the characteristic line segment $y = \eta = \text{const.}$ is drawn into the solution domain, and similarly the line segment $x = \xi = \text{const.}$ is drawn from B. Their intersection at the field point P has coordinates (ξ_0, η_0). The straight line characteristic AP is represented by varying x and fixing $y = \eta_0$, and the characteristic BP by varying y and fixing $x = \xi_0$. The

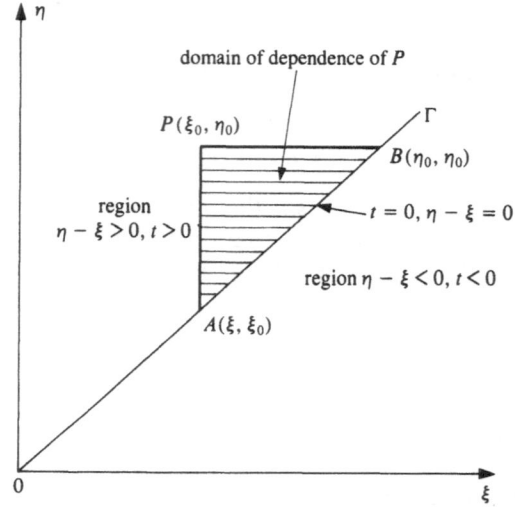

Fig. 2.11. Domain of dependence in the characteristic plane.

domain of dependence of P is the triangular region D bounded by the characteristics AP and BP and the *range of influence* which is the segment $[A, B]$ on Γ. The problem is to determine u_P from the initial data on $[A, B]$.

The essence of Riemann's method is to make use of the mathematical machinery described above. Specifically, we integrate the expression $vL(u) - uL^*(v)$ over the domain of dependence D, transform the integral by Green's identity so that u appears as a factor of the integrand, and attempt to determine the Riemann function $v(x, y)$ to satisfy the conditions of the problem. Using the divergence theorem in the form (2.74) we obtain

$$\iint_D [vL(u) - uL^*(v)] \, dx \, dy = \oint_C [(-uv_y + auv) \, dy - (vu_x + buv)] \, dx,$$

$$(2.87)$$

where use was made of (2.83). Referring to Fig. 2.12, the closed curve C is the path $AB + BP + PA$ bounding the region D. We shall define the Riemann function v to be the solution of

$$L^*(v) = 0, \qquad (2.88)$$

with additional conditions on v given by (2.90) below. These conditions arise from constructing v in such a way that the line integral in (2.87) will only involve the path AB on Γ, which means that this line integral only depends on the initial data.

Since $dx = 0$ on BP, $dy = 0$ on PA, and $L(u) = f$, (2.87) becomes

$$\iint_D vf \, dx \, dy = \oint_{AB} [(-uv_y + auv) \, dy - (vu_x - buv) \, dx]$$

$$+ \int_{BP} (-uv_y + auv) \, dy + \int_{PA} (vu_x + buv) \, dx.$$

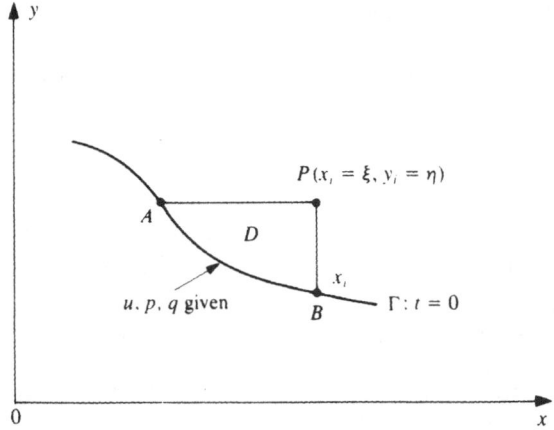

Fig. 2.12. Riemann's method.

Upon integrating the line integral along the path PA by parts in order to get vu at the end points, we obtain, after isolating $v_P u_P$:

$$v_P u_P = v_A u_A + \int_{AB} [v(u_x + bu)\, dx + u(v_y - av)\, dy] + \int_{BP} u(v_y - av)\, dy$$

$$- \int_{PA} u(v_x - bv)\, dx + \iint_D vf\, dx\, dy. \tag{2.89}$$

As indicated above we want u_P to depend on the integral of the region D and the initial data on Γ. This means we demand that the line integrals along the characteristics BP and PA vanish. v is partially defined by (2.88). We therefore complete its definition by imposing the following conditions compatible with these restrictions on the line integrals:

$$v_y - av = 0 \quad \text{on } BP \qquad \text{where} \quad x = \xi,$$

$$v_x - bv = 0 \quad \text{on } PA \qquad \text{where} \quad y = \eta, \tag{2.90}$$

$$v_P = 1 \quad \text{(for convenience).}$$

Inserting (2.90) into (2.89) yields

$$u_P = u_A v_A + \int_{AB} [v(u_x + bu)\, dx + u(v_y - av)\, dy + \iint_D vf\, dx\, dy. \tag{2.91}$$

We want a more symmetric form in the sense that it would also involve $v_B u_B$. To this end, we integrate the line integral in (2.91) by parts upon recognizing that $vu_x\, dx + uv_y\, dy = d(uv) - uv_x\, dx - vu_y\, dy$. Using this expression, (2.91) becomes

$$u_P = v_B u_B - \int_{AB} [u(v_x - bv)\, dx + v(u_y + au)\, dy] + \iint_D vf\, dx\, dy. \tag{2.92}$$

We then add (2.91) and (2.92) to obtain the desired symmetric expression for u_P

$$u_P = u(\xi, \eta) = \tfrac{1}{2}(v_A u_A + v_B u_B)$$

$$+ \frac{1}{2} \int_{AB} [(vp - uv_x + 2buv)\, dx + (uv_y - vq - 2auv)\, dy]$$

$$+ \iint_D vf\, dx\, dy. \tag{2.93}$$

(Recall that $p = u_x$, $q = u_y$.) Equation (2.93) is the solution for u_P in terms of the Riemann function v in the domain D and on the range of influence $[A, B]$. For the homogeneous case the forcing function $f = 0$. Then u_P can be obtained by integrating the data (u, p, q) and v along the range of influence. Otherwise, we have the additional integral of vf in D.

It remains to solve for v from (2.88) and (2.90). We observe that v is a continuous differentiable function of two sets of variables: (x, y), the coordinates of Γ; and (ξ, η), the coordinates of the field point P. We therefore rewrite the defining equations of v as

$$L_{(x, y)}[v(x, y; \xi, \eta) = 0,$$

$$v_y(\xi, y; \xi, \eta) - a(\xi, y)v(\xi, y; \xi, \eta) = 0 \quad \text{on } BP,$$

$$v_x(x, \eta; \xi, \eta) - b(x, \eta)v(x, \eta; \xi, \eta) = 0 \quad \text{on } PA,$$

$$v(\xi, \eta; \xi, \eta) = 1 \quad \text{at } P,$$

$$(2.94)$$

where the notation $L_{(x, y)}$ means differentiation with respect to (x, y). The second and third equations of (2.94) are ODEs along the respective characteristics. Their solutions are

$$v(\xi, y; \xi, \eta) = \exp\left(\int_\eta^y a(\lambda, \xi)\right) d\lambda \quad \text{along } BP,$$

$$v(x, \eta; \xi, \eta) = \exp\left(\int_\xi^x b(\lambda, \eta)\right) d\lambda \quad \text{along } PA.$$

$$(2.95)$$

In summary, the system (2.93) and (2.95) allows us to solve the Cauchy initial value problem in the characteristic coordinates (x, y) in terms of the Riemann function $v(x, y; \xi, \eta)$.

EXAMPLE. $L(u) = u_{xy} = 0$, the one-dimensional homogeneous wave equation in characteristic coordinates. It is clear that $L(u)$ is self-adjoint so that $L^*(v) = v_{xy} = 0$. Then v has the form

$$v(x, y; \xi, \eta) = F(x; \xi, \eta) + G(y; \xi, \eta).$$

However, we see from the defining equations for v that

$$v(x, y; \xi, \eta) = 1$$

for the one-dimensional wave equation. Then the solution for u in characteristic coordinates becomes

$$u(\xi, \eta) = \tfrac{1}{2}(u_A + u_B) + \frac{1}{2}\int_A^B (u_x \, dx - v_y \, dy). \qquad (2.96)$$

Equation (2.96) is the representation of the solution of the wave equation in characteristic coordinates. We can reformulate this solution to yield D'Alembert's solution in physical coordinates. To this end, we introduce the physical coordinates X-space and T-time, related to the characteristic coordinates (x, y) by the following transformation:

$$X = x + y, \qquad T = x - y. \qquad (2.97)$$

The inverse transformation is

$$x = \tfrac{1}{2}(X + T), \qquad y = \tfrac{1}{2}(X - T). \qquad (2.97')$$

Equation (2.97) transforms the wave equation from characteristic coordinates to physical coordinates, which is

$$u_{XX} - u_{TT} = 0. \tag{2.98}$$

The transformation (2.97) transforms (2.96) into

$$u_P = u(X, T) = \tfrac{1}{2}(u_A + u_B)$$

$$+ \frac{1}{4} \int_A^B [(u_T + u_X)(dX + dT) - (u_X - u_T)(dX - dT)]$$

$$= \tfrac{1}{2}(u_A + u_B) + \frac{1}{2} \int_A^B (u_T \, dX + u_X \, dT). \tag{2.99}$$

Suppose the initial curve Γ is on the X axis so that $T = 0$. Then the range of influence in (X, T) space is the interval $[A, B]$ where A, B have the coordinates $(X - T, 0)$, $(X + T, 0)$, and (2.99) becomes

$$u(X, T) = \tfrac{1}{2}(u(X - T, 0) + u(X + T, 0) + \frac{1}{2} \int_{X-T}^{X+T} u_T(\lambda, 0) \, d\lambda. \tag{2.100}$$

Equation (2.100) is D'Alembert's solution given by (2.13) where the initial conditions are $u(X, 0) = f(X)$ and $u_T(X, 0) = g(X)$. The general solution of (2.98) can also be written as

$$u(X, T) = F(X - T) + G(X + T), \tag{2.101}$$

where the solution is given in terms of traveling waves. F and G are arbitrary functions of their arguments and are obtained by prescribing initial conditions on u and u_T. $F(X - T)$ represents a progressing wave and $G(X + T)$ a regressing wave.

2.12. Nonlinear Hyperbolic Equations in Two Independent Variables

Introduction

Up to now we have treated the mathematical theory of one-dimensional wave propagation from the point of view of quasilinear hyperbolic PDEs in two variables. In this section we extend the treatment to more general equations of the form

$$F(x, y, p, q) = 0, \tag{2.102}$$

where $u_x = p$, $u_y = q$. In Section 2.3, (2.102) was specialized to be quasilinear. We saw that this means that (2.102) has the property that p and q occur linearly, where the coefficients are known functions of x, y, and u. In this section p and q occur nonlinearly meaning that their coefficients may also depend on p and q. We see that this is a more general type of nonlinearity. An example is

$$(\mathbf{grad} \ u)^2 = p^2 + q^2 = 1.$$

This is the PDE of straight line rays in the (x, y) plane that occurs in geometric optics.

As for the quasilinear case, characteristic theory for the fully nonlinear case will be developed in this section. We shall see that the theory reduces to the method of characteristics for a quasilinear system, which was already treated.

Geometric Considerations

We interpret (2.102) geometrically as a generalization of the geometric interpretation of the quasilinear system given in Section 2.3 where the Monge axis and pencil were introduced. We shall see that, for the fully nonlinear case, the Monge pencil is generalized to a Monge cone. We first observe that, by solving (2.102) for $u(x, y)$, we obtain a family of integral surfaces in (x, y, u) space. We also know that p and q are the direction coefficients of the tangent plane to an integral surface, and that they must satisfy the equation $F = 0$ at every field point P: (x, y, u). Since this equation is not linear in p and q, in general, the tangent planes to an integral surface do not form a pencil of planes through the Monge axis (as they do for the quasilinear case), but they form a one-parameter family enveloping a conical surface with P as its vertex. This conical surface is called the *Monge cone*. (As stated, it is a generalization of the Monge pencil for a fully nonlinear system.) Each field point P in the domain of the solution is the vertex of a Monge cone. We therefore want to determine the field of cones which form a dense set in the solution domain. Therefore the problem of integrating $F = 0$ for $u(x, y)$ or finding the one parameter of integral surfaces consists in that of finding surfaces which fit this field of cones, i.e., surfaces which are tangent to the corresponding cones at each field point. We point out that the generator of the Monge cone (along which the integral surface is tangent) is a characteristic. In this fully nonlinear case there is a whole cone of characteristics through each field point.

Since the problem of solving (2.102) for $u(x, y)$ is reduced to the problem of finding the Monge cone through each field point, our task now is to find the generators of the Monge cone. This will yield the characteristic field.

Since the Monge cone is the envelope of a one-parameter family of tangent planes through each field point, it would not be amiss at this point to give an example of how to find an envelope. Suppose $F(x, y, p, q)$ is such that its solution is represented by

$$f(x, y, u, s) = x^2 + y^2 + (u - s)^2 - 1 = 0.$$

This equation defines a one-parameter family of integral surfaces in (x, y, u) space, namely, a family of spheres of unit radius whose centers are on the u axis. Each integral curve is defined by a value of the parameter s. It is obvious, intuitively, that the envelope of this family of spheres is a cylinder of unit radius whose axis is the u axis. To obtain this envelope analytically we eliminate s between $f = 0$ and $f_s = 0$. The latter equation yields $u - s = 0$. Eliminating s yields the equation of the envelope which is the cylindrical surface $x^2 + y^2 - 1 = 0$, which is our result.

We now return to the general problem of finding the generator of the Monge cone. Consider a particular field point P_0: (x, y, u). The equation of the tangent plane to the integral surface $u(x, y)$ at P is

$$p(\xi - x) + q(\eta - y) = \zeta - u, \qquad (2.103)$$

where (ξ, η, ζ) are the running coordinates of the tangent plane. p and q are connected by $F = 0$ (given by (2.102)) from which we can eliminate q and thus obtain $q = q(p)$ which we insert in (2.103). Since (2.103) depends on the parameter p we differentiate this equation partially with respect to p and obtain

$$(\xi - x) + \frac{dq}{dp}(\eta - y) = 0.$$

Since the coordinates of the tangent plane (ξ, η) are near P we replace $\xi - x$, $\eta - y$ by dx, dy, since we are interested in what happens locally in the neighborhood of P ("in the small"). We therefore replace the above equation by

$$dx + \frac{dq}{dp} dy = 0. \qquad (2.104)$$

We also replace (2.103) by

$$p \, dx + q \, dy = du. \qquad (2.105)$$

Differentiating $F = 0$ with respect to p gives

$$F_p + \left(\frac{dq}{dp}\right) F_q = 0. \qquad (2.106)$$

Let (x, y, u) depend on the parameter t (which we may take as time). From (2.104) and (2.106) we get

$$\frac{dx}{F_p} = \frac{dy}{F_q} = dt,$$

which gives $dx/dt = F_p$, $dy/dt = F_q$. Using these equations in (2.105) yields $du/dt = pF_p + qF_q$. Summarizing, we get the following system of equations for the generators of the Monge cone:

$$\frac{dx}{dt} = F_p, \qquad \frac{dy}{dt} = F_q, \qquad \frac{du}{dt} = pF_p + qF_q. \qquad (2.107)$$

These are the ODEs whose solutions give the family of characteristic curves.

Let us apply (2.107) to the quasilinear case. Set $F = ap + bq - cu - d = 0$, which is (2.16). Then (2.107) becomes

$$\frac{dx}{dt} = F_p = a, \qquad \frac{dy}{dt} = F_q = b, \qquad \frac{du}{dt} = pF_p + qF_q = ap + bq = cu + d,$$

which are the characteristic equations given by (2.27) and (2.28). It is therefore clear that the system of three equations given by (2.107) is sufficient to yield the family of characteristic equations for the quasilinear case, since we have three unknowns (x, y, u) to be solved for as functions of t.

For the fully nonlinear case the system (2.107) depends on p and q as well as (x, y, u), since $F(x, y, u, p, q)$ is not linear in p and q. For example: Suppose $F = p^2 + q^2 - 1 = 0$ (which we used previously as an example in optics). Then $dx/dt = 2p$ and $dy/dt = 2q$, which is sufficient to show that the three equations of (2.107) are not sufficient to yield the characteristic curves. In fact, we have five unknowns: x, y, u, p, q, to be solved for as functions of t. We therefore need two more equations. These are obtained by expanding dp/dt and dq/dt and using the first two equations of (2.107). We get

$$\frac{dp}{dt} = p_x \left(\frac{dx}{dt}\right) + p_y \left(\frac{dy}{dt}\right) = p_x F_p + p_y F_y,$$

$$\frac{dq}{dt} = \left(\frac{dx}{dt}\right) + q_y \left(\frac{dy}{dt}\right) = q_x F_p + q_y F_q. \tag{2.108}$$

Differentiating $F = 0$ first with respect to x, then with respect to y, yields

$$F_x + pF_u + p_x F_p + q_x F_q = 0,$$

$$F_y + qF_u + p_y F_p + q_y F_q = 0. \tag{2.109}$$

Using the fact that $q_x = p_y$ and inserting (2.108) into (2.109) yields the two additional equations:

$$\frac{dp}{dt} = -(F_x + pF_u), \qquad \frac{dq}{dt} = -(F_y + qF_u). \tag{2.110}$$

For the convenience of the reader we restate the characteristic ODEs for the fully nonlinear case.

$$\frac{dx}{dt} = F_p, \qquad \frac{dy}{dt} = F_q, \qquad \frac{du}{dt} = pF_p + qF_q,$$

$$\frac{dp}{dt} = -(F_x + pF_u), \qquad \frac{dq}{dt} = -(F_y + qF_u). \tag{2.111}$$

We now give some examples.

EXAMPLE 1. We illustrate the use of (2.111) by considering the set of spheres of unit radius whose centers have the coordinates $(a, b, 0)$ in the (x, y) plane. We represent these spheres by the two-parameter family of surfaces in (x, y, u) space.

$$f(x, y, u, a, b) = (x - a)^2 + (y - b)^2 + u^2 - 1 = 0. \tag{2.112}$$

The two parameters are, of course (a, b). We now obtain the PDE whose solution is the above equation by first differentiating (2.112) with respect to x, then with respect to y, and then eliminating a and b. We get

$$x - a = -up, \qquad y - b = -uq.$$

Inserting these expressions into (2.112) eliminates a and b and gives the required equation, namely

$$F(u, p, q) = u^2(1 + p^2 + q^2) - 1 = 0. \tag{2.113}$$

Clearly (2.113) is fully nonlinear and, in this case, F does not depend on (x, y). Using (2.113) the characteristic equations (2.111) become

$$\frac{dx}{dt} = 2pu^2,$$

$$\frac{dy}{dt} = 2qu^2,$$

$$\frac{du}{dt} = 2(1 - u^2), \tag{2.114}$$

$$\frac{dp}{dt} = -\frac{2p}{u},$$

$$\frac{dp}{dt} = -\frac{2q}{u}.$$

Dividing the fifth equation of (2.114) into the fourth and integrating gives $p = kq$, where k is a constant. Multiplying the third equation by p, the fourth by u, and adding the results to the first gives $d(x + pu) = 0$. A similar manipulation on the second, third, and fifth gives $d(y + qu) = 0$. Also $q\,dp - p\,dq = q^2 d(p/q) = 0$ which again yields $p = kq$. We immediately get

$$x + pu = a, \qquad y + qu = b, \qquad p = kq.$$

Combining these expressions with (2.113) we get the equations

$$(x - a)^2 + (y - b)^2 + u^2 - 1 = 0, \qquad \frac{x - a}{y - b} = k. \tag{2.115}$$

The first equation of (2.115) is the family of unit spheres whose centers lie in the (x, y) plane. If we then single out those spheres whose centers are on a given curve represented by $b = g(a)$, then it reduces to a one-parameter family of surfaces. The second equation arises from the proportionality of p and q. Combining these two yields

$$(x - a)^2 + (y - g(a))^2 - 1 = 0. \tag{2.116}$$

The integrals of the characteristic equations tell us that the characteristic curves corresponding to the PDE (2.113) are the great circles of unit spheres parallel to the u axis. The envelope is therefore a tubular surface.

EXAMPLE 2 (Geometric Optics). We revisit the example in two-dimensional geometric optics given in the introduction to this section by reconsidering the fully nonlinear equation

$$p^2 + q^2 - 1 = 0, \tag{2.117}$$

which, we recall, is the PDE of straight line light rays in the (x, y) plane. We note that this example is most important in giving us a mathematical description of the properties of light rays vis-à-vis characteristic theory. The extension to three dimensions yields the more general PDE

$$\mathbf{grad}\, u(x, y, z) = n(x, y, z), \quad \text{or} \quad p^2 + q^2 + r^2 = n(x, y, z), \quad r = u_z, \tag{2.118}$$

where the index of refraction n in an inhomogeneous medium is a function of x, y, z. In three dimensions the surfaces $u(x, y, z) = $ const. represent the *wave fronts*; and in two dimensions, the curves $u(x, y) = $ const. (solutions of $p^2 + q^2 = 1$ for the case of unit index of refraction) represent the projection of these wave fronts in the (x, y) plane. The wave fronts are planar surfaces normal to the direction of wave propagation (the direction of the light rays). This fact stems from the nature of the solution of the three-dimensional PDE. The characteristics in three dimensions represent the *light rays*. In two dimensions, they represent the projection of these light rays in the (x, y) plane. This will be seen below for the two-dimensional case.

We return to the two-dimensional case and first show how the solution to (2.117) can be obtained by elementary means. This is most easily obtained by assuming that u is the sum of a function of x and a function of y, i.e.,

$$u(x, y) = f(x) + g(y).$$

Using this expression in the differential equation we get

$$[f'(x)]^2 = 1 - [g'(y)]^2.$$

Since the left-hand side is independent of y and the right-hand side is independent of x, both sides must be independent of both x and y or $f'(x) = $ const. $= a$. This yields the solution (the complete integral) in the form

$$u(x, y) = ax + \sqrt{(1 - a^2)}y + b, \tag{2.119}$$

where a, b are two parameters. Differentiating (2.119) with respect to a yields

$$0 = x - \left(\frac{a}{\sqrt{1 - a^2}} \right) y + c, \tag{2.120}$$

where c is another parameter. We also have the relations

$$u_x = p = a, \quad u_y = q = \sqrt{1 - a^2}. \tag{2.121}$$

It is obvious that the direction cosines of the tangent plane u_x and u_y are constant on each planar surface (2.119) defined by a particular value of the parameter p. Let the field point P_0: (x_0, y_0, u_0) be the vertex of the Monge cone. Then the parameters b in (2.119) and c in (2.120) can be defined by using the coordinates of P_0 in (x, y, u) space. Using (2.121), (2.119) and (2.120) become, respectively

$$u - u_0 = p(x - x_0) + q(y - y_0), \tag{2.122}$$

$$p(y - y_0) = q(x - x_0), \tag{2.123}$$

where

$$b = u_0 - px_0 - qy_0, \qquad c = \left(\frac{p}{q}\right)y_0 - x_0.$$

By using (2.117) and the expressions $p = (u - u_0)/(x - x_0)$, $q = (u - u_0)/(y - y_0)$, we obtain the following equation for the Monge cone whose vertex is at P_0:

$$(x - x_0)^2 + (y - y_0)^2 = (u - u_0)^2. \tag{2.124}$$

This cone has its axis parallel to the u axis (in (x, y, u) space). The generators, which lie on the surface of the cone, are the characteristic lines; and they form a $45°$ angle with the (x, y) plane. There is a given direction to each characteristic line which is determined by a given value of p and $q = \sqrt{1 - p^2}$.

The above results can be obtained by setting down and solving the characteristic ODEs (2.111). From $F = p^2 + q^2 - 1 = 0$ we get $F_p = 2p$, $F_q = 2q$, $F_x = F_y = F_u = 0$, so that (2.111) becomes

$$\frac{dx}{dt} = 2p, \qquad \frac{dy}{dt} = 2q, \qquad \frac{du}{dt} = 2, \qquad \frac{dp}{dt} = 0, \qquad \frac{dq}{dt} = 0.$$

From these characteristic equations and $F = p^2 + q^2 - 1 = 0$ we get

$$\frac{dy}{dx} = \frac{q}{p}, \qquad \frac{du}{dt} = \frac{1}{p},$$

whose integrals yield the required results. Furthermore, by eliminating p and q from the above equations and $p^2 + q^2 = 1$, we get the following differential equation for the functions $u(x)$ and $y(x)$:

$$\left(\frac{du}{dx}\right)^2 - \left(\frac{dy}{dx}\right)^2 = 1. \tag{2.125}$$

Equation (2.125) is called the *Monge equation*. It is a necessary consequence of the equation $F = 0$ and the characteristic equations. Its solutions are those curves whose tangents from a $45°$ angle with the (x, y) plane. These curves are sometimes called "caustics"; they are the involutes of the curves $u = $ const.

We now look at this problem from the point of view of the Cauchy initial value problem, which is defined as follows: Find the solution $u = u(x, y)$ which satisfies

$$F(p, q) = p^2 + q^2 - 1 = 0,$$

on Γ: $x(s, 0) = s$, $\qquad y(s, 0) = f(s)$, $\qquad t = 0$; $\qquad$ initial datum $u(s) = 0$.

We recall the parametric representation of the Cauchy problem: In the (x, y)

plane, t is the parameter that varies along a characteristic curve for which s is constant (defining that curve) and s varies along the initial value curve Γ on which $t = 0$. The initial datum is: $u = u(s) = 0$ on Γ. The characteristic equations were given above. The family of characteristic curves is $\Delta y/\Delta x = -p/q$, where $\Delta x = x - x_0$, $\Delta y = y - y_0$, and (x_0, y_0) are the coordinates of a point on Γ defined by a particular value of s. Since the slope of a characteristic equation that passes through (x_0, y_0) on Γ is $dy/dx = q/p$, it follows that the characteristic curve is a straight line which is orthogonal to Γ at (x_0, y_0). This means that there is a one-parameter family of parallel straight line characteristics of slope $dy/dx = \sqrt{(1 - p^2)}/p$ which intersects the initial value curve orthogonally. Figure 2.13 shows the geometry of this problem. Each characteristic curve C_s is defined by the equations $x(s, t) = pt + s$, $y(s, t) = qt + f(s)$. The family of characteristics is the set of straight lines that arise from Γ, are orthogonal to it, and enter the region $t > 0$. It is clear that for $t = 0$, we have $x(s, 0) = s$, $y(s, 0) = f(s)$.

From a physical point of view: the one-parameter family of characteristic curves represent the projection of the light rays in the (x, y) plane; and these trajectories (the paths of the light rays) are orthogonal to the wave fronts. The initial value curve is the projection of the initial position ($t = 0$) of the wave front in the (x, y) plane. As the time t increases, the wave fronts move with a unit velocity normal to the light rays. We recall that this is the case where the index of refraction $n(x, y) = 1$, where the medium is homogeneous. For a nonhomogeneous medium the index of refraction is a given nonconstant function of (x, y), so that the wave fronts move with a nonuniform velocity (the wave speed is spatially dependent).

A final note on our example. We may start with the following problem:

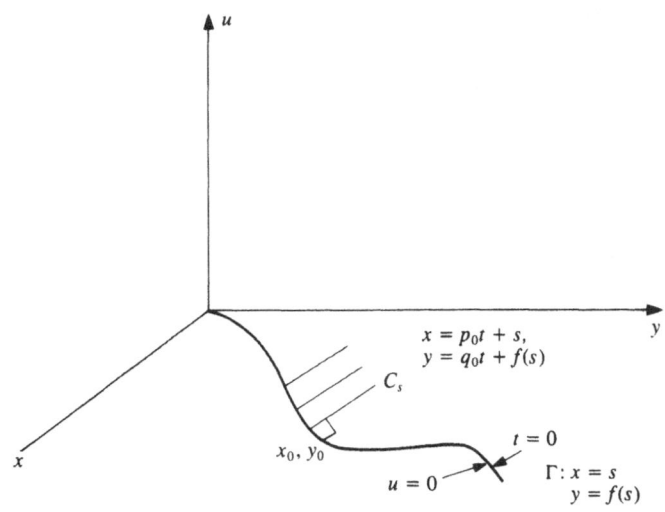

Fig. 2.13. Geometry of C_s characteristics in (x, y, u) space.

Find the orthogonal trajectories to a given curve $u = \text{const.}$ On this curve we have

$$\frac{du}{dt} = u_x \frac{dx}{dt} + u_y \frac{dy}{dt} = p \frac{dx}{dt} + q \frac{dy}{dt} = 0.$$

The orthogonal trajectories are characterized by the differential equations

$$\frac{dx}{dt} = p, \qquad \frac{dy}{dt} = q.$$

Squaring and adding these equations yields

$$\left(\frac{dx}{dt}\right)^2 + \left(\frac{dy}{dt}\right)^2 = 1,$$

which tells us that t represents the arc length along each of the orthogonal trajectories. From $dx/dt = u_x$ and $dy/dt = u_y$ we obtain

$$\frac{d^2x}{dt^2} = u_{xx}u_x + u_{xy}u_y = p_x p + q_x q = 0,$$

$$\frac{d^2y}{dt^2} = u_{xy}u_x + u_{yy}u_y = p_y p + q_y q = 0,$$

upon differentiating $p^2 + q^2 - 1 = 0$ with respect to x and y, respectively. This tells us that the trajectories are straight lines.

For the case of three independent variables we may consider the hyperspace (x, y, z, u). The same method shows that the solutions of the PDE (2.118) are given by a family of surfaces $u(x, y, z) = \text{const.}$ (embedded in this space) which are parallel to a prescribed initial surface $f(x, y, z) = 0$. These surfaces possess rectilinear orthogonal trajectories. The straight line characteristics which lie between the surfaces $u = c_1$ and $u = c_2$ have the constant length $c_2 - c_1$, and $u(x, y, z)$ is the distance of the field point (x, y, z) from the initial surface.

CHAPTER 3

Hyperbolic Partial Differential Equations in More Than Two Independent Variables

Introduction

In Chapter 2 we gave a treatment of the partial differential equations (PDEs) of hyperbolic type in two independent variables, as a mathematical description of electromagnetic wave propagation in the (x, t) plane. We first discussed the one-dimensional wave equation, then the theory of quasilinear hyperbolic equations in two independent variables, and finally the theory of fully non-linear equations in two variables. In accordance with our plan of going from simple to more complex situations, we extend the treatment of PDEs of hyperbolic type to more than two independent variables. This will essentially supply the mathematical description for wave propagation phenomena in electromagnetic media in more than one dimension vis-à-vis the Cauchy initial value problem and related problems. We shall show that the extension of characteristic curves to characteristic surfaces, and the rays along which disturbances are propagated, will play a central role in our generalization to more than two dimensions. We showed in Chapter 2 that we can transform from a single second-order to a pair of first-order PDEs.

In this chapter we extend the treatment by starting with a system of first-order PDEs in n independent variables. We shall then treat the equivalent system of second-order equations in n independent variables. For $n > 2$ there is no essential change in the theory. If we have one dependent variable $u(x_1, x_2, \ldots, x_n)$ then we have an $(n + 1)$-dimensional hyperspace: $(x_1, \ldots, x_n, u)$. For m dependent variables $(u_1, \ldots, u_m)$ we have an $(n + m)$-dimensional hyperspace: $(x_1, \ldots, x_n, u_1, \ldots, u_m)$.

3.1. First-Order Quasilinear Equations in n Independent Variables

We start this chapter by considering a single variable $u = u(\mathbf{x})$, where the vector $\mathbf{x} = (x_1, x_2, \ldots, x_n)$. u satisfies the quasilinear PDEs

$$\sum_{i=1}^{n} a_i u_{x_i} = a_{n+1}, \qquad u_{x_i} \equiv \frac{\partial u}{\partial x_i}, \tag{3.1}$$

where the coefficients a_i are known functions of (x, u) with continuous derivatives, such that $\sum_{i=1}^{n} a_i^2 \neq 0$. If $a_{n+1} = 0$ then (3.1) is homogeneous, otherwise it is nonhomogeneous. Extending characteristic theory to $\mathbf{x}$ we know that along a characteristic curve we have

$$\frac{du}{dt} = \sum_{i=1}^{n} u_{x_i}\left(\frac{dx_i}{dt}\right) = a_{n+1}. \tag{3.2}$$

Comparing (3.2) with (3.1) leads to the system of $(n + 1)$ ordinary differential equations, the *characteristic equations* for the quasilinear equation (3.1).

$$\frac{dx_i}{dt} = a_i, \qquad \frac{du}{dt} = a, \qquad i = 1, 2, \ldots, n, \tag{3.3}$$

where we set $a_{n+1} = a$. As for the case of two independent variables x, y, we define the n-parameter family of curves obtained by integrating the system (3.2) as the *family of characteristic curves* belonging to (3.1). We define a *characteristic base curve* as the projection of a characteristic curve in $(\mathbf{x}, u)$ space on the $\mathbf{x}$ subspace. The $(n + 1)$ characteristic differential equations given by (3.3) define an n-parameter family of curves (rather than an $(n + 1)$ family) because the variable t does not occur explicitly in these equations (u depends only on the n-dimensional space $\mathbf{x}$). The quasilinear system (3.3) is coupled in the sense that the characteristic ordinary differential equations (ODEs) depend on u as well as $\mathbf{x}$. For the linear case the coefficients do not depend on u so that the characteristic equations $dx_i/dt = a_i$, $(i = 1, 2, \ldots, n)$ form a determined system. This means the first n equations of (3.3) are uncoupled, so that the characteristic base curves consist of an $(n - 1)$-parameter family in the $\mathbf{x}$ space, and are independent of the other characteristic equation $du/dt = a$. This is not true if (3.1) is quasilinear where the system (3.3) is coupled so that an n-parameter family is obtained and the first n characteristic equations depend on $du/dt = a$.

Suppose we have solved (3.1) for a given set of coefficients. Then we know $u = u(\mathbf{x})$. Substituting this expression for u in the right-hand side of the equations

$$\frac{dx_i}{dt} = a(\mathbf{x}, u(\mathbf{x})), \qquad i = 1, 2, \ldots, n,$$

we are able, in principle, to solve for the trajectories $x_i = x_i(t)$, which are the characteristic base curves. Another way of looking at the situation is: The integral surface $u = u(\mathbf{x})$ is generated by an $(n - 1)$-parameter family of characteristic curves C obtained by integrating the characteristic equations. Recall that, in parametric form, each characteristic curve C of this family is generated by the parameter t. Each $C = C_s$ is defined by the parameter s which represents the position on the initial value curve which C_s crosses. (We shall discuss the Cauchy initial value problem below.) Along each C_s we must have

$$\frac{du}{dt} = \sum_{i=1}^{n} u_{x_i}\left(\frac{dx_i}{dt}\right) = \sum_{i=1}^{n} a_i u_x,$$

using $dx_i/dt = a_i$. Since u must satisfy (3.1) we must have $du/dt = a$, which clearly means that each C_s is a characteristic curve. It follows from the definition leading to each C_s, that every surface $u = u(\mathbf{x})$ generated by such a family of C_s's is an integral surface. Furthermore, from the uniqueness theorem of ODEs under a wide range of conditions there is one and only one C_s through a field point whose coordinates are $(\mathbf{x}, u)$. This means that, if a C_s has a point in common with an integral surface then it must lie entirely on that surface. We have therefore proven the following facts which may be stated as a theorem:

On every integral surface defined by $u = u(\mathbf{x})$ of (3.1), there exists an $(n - 1)$-parameter family of characteristic curves C_s (defined by the parameter s, and generated by the parameter t) which generate the integral surface. Conversely, every surface $u = u(\mathbf{x})$ generated by such a family is an integral surface. In addition, if C_s has a point in common with an integral surface, then it lies entirely on the surface.

Cauchy Initial Value Problem

We define the Cauchy initial value problem as follows: Find the solution $u = u(\mathbf{x})$ of the quasilinear PDE (3.1) which satisfies the initial conditions that u is prescribed on a given surface in the $(\mathbf{x}, u)$, $(n + 1)$-dimensional space. Specifically, the initial surface S_0, which is an $(n - 1)$-dimensional manifold, is prescribed and is defined by

$$S_0: \quad x_i = x_i(s_1, s_2, \ldots, s_{n-1}, 0), \qquad t = 0 \quad i = 1, 2, \ldots, n,$$

The initial data are

$$u = f(s_1, s_2, \ldots, s_{n-1}, 0) \quad \text{on } S_0 \qquad \text{for} \quad t = 0, \tag{3.4}$$

which means that u is a prescribed function f of the $(n - 1)$ parameters $s_1, \ldots, s_{n-1}$ on the initial surface S_0. We want to find the integral surface $u = u(\mathbf{x})$, which is the value of u that satisfies (3.1) and has the given value on the initial surface S_0. Our theorem tells us that this integral surface is generated by the trajectories $\mathbf{x} = \mathbf{x}(t)$ which arise as solutions of the characteristic equations.

We solve the Cauchy initial value problem in the following way: For a prescribed set of values of the parameters $s_1, s_2, \ldots, s_{n-1}$ we determine the solutions

$$x_i = x_i(t, s_1, s_2, \ldots, s_{n-1}), \qquad u = u(t, s_1, s_2, \ldots, s_{n-1}), \tag{3.5}$$

of the system of n characteristic ODEs (3.3). These solutions must satisfy the initial conditions at $t = 0$ given by (3.4). The $(n + 1)$ functions given by (3.5) depend in a smooth way (continuously differentiably) on $s_1, \ldots, s_{n-1}$ as well as t. If we now invert the functional relationships by considering $t, s_1, \ldots, s_{n-1}$ to be expressed in terms of $x_1, x_2, \ldots, x_n$ and substitute them into $u = u(t, s_1, \ldots, s_{n-1})$, we obtain $u = u(x_1, x_2, \ldots, x_n)$. We have thus introduced $\mathbf{x} = (x_1, \ldots, x_n)$ as the new independent variables. The transformation from

$(t, s_1, s_2, \ldots, s_{n-1})$ to $(x_1, x_2, \ldots, x_n)$ space has a unique inverse if and only if the Jacobian of the mapping J is nonsingular (has a nonzero determinant), where

$$J = \frac{\partial(x_1, x_2, \ldots, x_n)}{\partial(t, s_1, s_2, \ldots, s_{n-1})} = \begin{pmatrix} \partial x_1/\partial t & \cdots & \partial x_n/\partial t \\ \partial x_1/\partial s_1 & \cdots & \partial x_n/\partial s_1 \\ \cdots\cdots\cdots\cdots\cdots\cdots\cdots \\ \partial x_1/\partial s_{n-1} & \cdots & \partial x_n/\partial s_{n-1} \end{pmatrix}. \quad (3.6)$$

By using the characteristic ODEs we can express the elements of the first row of J by the elements $a_1, a_2, \ldots, a_n$. Then J is valid along a characteristic curve C and $\det(J) \neq 0$. As long as this condition (of the nonvanishing of the determinant of the Jacobian) is valid we obtain $u = u(\mathbf{x})$ from $u = u(t, s_1, s_2, \ldots, s_{n-1})$. The relation $du/dt = a$ (for this functional relationship) goes over into

$$\sum_{i=1}^{n} u_{x_i}\left(\frac{dx_i}{dt}\right) = \sum_{i=1}^{n} a_i u_{x_i} = a,$$

so that $u = u(\mathbf{x})$ is the solution of the quasilinear equation (3.1). Thus the Cauchy initial value problem for our first-order linear equation in n independent variables has a unique solution as long as we demand that $\det(J) \neq 0$.

3.2. First-Order Fully Nonlinear Equations in n Independent Variables

We generalize the previous section to include a fully nonlinear PDE in n independent variables which we may put in the form

$$F(x_1, x_2, \ldots, x_n, u, p_1, p_2, \ldots, p_n) = 0, \quad (3.7)$$

where $p_i = \partial u/\partial x_i$.

The case $n = 2$ was considered in Section 2.12 where we treated the fully nonlinear hyperbolic PDE in two independent variables. In a geometric approach we introduced the Monge cone as a conical surface composed of a one-parameter family of characteristics. It was shown that each field point is the vertex of a Monge cone, and a field of cones was thus determined. The generators of each Monge cone was obtained by solving the characteristic ODEs (2.107).

We shall not repeat the arguments given in Chapter 2 leading to the characteristic equations for the case $n > 2$. However, we shall discuss the role of the *characteristic strips*, defined below, for the PDE given by (3.7). By analogy with the treatment in Chapter 2 we associate the following system of ODEs with (3.7):

$$\frac{dx_i}{dt} = F_{p_i}, \qquad \frac{du}{dt} = \sum_i p_i F_{p_i}, \qquad \frac{dp_i}{dt} = -(F_u p_i + F_{x_i}). \quad (3.8)$$

Equation (3.8) is called the system of *characteristic differential equations* belonging to the PDE (3.7). It consists of a set of $(2n + 1)$ equations for the $(2n + 1)$ functions (x_i, u, p_i) in terms of the parameter t. This system is the generalization of (2.107) for n. Clearly, we have $x_1 = x$, $x_2 = y$, $p_1 = p$, $p_2 = q$, for the case $n = 2$.

If we expand dF/dt by the chain rule and use (3.8) we obtain

$$\frac{dF}{dt} = \sum_{i=1}^{n} F_{x_i}\left(\frac{dx_i}{dt}\right) + \sum_{i} F_{p_i}\left(\frac{dp_i}{dt}\right) + F_u\left(\frac{du}{dt}\right) = 0. \tag{3.9}$$

Since (3.9) must hold for every solution of (3.8) it follows that $F = $ const. (which we may set equal to zero), which means that F is an integral of the system of ODEs (3.8), so that (3.8) is indeed the characteristic system of equations associated with $F = 0$.

The *characteristic strips* are defined as all the solutions of (3.8) which satisfy (3.7). We define an *element* of the characteristic strip as a set of values of the $(2n + 1)$ variables $(x_1, x_2, \ldots, x_n, u, p_1, p_2, \ldots, p_n)$. As in the case of $n = 2$, infinitely many characteristic strips lie on every integral surface which is a solution of (3.7) such that $u = u(x_1, x_2, \ldots, x_n)$. Also, as for the case $n = 2$, every characteristic strip which has an element in common with an integral surface must lie wholly on that surface.

3.3. Directional Derivatives in n Dimensions

In order to get a deeper understanding of the geometry of the characteristic surfaces (and also noncharacteristic or initial value surfaces) associated with the first-order PDE in n-dimensional space given by (3.7), we need to generalize the concept of the directional derivative for curves in two dimensions for curves to surfaces embedded in an n-dimensional space.

We start with the concept of a directional derivative in two-dimensional or (x, y) space. Recall that in Chapter 2, Section 2.4, we studied a pair of first-order quasilinear PDEs. We introduced the idea of a directional derivative by finding that linear combination of the derivatives of $p = u_x$ and $q = u_y$ (compatible with the two given PDEs) which would yield the directional derivative, which turned out to be in the direction tangent to the two families of characteristic curves. We turn now to the simpler problem of defining the directional derivative of a single function $u(x, y)$ defined in an appropriate domain in (x, y) space. Let the point P have coordinates (x, y). Associate with P the vector $\mathbf{a} = (\alpha, \beta)$ such that $\alpha^2 + \beta^2 \neq 0$ or $\mathbf{a} \neq 0$. Let t be a parameter generating a straight line which goes through P. Then the coordinates of any point on this line are $(x + \alpha t, y + \beta t)$. Suppose $\bar{x} = x + \alpha t$, $\bar{y} = y + \alpha t$. Then eliminating t yields $\beta(\bar{y} - y) = \alpha(\bar{x} - x)$ which clearly is a straight line going through P. Expanding du/dt by the chain rule gives

$$\frac{du}{dt} = u_x \frac{dx}{dt} + u_y \frac{dy}{dt} = \alpha u_x + \beta u_y, \tag{3.10}$$

by using the parametric representation of the straight line. This tells us that the operator

$$\frac{d}{dt} = \alpha \frac{\partial}{\partial x} + \beta \frac{\partial}{\partial y} \tag{3.11}$$

represents differentiation in the direction of the vector **a**, which may depend continuously on x and y. We note that the direction defined by **a** at each point in (x, y) space forms a *direction field* whose trajectories are uniquely determined by the system of differential equations

$$\frac{dx}{dt} = \alpha, \qquad \frac{dy}{dt} = \beta. \tag{3.12}$$

Consider a characteristic curve C represented by $\phi(x, y) = 0$.† Since $u(x, y)$ is defined in the (x, y) plane it is certainly defined on C. From the equation for C we get

$$\frac{d\phi}{dt} = \phi_x \frac{dx}{dt} + \phi_y \frac{dy}{dt} = 0, \qquad \text{or} \qquad \frac{dy}{dx} = -\frac{\phi_x}{\phi_y}, \tag{3.13}$$

which holds on C. In order for (3.10) to represent the derivative of u on C in the direction *normal* to C we must have the following conditions:

$$\alpha = \lambda \phi_x, \qquad \beta = \lambda \phi_y, \tag{3.14}$$

where λ is an arbitrary parameter that is constant at each point P on C. It follows that du/dt, the normal derivative of u, is given by

$$\frac{du}{dt} = \left(\frac{1}{\sqrt{\phi_x^2 + \phi_y^2}} \right) (\phi_x u_x + \phi_y u_y), \tag{3.15}$$

where **a** is normalized by setting $\alpha^2 + \beta^2 = 1$.

If **a** is tangent to C at P (and therefore perpendicular to the normal at P), then the scalar product of **a** and **grad** $\phi = (\phi_x, \phi_y)$ must be zero. This is clear, since **grad** ϕ is normal to C. We have

$$\alpha \phi_x + \beta \phi_y = 0. \tag{3.16}$$

Then $du/dt = \alpha u_x + \beta u_y$ is the tangential derivative of u at P on C. du/dt is sometimes called the *inner derivative* on C and clearly lies in the direction tangent to C. From the above relations it follows that the condition for inner differentiation on C is given by

$$\alpha \frac{dy}{dt} - \beta \frac{dx}{dt} = 0, \tag{3.17}$$

where α and β satisfy the conditions given by (3.12). The inner derivative for u becomes

$$\frac{du}{dt} = \phi_y u_x - \phi_x u_y. \tag{3.18}$$

---

† C doesn't have to be c characteristic curve but may be an initial value curve having nowhere a characteristic direction.

We note that the inner derivative of u on C depends only on the distribution of the values of u on C. Suppose C is an *initial value curve* instead of a characteristic curve. If the initial data u is known on C, then the inner derivative du/dt is also known on C. This is easily seen.

Case $n > 2$. We are now in a position to investigate the nature of inner derivatives (we will have more than one for $n > 2$) and characteristic surfaces for $n > 2$. Clearly, for this more general case, the family of characteristic curves C goes over into a family of characteristic surfaces. For an n-dimensional space given by $\mathbf{x} = (x_1, x_2, \ldots, x_n)$ have an $(n - 1)$-dimensional surface S expressed by the equation

$$\phi(\mathbf{x}) = 0 \tag{3.19}$$

embedded in this space. We also consider a function $u = u(\mathbf{x})$ defined in a specified domain in $\mathbf{x}$ space which is smooth in the neighborhood of S (continuous derivatives with respect to $\mathbf{x}$). At each P whose coordinates are $\mathbf{x}$ we define a vector $\mathbf{a} = (a_1, a_2, \ldots, a_n)$, which is a generalization to n space of the vector (α, β) for $n = 2$. We demand that $\sum_{i=1}^{n} a_i^2 \neq 0$ or $\mathbf{a} \neq 0$. At the point $P: \mathbf{x}$ we construct a straight line L. The coordinates of L in parametric form are

$$x_i + a_i t, \qquad i = 1, 2, \ldots, n,$$

where, as before, t is the parameter producing L. We define the derivative of $u(\mathbf{x})$ with respect to t, in the direction given by the vector $\mathbf{a}$, by the expression

$$\frac{du}{dt} = \sum_{i=1}^{n} a_i u_{x_i}. \tag{3.20}$$

Note that the right-hand side of (3.20) is the scalar product of the vectors $\mathbf{a}$ and $\mathbf{grad}\ u$. Since the function u is defined in the n-dimensional domain containing the surface S it is clearly defined on S. Therefore the expression (3.20) is valid on S. Consider any point P on S. Suppose we want du/dt to be *in the direction normal to S at P.* It is clear that, if we evaluate $\mathbf{grad}\ \phi$ at P, then $\mathbf{grad}\ \phi$ is in the direction normal to S at P. Since $\mathbf{a}$ specifies the direction of du/dt and we want this derivative to be in the direction normal to S, we must choose $\mathbf{a}$ to be in the same direction as $\mathbf{grad}\ \phi$. We then have the relations

$$a_i = \lambda \phi_{x_i}, \qquad i = 1, 2, \ldots, n, \tag{3.21}$$

where λ is an arbitrary parameter which is constant at each point P on S, but may vary along S. The equations (3.21) tell us that each component of $\mathbf{a}$ is in the same proportion to the corresponding component of $\mathbf{grad}\ \phi$. If we further normalize $\mathbf{a}$ (for convenience) by setting $\sum_{i=1}^{n} a_i^2 = 1$, the derivative of u in the direction normal to S becomes

$$\frac{du}{dt} = \frac{1}{\sum\limits_{i=1}^{n} \sqrt{\phi_{x_i}^2}} \sum_{i=1}^{n} \phi_{x_i} u_{x_i}, \tag{3.22}$$

where du/dt, as given by (3.22), is the expression for the normal derivative of u at P.

Suppose we want the derivative du/dt to be in the direction tangent to S at P. Then **a** must be in that direction, or **a** is normal to **grad** ϕ. This, of course, means that the scalar product of **a** and **grad** ϕ must be zero, or

$$\sum_{i=1}^{n} a_i \phi_{x_i} = 0. \tag{3.23}$$

If the condition given by (3.23) is valid then $du/dt = \sum_i a_i u_{x_i}$ is the *tangential derivative* of u with respect to t, which, as mentioned, is sometimes called the *inner derivative*. Therefore du/dt lies on the surface S at each point P. Suppose (3.23) does not hold so that $\sum_i a_i \phi_{x_i} \neq 0$. Then du/dt has a direction either into or out of S and is therefore no longer the inner derivative. It is sometimes called the "outer derivative". The terms "inner" and "outer" derivatives are somewhat misleading. We prefer the terms "tangential" and "normal", instead. (For the case where du/dt has a nonzero normal component but is not normal to S we shall say that du/dt is a "nontangential derivative".)

Consider the tangent plane through P on S. If we cut the surface $\phi = 0$ at P by a plane whose orthogonal coordinates are (x_i, x_j), then the projection of the tangential derivative du/dt on this plane is easily seen to be

$$\phi_{x_i} u_{x_j} - \phi_{x_j} u_{x_i}, \quad \text{for } i \neq j, \quad i, j = 1, 2, \ldots, n. \tag{3.24}$$

As an example, let $n = 3$ and let $\mathbf{x} = (x, y, z)$. Then the three orthogonal tangential derivatives are

$$\phi_y u_x - \phi_x u_y,$$
$$\phi_z u_y - \phi_y u_z, \tag{3.25}$$
$$\phi_x u_z - \phi_z u_x.$$

Suppose S is a noncharacteristic surface. Then the n-dimensional generalization of the Cauchy initial value problem tells us that we can obtain a solution for $u(\mathbf{x})$ of a first-order quasilinear PDE in **x** space by expanding in a Taylor series off S, provided we prescribe initial data for u on S. The following question now arises: Is a knowledge of u on S sufficient to determine the tangential derivative du/dt on S? We shall attempt to answer this question in three dimensions and leave the generalization to the reader. Let the equation of the noncharacteristic surface S be $\phi(x, y, z) = 0$. Then we can calculate **grad** ϕ or (ϕ_x, ϕ_y, ϕ_z) at any point P on S. Turning to the tangential derivatives given by the system (3.25) we see that we cannot calculate **grad** u since we cannot calculate partial derivatives in directions off the tangent plane at P on S. This suggests that we introduce new independent variables, one of which is ϕ, the other two being variables on the tangent plane of S at P. We therefore introduce the vector $\mathbf{z} = (\xi, \eta, \zeta)$ where $\zeta = \phi$ and (ξ, η) are in the tangent plane of S. Let $\mathbf{a} = (\alpha, \beta, \gamma)$. Expanding the tangential derivative by the chain rule gives

$$\frac{du}{dt} = \alpha u_x + \beta u_y + \gamma u_z$$

$$= u_\phi(\alpha\phi_x + \beta\phi_y + \gamma\phi_z) + \alpha(u_\xi\xi_x + u_\eta\eta_x) \qquad (3.26)$$

$$+ \beta(u_\xi\xi_y + u_\eta\eta_y) + \gamma(u_\xi\xi_z + u_\eta\eta_z).$$

The condition that du/dt be a tangential derivative (in the tangent plane of S) is that the following expression must hold:

$$\alpha\phi_x + \beta\phi_y + \gamma\phi_z = 0. \qquad (3.27)$$

This clearly means that all the terms on the right-hand side of (3.26) can be calculated, since $u = u(\xi, \eta, 0)$ is prescribed initial data on the surface $\phi = 0$. Therefore, the answer is "yes", the tangential derivative in any direction in the tangent plane of S at any point P on S can be calculated if we know the equation of S and the initial data for u on S. By taking appropriate linear combinations we may obtain all the derivatives on S, for example, (u_x, u_y, u_z), and then the higher-order derivatives. This will give us sufficient information on S to allow us to construct solutions for $u(\mathbf{x})$ off the initial value surface by Taylor series expansions in terms of the data on S (the "initial data"). We repeat that S must nowhere have a direction tangent to a characteristic surface.

3.4. Characteristic Surfaces in n Dimensions†

We are now in a position to turn to a deeper approach to the method of characteristics which we develop for a system of first-order PDES in n dimensions (in $\mathbf{x}$ space). Instead of merely considering a single dependent variable $u(\mathbf{x})$ we generalize to a system of m dependent variables in n space. We can write a first-order system of quasilinear PDES in m unknowns in n-dimensional space as follows:

$$a_{ijk}u_{i,j} + b_k = 0, \qquad i, k = 1, 2, \ldots, m, \quad j = 1, 2, \ldots, n, \qquad (3.28)$$

where the tensor summation convention is used, so that the kth equation of (3.28) is summed over i for 1 to m and j from 1 to n. There is no sum over k. There are m functions $u_i = u_i(x_1, x_2, \ldots, x_n)$, where we use the notation $u_{i,j} \equiv \partial u_i/\partial x_j$. The coefficients a_{ijk} and b_k are prescribed functions of the x_j's and the u_i's. (They depend only on the x_j's for the linear case.) The system (3.28) is a set of m equations for the m unknown functions u_i. Clearly for $m = 1$ this system reduces to a single equation for $u(x_1, x_2, \ldots, x_n)$. We can recast (3.28) into a more compact matrix representation by introducing the following notation: Let $\mathbf{u}$ and $\mathbf{b}$ be m-dimensional vectors such that $\mathbf{u} = (u_1, u_2, \ldots, u_m)$,

---

† See, for example, [11, Vol. II] and [7].

$\mathbf{b} = (b_1, b_2, \ldots, b_m)$. Let the matrix $\mathbf{A}_j$ consist of the elements a_{ijk} for a given x_j. Also let the set of derivatives of $\mathbf{u}$ with respect to each x_j be given by $\mathbf{u}_j \equiv \partial \mathbf{u}/\partial x_j$. Then the matrix representation of (3.28) is

$$\mathbf{A}_j\mathbf{u}_j + \mathbf{b} = 0 \qquad \text{(summed over } j\text{)}. \qquad (3.28a)$$

Case $n = 2$. We first study the two-dimensional case setting $x_1 = x$, $x_2 = y$, $x_j = 0$ for $j = 3, 4, \ldots, n$. The system (3.28) reduces to

$$a_{i1k}u_{i,x} + a_{i2k}u_{i,y} + b_k = 0, \qquad k = 1, 2, \ldots, m. \qquad (3.29)$$

The system (3.29) of m equations is summed over i from 1 to m. We introduce the set of differential operators D_{ik} defined by

$$D_{i1k} = a_{i1k}\frac{\partial}{\partial x} + a_{i2k}\frac{\partial}{\partial y}. \qquad (3.30)$$

Equation (3.30) represents the set of m^2-directional derivative operators. Using (3.30), we put (3.29) in the following form:

$$\sum_{i=1}^{m} D_{ik}u_i + b_k = 0, \qquad k = 1, 2, \ldots, m. \qquad (3.31)$$

From the definition of the D_{ik}'s we see in (3.31) that, in general, each u_i is differentiated in a different direction.

We recall from Chapter 2, Section 2.4, that we constructed a linear combination of two first-order quasilinear PDEs that would yield the same directions for the directional derivatives of p and q. In the treatment given here we extend this approach to the m functions u_i. To this end, we return to the system (3.29), multiplying the kth equation by λ_k, and sum over k from 1 to m, obtaining

$$\sum_{i,k=1}^{m} \lambda_k a_{i1k}u_{i,x} + \sum_{i,k} \lambda_k a_{i2k}u_{i,y} = -\sum_{k} \lambda_k b_k. \qquad (3.32)$$

The set of m λ_k's may vary from point to point in the (x, y) plane, but are constant at each field point. In (3.32) each u_i is differentiated in the direction given by the direction numbers

$$\sum_{k=1}^{m} \lambda_k a_{i1k} = l_i \qquad \text{and} \qquad \sum_{k=1}^{m} \lambda_k a_{i2k} = m_i.$$

We want the directions given by the vector (l_k, m_k) to be the same for all $k = 1, \ldots, m$. This means that the direction numbers or components of this vector must be proportional, so that

$$\sum_{k=1}^{m} \lambda_k(a_{i1k} - \mu a_{12k}) = 0, \qquad i = 1, 2, \ldots, m, \qquad (3.33)$$

where the parameter μ is the proportionality constant which is constant at each field point. In order that the set of λ_k's be nontrivially zero, the determinant of the left-hand side of (3.33) must vanish. This means that μ must be the roots of the equation

$$\det(a_{i1k} - \mu a_{i2k}) = 0. \tag{3.34}$$

Since the left-hand side of (3.34) is an mth degree polynomial in μ there are m roots for μ of (3.34). Inserting (3.33) into (3.32) yields

$$\sum_{i,k=1}^{m} \lambda_k a_{i2k}(\mu u_{i,x} + u_{i,y}) = \sum_{i=1}^{m} m_i(\mu u_{i,x} + u_{i,y}) = -\sum_{k=1}^{m} \lambda_k b_k. \tag{3.35}$$

The left-hand side of (3.35) can be written in terms of the directional derivative D whose direction numbers are $(\mu, 1)$. Then (3.35) can be written as

$$\sum_i m_i(\mu u_{i,x} + u_{i,y}) = \sum_i m_i D u_i = D \sum_i m_i u_i - \sum_i (Dm_i)u_i = -\sum_k \lambda_k b_k. \tag{3.36}$$

We are interested in recasting (3.36) into an expression analogous to (3.31), where the operator D_{ik} is replaced by

$$D_l = \mu_l \frac{\partial}{\partial x} + \frac{\partial}{\partial y}. \tag{3.37}$$

The significance of D_l will be seen below. Using the following notation:

$$v = \sum_i m_i u_i, \qquad w = \sum_i [-\lambda_i b_i + (Dm_i)u_i],$$

(3.35) becomes

$$Dv = w. \tag{3.38}$$

The significance of (3.38) is seen by assuming that there are r real distinct roots $\mu_1, \mu_2, \ldots, \mu_r$ of (3.34). Then (3.38) is *uncoupled* in the sense that, for each distinct real root μ_l of (3.34) there exists a function v_l such that

$$D_l v_l = w_l, \qquad l = 1, 2, \ldots, r, \tag{3.38a}$$

where

$$v_l = \sum_i m_{il} u_i, \qquad m_{il} = \sum_k \lambda_{kl} a_{i2k}. \tag{3.39}$$

As mentioned, (3.38) is uncoupled, as seen from (3.38a). Equation (3.38) or (3.38a) is therefore the analogue of (3.31). It is much simpler than the original system (3.31) since it contains only one of the v_i's corresponding to the root μ_l. We note that, if $r < m$ then there are not enough equations to solve for the m values of $\mathbf{u}$. We shall not treat this rather difficult case, but shall assume that $r = m$, so that the system of PDEs is *totally hyperbolic*. For the other extreme case of $r = 0$ then there are no real roots and the system is elliptic, having no real characteristics.

EXAMPLE (Wave Equation). As an example we reinvestigate the one-dimensional wave equation $c^2 w_{xx} - w_{yy}$ (for constant wave speed c). Setting

$$cw_x = u_1 = u, \qquad w_y = u_2 = v,$$

the wave equation is thereby transformed into the pair of first-order PDEs

$$cu_x - v_y = 0, \qquad u_y - cv_x = 0,$$

which may be put into the matrix equation

$$\begin{pmatrix} c & 0 \\ 0 & -c \end{pmatrix}\begin{pmatrix} u_x \\ v_x \end{pmatrix} + \begin{pmatrix} 0 & -1 \\ 1 & 0 \end{pmatrix}\begin{pmatrix} u_y \\ v_y \end{pmatrix} = 0. \tag{3.40}$$

Equation (3.34) becomes

$$\det\left[\begin{pmatrix} c & 0 \\ 0 & -c \end{pmatrix} - \mu\begin{pmatrix} 0 & -1 \\ 1 & 0 \end{pmatrix}\right] = \det\begin{pmatrix} c & \mu \\ -\mu & -c \end{pmatrix} = -c^2 + \mu^2 = 0, \tag{3.41}$$

which yields

$$\mu = \pm c, \tag{3.42}$$

for the roots of (3.34). The physical interpretation of μ for this example is obtained by referring to the directional derivative given by (3.37), which may be written as

$$D_{1,2} = (\pm c)\frac{\partial}{\partial x} + \frac{\partial}{\partial y}, \tag{3.43}$$

where D_1 refers to $\mu = +c$ and D_2 to $\mu = -c$. Since

$$D = \frac{dx}{dy}\frac{\partial}{\partial x} + \frac{\partial}{\partial y},$$

it follows that

$$\mu_1 = c = \frac{dx}{dy}: \qquad \text{for } C_+ \text{ characteristic,}$$

$$\tag{3.44}$$

$$\mu_2 = -c = \frac{dx}{dy}: \qquad \text{for } C_- \text{ characteristic.}$$

EXAMPLE (Elliptic System). We now study the Cauchy–Riemann equations from the viewpoint of the above theory. They are given by the following system of first-order equations:

$$u_x - v_y = 0, \qquad u_y + v_x = 0,$$

which may be written as the matrix equation

$$\begin{pmatrix} 1 & 0 \\ 0 & 1 \end{pmatrix}\begin{pmatrix} u_x \\ v_x \end{pmatrix} + \begin{pmatrix} 0 & -1 \\ 1 & 0 \end{pmatrix}\begin{pmatrix} u_y \\ v_y \end{pmatrix} = 0. \tag{3.45}$$

Equation (3.34) becomes

$$\det\begin{pmatrix} 1 & \mu \\ -\mu & 1 \end{pmatrix} = \mu^2 + 1 = 0. \tag{3.46}$$

The roots are $\mu_{1,2} = \pm i$ which are imaginary, thus yielding no real characteristics in the (x, y) plane. Therefore the system (3.45) is elliptic. But we know the Cauchy–Riemann equations yield harmonic functions for u and v in the (x, y) plane, and the Laplace equation is elliptic.

We see that for the special case of two dimensions $(n = 2)$ the characteristic surfaces reduce to characteristic curves in the (x, y) plane. We now generalize to higher-dimensional spaces.

Case $n > 2$. For this more general case we must be concerned with characteristic surfaces in $\mathbf{x} = (x_1, x_2, \ldots, x_n)$ space or "n space". The first-order quasi-linear system of PDEs in n space is given by (3.28), or in matrix form, by (3.28a). We now study this equation in the setting of the nth-dimensional generalization of the Cauchy initial value problem described in Section 3.3. To this end, we again consider a surface $S: \phi(\mathbf{x}) = 0$ such that $\mathbf{grad}\ \phi \neq 0$.

Case $m = 1$. In order to get a firmer understanding of the general theory for m components of $\mathbf{u}$ in n space, we shall first review the special case $m = 1$, so that the vector $\mathbf{u} = u(\mathbf{x})$, a scalar function of $\mathbf{x}$. For this case the system of m equations (3.28) reduces to a single first-order PDE

$$a_j u_{,j} + b = Du = 0, \qquad \text{where} \quad u_{,j} \equiv \frac{\partial u}{\partial x_j}, \tag{3.47}$$

where the tensor summation convention is used (the first term is summed over j from 1 to n). The n coefficients a_j and b are prescribed functions of $\mathbf{x}$ and u for the quasilinear case. The directional derivative operator is $D = a_j\, \partial/\partial x_j$, where the coefficients a_j are the direction numbers of D. Since (3.47) is a single equation, this case is a little too specialized: we cannot take a linear combination of equations as we did in obtaining (3.32). Thus, there is a single directional derivative, not a combination giving a certain direction as in the case of $\mathbf{u}$. In any event, the reason for considering this case is to show the relationship between the direction numbers and the *characteristic ODEs*. This is given by the system

$$\frac{dx_j}{dt} = a_j, \qquad \frac{du}{dt} = -b, \qquad j = 1, 2, \ldots, n, \tag{3.48}$$

where, as usual, (3.48) is the parametric representation of the characteristic equations in terms of the parameter t. This means that the solution of the system (3.48) yields a one-parameter family of characteristic surfaces which, for the quasilinear case, is coupled in the sense that the surfaces depend on $u(\mathbf{x}(t))$. We prove (3.48) by the following argument: Consider a characteristic surface S given by

$$\phi(\mathbf{x}) = 0 \quad \text{on } S.$$

Using the chain rule, the following equation holds:

$$\text{on } S: \quad \frac{d\phi}{dt} = \sum_{i=1}^{n} \left(\frac{dx_i}{dt}\right)\phi_i = 0, \qquad \text{where} \quad \phi_i \equiv \frac{\partial \phi}{\partial x_i}. \tag{3.49}$$

Equation (3.49) is called *the characteristic condition*; it holds on S. To ensure that both (3.49) and the PDE (3.47) be valid, we deduce that the characteristic equations (3.48) must also hold, thus proving that the PDE along with the characteristic condition yield the characteristic equations. As we know, the characteristic condition (3.49) may also be written in vector form in terms of the scalar product of **a** and **grad** ϕ or

$$\mathbf{a} \cdot \mathbf{grad}\ \phi = 0, \tag{3.49a}$$

where the direction numbers are given by the vector $\mathbf{a} = (a_1, a_2, \ldots, a_n)$. Equation (3.49a) clearly tells us that **grad** ϕ is normal to the surface S and the vector **a** lies in the tangent plane of S.

General Case. We are now in a position to generalize to $m > 1, n > 2$.

EXAMPLE ($m = 2, n = 3$). In order to fully understand the notation in (3.28) or (3.28a) we first study the special case $m = 2, n = 3$. To this end, we let the vector $\mathbf{u} = (u, v)$ and $\mathbf{x} = (x, y, z)$. Then (3.28) becomes the following system of two first-order quasilinear equations in (x, y, z) space:

$$\begin{aligned} a_{111}u_x + a_{121}u_y + a_{131}u_z + a_{211}v_x + a_{221}v_y + a_{231}v_z + b_1 = 0, \\ a_{112}u_x + a_{122}u_y + a_{132}u_z + a_{212}v_x + a_{222}v_y + a_{232}v_z + b_2 = 0. \end{aligned} \tag{3.50}$$

Note that the subscripts in a_{ijk} refer to $u, v; x, y, z$, first, or second equation, respectively. It may also be written in the following matrix form, analogous to (3.28a):

$$\mathbf{A}_1 \mathbf{u}_x + \mathbf{A}_2 \mathbf{u}_y + \mathbf{A}_3 \mathbf{u}_z + \mathbf{b} = 0, \tag{3.50a}$$

We consider the surface S defined by

$$S: \quad \phi(x, y, z) = 0.$$

The characteristic condition given by (3.49) is generalized to

$$\mathbf{A} = (\mathbf{A}_1 + \mathbf{A}_2 + \mathbf{A}_3) \cdot (\phi_x, \phi_y, \phi_z). \tag{3.51}$$

Equation (3.51) is valid on S. **A** is called the *characteristic matrix*, and the right-hand side of (3.51) is interpreted as the scalar product of the matrices $\mathbf{A}_1, \mathbf{A}_2, \mathbf{A}_3$ with **grad** $\phi = (\phi_x, \phi_y, \phi_z)$.

For the general case the characteristic matrix **A** is defined by

$$\mathbf{A} = \mathbf{A}_j \phi_j = \mathbf{A}_1 \phi_1 + \mathbf{A}_2 \phi_2 + \cdots + \mathbf{A}_n \phi_n, \qquad \phi_j \equiv \frac{\partial \phi}{\partial x_j}. \tag{3.52}$$

The determinant of **A** is defined by $Q(\phi_1, \phi_2, \ldots, \phi_n)$, called the *characteristic determinant*. We have

$$Q(\phi_1, \phi_2, \ldots, \phi_n) = \det(\mathbf{A}). \tag{3.53}$$

EXAMPLE (Wave Equation). We again revisit the one-dimensional wave equation represented by the pair of first-order equations in (u, v) given in matrix form by (3.40), from which we get $\mathbf{A}_1$ and $\mathbf{A}_2$ by inspection. The characteristic determinant, given by (3.53), becomes

$$\det\left[\begin{pmatrix} c & 0 \\ 0 & -c \end{pmatrix}\phi_x + \begin{pmatrix} 0 & -1 \\ 1 & 0 \end{pmatrix}\phi_y\right] = -c^2\phi_x^2 + \phi_y^2 = 0. \qquad (3.54)$$

This yields

$$c^2 = \left(\frac{\phi_y}{\phi_x}\right)^2.$$

On the surfae $\phi(x, y) = $ const. we have $d\phi/dy = (dx/dy)\phi_x + \phi_y = 0$. (As usual, y stands for time.) This gives $dx/dy = -\phi_y/\phi_x$, which yields $c = \pm dx/dy$, the slopes of the $C_\pm$ characteristic curves in the (x, y) plane. The reader should compare this treatment of the wave equation with that given in the same example which led to (3.41). Recall that this equation arose from constructing the proper linear combination of the PDEs giving the directional derivative in the direction characterized by the parameter μ.

3.5. Maxwell's Equations

The previous sections in this chapter give us sufficient information to treat Maxwell's equations as a system of first-order PDEs, so that we may directly consider the subject matter of this book, namely, wave propagation in electro-magnetic media. A more detailed treatment of these electromagnetic equations is reserved for the section on second-order PDEs where Maxwell's equations will be discussed more thoroughly and efficiently.

The propagation of plane waves in an electromagnetically conducting medium was discussed in Section 1.5 of Chapter 1. Maxwell's equations for a three-dimensional conducting medium was given by the system (1.71) and (1.72). We use the notation more in keeping with our current investigations: Let the electric field vector $\mathbf{E} = (u_1, u_2, u_3)$, the magnetic field vector $\mathbf{H} = (u_4, u_5, u_6)$, and $\mathbf{x} = (x, y, z, t)$. We introduce the vector $\mathbf{u} = (u_1, u_2, \ldots, u_6)$. $\mathbf{u}$ is sometimes called a "super vector". Its six components are the three components of $\mathbf{E}$ and $\mathbf{H}$. We need not be concerned with the continuity equations for $\mathbf{E}$ and $\mathbf{H}$ in (1.72). Writing out the other two equations of (1.72) in extended form we get:

$$u_{3,y} - u_{2,z} + \alpha u_{4,t} = 0,$$

$$u_{1,z} - u_{3,x} + \alpha u_{5,t} = 0,$$

$$u_{2,x} - u_{1,y} + \alpha u_{6,t} = 0,$$

$$u_{6,y} - u_{5,z} - \beta u_{1,t} - \gamma u_1 = 0, \qquad (3.55)$$

$$u_{4,z} - u_{6,x} - u_{2,t} - \alpha u_2 = 0,$$

$$u_{5,x} - u_{4,y} - u_{3,t} - \alpha u_3 = 0,$$

$$\alpha = \frac{\mu}{c}, \qquad \beta = \frac{\varepsilon}{c}, \qquad \gamma = \frac{4\pi\sigma}{c}.$$

The system (3.55) is a set of six linear first-order PDEs for the components of **u** in the four-dimensional **x** space. The matrix representation of (3.55) is given by (3.50a) where each of the (6×6) matrices $\mathbf{A}_i$ $(i = 1, 2, 3)$ and the vector **b** can easily be written down by inspection from (3.55). From the characteristic condition (3.51) we can construct the characteristic matrix **A**. Then the characteristic determinant $Q = \det(\mathbf{A})$ becomes

$$Q = \begin{vmatrix} 0 & -\phi_z & \phi_y & \alpha\phi_t & 0 & 0 \\ \phi_z & 0 & -\phi_x & 0 & \alpha\phi_t & 0 \\ -\phi_y & \phi_x & 0 & 0 & 0 & \alpha\phi_t \\ -\beta\phi_t & 0 & 0 & 0 & -\phi_z & \phi_y \\ 0 & -\beta\phi_t & 0 & \phi_z & 0 & -\phi_x \\ 0 & 0 & -\beta\phi_t & -\phi_y & \phi_x & 0 \end{vmatrix}, \tag{3.56}$$

which is a six-degree polynomial in $(\phi_x, \phi_y, \phi_z, \tau)$, where $\tau = \phi_t$. The surface S is given by the equation $\phi(\mathbf{x}) = \phi(x, y, z, t) = 0$, so that

$$\frac{d\phi}{dt} = \left(\frac{dx}{dt}\right)\phi_x + \left(\frac{dy}{dt}\right)\phi_y + \left(\frac{dz}{dt}\right)\phi_z + \tau = 0 \quad \text{on } S.$$

If S is a characteristic surface then $Q = 0$. The roots are obtained from

$$Q = \tau^2(\alpha\beta\tau^2 - \phi_x^2 - \phi_y^2 - \phi_z^2)^2 = 0. \tag{3.57}$$

We neglect the roots $\tau^2 = 0$ (which correspond to a rigid body motion and therefore does not directly involve wave propagation).

Relation to Second-Order Equations. The factor in the parentheses of (3.57), namely, $\alpha\beta\tau^2 - \phi_x^2 - \phi_y^2 - \phi_z^2 = 0$, yields the roots of the characteristic relation $Q = 0$ for the three-dimensional wave equation which is satisfied by each component of **u**. This was shown in Chapter 1 for the components of **E** and **H**. The characteristic determinant Q for the wave equation for each component of **u** is a factor of Q for the system (3.55).

We state this result for the more general case of a second-order quasilinear PDE for w in $\mathbf{x} = (x_1 x_2, \ldots, x_n))$ space:

$$\sum_{i,j=1}^{n} a_{ij}w_{x_i x_j} + b = 0, \tag{3.58}$$

where the coefficients a_{ij} and b are known functions of $\mathbf{x}$ and $w_{\mathbf{x}}$. (We are only interested in the principle part of (3.58).)

The rule for constructing the characteristic determinant Q is: Replace each second-order partial derivative by the corresponding first-order derivative of ϕ. This means $w_{x_i x_j}$ is replaced by $\phi_{x_i}\phi_{x_j}$. The same rule applies to the principle part of *any* nth-order PDE. For example, for the principle part of a fourth-order equation a term such as w_{xxyz} is replaced by $(\phi_x)^2\phi_y\phi_z$. Therefore, the characteristic condition (which is obtained by setting $Q = 0$) for the above PDE for w becomes

$$Q = \sum_{i,j=1}^{n} a_{ij}\phi_i\phi_j = 0. \tag{3.59}$$

We may replace (3.58) by the system of first-order PDEs

$$\sum_{i,j=1}^{n} a_{ij} u_{i,x_j} + \cdots = 0,$$

$$u_{i,x_n} - u_{n,x_i} = 0, \qquad i = 1, 2, \ldots, n - 1,$$

(3.60)

where we introduced the n variables

$$u_i = w_{x_i}, \qquad i = 1, 2, \ldots, n - 1.$$

(3.61)

The system (3.60) is an example of the set of first-order PDEs given by (3.28) for $b_k = 0$ (in slightly different notation). The matrix representation is a special case of (3.28a), which we write as

$$\sum_{i=1}^{n} \mathbf{A}_i \mathbf{u}_{x_i} = 0.$$

(3.60a)

A word about the last $(n - 1)$ equations of (3.60): The continuity of the partial derivatives $\mathbf{u}_x$ imply that $u_{i,x_j} - u_{j,x_i} = 0$ for $i, j = 1, 2, \ldots, n, i \neq j$. Clearly, there are more than $n - 1$ of these equations. We must only select $(n - 1)$ equations to get a linearly independent set of n PDEs for $\mathbf{u}$.

The characteristic condition is obtained from (3.60a) by replacing each $u_{i,x}$ by $\phi_{x_i} \equiv \phi_i$. We thereby obtain (3.52). Picking off the $\mathbf{A}_i$'s from (3.60) and evaluating $Q = \det(\mathbf{A})$ yields the characteristic condition

$$(-1)^{n-2} (\phi_n)^{n-2} \sum_{i,j=1}^{n} a_{ij} \phi_i \phi_j = 0,$$

(3.62)

which is an nth degree polynomial in the ϕ_i's. Neglecting the $(n - 2)$ roots $(\phi_n)^{n-2} = 0$ we see that the roots of (3.62) agree with those of the original second-order equation given by (3.59).

We pursue this topic in more detail below, from the point of view of an investigation of second-order equations.

3.6. Second-Order Quasilinear Equation in n Independent Variables

The equations of mathematical physics that represent wave propagation phenomena, such as Maxwell's equations for electromagnetic media, are most conveniently couched in the form of systems of second-order PDEs, as we learned in Chapter 1 when treating Maxwell's equations as systems of coupled wave equations for $\mathbf{E}$ and $\mathbf{H}$.

In this section we treat a single second-order quasilinear PDE for the scalar u in $\mathbf{x} = (x_1, x_2, \ldots, x_n)$ space. We write it as

$$\sum_{i,j=1}^{n} a_{ij} u_{ij} + b = 0,$$

(3.63)

where we use the notation

$$u_i \equiv \frac{\partial u}{\partial x_i}, \qquad u_{ij} \equiv \frac{\partial^2 u}{\partial x_i \, \partial x_j}.$$

(A note on notation: In general, subscripts on functions that are differentiable will denote partial differentiation with respect to the subscript, as, for example, $\phi_i \equiv \partial \phi / \partial x_i$, while subscripts on coefficients such as a_{ij} are merely indices.) In (3.63) b and the coefficients a_{ij} are prescribed functions of $\mathbf{x}$, u, and u_i ($i = 1, 2, \ldots, n$). Clearly, for the linear case, b and the a_{ij}'s depend at most on $\mathbf{x}$.

We recall that the Cauchy initial value problem for a first-order system of PDEs involves initial data on u on an initial value curve S given by $\phi(x) = 0$. For the second-order system given by (3.63) the *Cauchy initial value problem* is formulated in terms of prescribed initial data for u and the u_i's on the initial value curve. For this system we shall obtain a two-parameter family of characteristic surfaces $\phi(x) = 0$ such that: if u and the u_i's are prescribed on these characteristic surfaces we must obtain $Q(\phi_1, \phi_2, \ldots, \phi_n) = 0$, so that we cannot obtain a solution for (3.63) in the characteristic strip surrounding the characteristic surface. This means that if the characteristic condition $Q = 0$ is satisfied on a surface, then that surface is characteristic, so that we cannot calculate the higher-order partial derivatives u_{ij}, u_{ijk}, etc., on it, which is necessary for a Taylor series expansion to obtain solutions in the strip off the surface. Clearly then, an initial value surface can nowhere be tangent to a characteristic surface. We saw that there is a one-parameter family of characteristic surfaces for a first-order system of PDEs. Generalizing, for an nth-order hyperbolic PDE in three- or higher-dimensional space there exists an n-parameter family of characteristic surfaces embedded in the space. This is a generalization for an nth-order hyperbolic equation in two-dimensional space, where there is an n-parameter family of characteristic curves.

Classification of Second-Order Equations in n Space. We recall that in Chapter 2, Section 2.10, we classified second-order PDEs in two independent variables into three types:

(1) *Hyperbolic*, where there are two distinct families of characteristics.
(2) *Parabolic*, where there is only one family of characteristics.
(3) *Elliptic*, where there are no real characteristics.

As examples, we recall that the one-dimensional wave equation is hyperbolic, the one-dimensional unsteady heat conduction equation (Fourier's equation) is parabolic, and the two-dimensional Laplace equation is elliptic.

Continuing with a review of second-order equations in the (x, y) space, as described in Chapter 2, the *canonical form* of the principal part was given in (2.63) in terms of the transformed variables (ξ, η) given by (2.60). The coefficient of $u_{\xi\xi}$ in (2.65) is A and is given by the first equation of (2.64) which tells us that $A = Q(\phi_x, \phi_y)$ which is the two-dimensional form of the characteristic determinant $Q(\phi_1, \phi_2, \ldots, \phi_n)$ given by (3.52). We saw in Chapter 2 that the canonical or normal form for the principal part of the hyperbolic second-order equation in two dimensions is given by (2.70), which arose from setting $A = Q(\phi_x, \phi_y) = 0$, thus yielding the characteristic condition for the two-dimensional case. In this section, we extend these ideas to the general case of n-dimensional space.

Let the principal part of (3.63) be given by

$$L(u) = \sum_{i,j=1}^{n} a_{ij} u_{ij},$$ (3.64)

so that the differential equation (3.63) becomes $L(u) + b = 0$. To extend the concepts in Chapter 2 concerning canonical forms we introduce a transformation of independent variables from x space to ξ space given by

$$\xi_i = \xi_i(x_1, x_2, \ldots, x_n), \qquad i = 1, 2, \ldots, n.$$ (3.65)

The derivatives u_i and u_{ij} transform as follows:

$$u_i \equiv u_{x_i} = \sum_{k=1}^{n} \xi_{ki} u_{\xi_k} \quad \text{and} \quad u_{ij} \equiv u_{x_i x_j} = \sum_{k,l=1}^{n} \xi_{ki} \xi_{lj} u_{\xi_k \xi_l} + \cdots, \qquad \xi_{rs} \equiv \frac{\partial \xi_r}{\partial x_s},$$

where the dots denote expressions containing the first-order derivatives of u. The transformation (3.65) transforms the principal part given by (3.64) into the following form in n space:

$$\Lambda(u) = \sum_{i,j=1}^{n} \alpha_{ij} u_{\xi_i \xi_j},$$ (3.66)

where the coefficients α_{ij} are given by the transformation

$$\alpha_{ij} = \sum_{r,s=1}^{n} \xi_{ir} \xi_{js} a_{rs}.$$ (3.67)

The reduction to (x, y) space discussed in Section 2.10 is as follows: $n = 2$. The principal part (3.64) reduces to (2.58), the transformation (3.65) reduces to (2.60), the transformed principal part or canonical form (3.66) reduces to (2.63), and the transformed coefficients given by (3.67) reduce to (2.64). Specifically, x reduces to (x, y), $(\xi_1, \xi_2, \ldots, \xi_n)$ to (ξ, η), a_{ij} to a, b, c, and α_{ij} to A, B, C.

Coming back to n space: In a small enough neighborhood of a field point P, the coefficients a_{ij} are approximately constant so that we can approximate the operator $L(u)$ as a linear operator with constant coefficients, thus considering an equation with constant coefficients instead of a quasilinear equation. This allows us to investigate the simpler case of characteristic theory for the linear case. For an extended domain, the quasilinear equation is not easily classified into one of the three types, but may vary in type according to the position of the field point, making the situation overly complicated. It is sufficient for our purposes to study only what happens locally in the neighborhood of P.

Relation to Quadratic Forms. If the principal part of the second-order equation given by (3.64) involves a *hyperbolic* PDE, then we saw that the characteristic condition was give by (3.59). In any case, the left-hand side of (3.59) is a quadratic form Q in the ξ_i's, and a rule was given enabling us to construct Q. For convenience we let $\mathbf{y}$ be the n-dimensional vector whose ith component is ξ_i, and let the $(n \times n)$ matrix $\mathbf{A}$ have the elements a_{ij}. A study of (3.64) shows us that $\mathbf{A}$ is symmetric so that $a_{ij} = a_{ji}$. This fact will be necessary when we want to find a transformation of coordinates that will yield a "pure" quadratic form ($\xi_i\xi_j = 0$ for $i \neq j$). We shall show that, in order to get Q as a pure quadratic with respect to a certain coordinate system, we must diagonalize $\mathbf{A}$ which can only be done if $\mathbf{A}$ is a nonsingular symmetric matrix. This process of diagonalizing a matrix involves a *similarity transformation* discussed below.† If $\mathbf{y}$ is represented by a column matrix, then its transpose $\mathbf{y}^*$ is a row matrix. Q can then be described by the matrix equation

$$Q(\mathbf{y}) = \mathbf{y}^*\mathbf{A}\mathbf{y} = \sum_{i,j} a_{ij}y_iy_j. \tag{3.68}$$

To obtain the canonical form which enables us to classify the second-order equations into the three above-mentioned types, we need to find a transformation from ξ to η space (which is also n dimensional) such that the quadratic form $Q(\eta)$, with respect to this new vector space, has only pure quadratic terms such as η_i^2, ($\eta_i\eta_j = 0$ for $i \neq j$). As mentioned above, in order for $Q(\eta)$ to be a pure quadratic form, it is clear that such a transformation must *diagonalize* the matrix $\mathbf{A}$. Since $\mathbf{A}$ is a nonsingular symmetric matrix, such a transformation exists; it is called a *similarity transformation*. To this end, we look for an orthogonal transformation given by the *orthonormal matrix* $\mathbf{R}$ such that

$$\mathbf{y} = \mathbf{R}\eta \quad \text{or} \quad \eta = \mathbf{R}^{-1}\mathbf{y} = \mathbf{R}^*\mathbf{y}, \tag{3.69}$$

where we used the fact that the inverse of an orthogonal matrix equals its transpose so that $\mathbf{R}^{-1} = \mathbf{R}^*$. The vector η is a column matrix whose jth element is η_j. An orthonormal matrix is one where each row (and column) has the property that $\mathbf{r}_i\mathbf{r}_j = \delta_{ij}$ the Kronecker delta. If we now insert (3.69) into (3.68) we get

$$Q(\mathbf{y}) = \mathbf{y}^*\mathbf{A}\mathbf{y} = \eta^*\mathbf{R}^*\mathbf{A}\mathbf{R}\eta = \eta^*\mathbf{A}'\eta = Q(\eta), \tag{3.70}$$

where

$$\mathbf{A}' = \mathbf{R}^{-1}\mathbf{A}\mathbf{R} = \mathbf{R}^*\mathbf{A}\mathbf{R}. \tag{3.71}$$

The transformation given by (3.71) is called a *similarity transformation*. $\mathbf{R}$ is constructed such that $\mathbf{A}'$ is a diagonal matrix whose diagonal elements are the eigenvalues λ_i which are the roots of the algebraic equation

$$\det(\mathbf{A} - \lambda\mathbf{I}) = 0, \tag{3.72}$$

where $\mathbf{I}$ is the $(n \times n)$ identity matrix. $\mathbf{R}$ is made up of the eigenvectors corre-

---

† Since similarity transformations are pivotal to wave phenomena and many other applications of mathematical physics and engineering, we reserve the appendix of this chapter for a complete discussion of this important topic.

sponding to the eigenvalues λ_i. The appendix to this chapter gives the details. It is clear that the diagonal matrix $\mathbf{A}'$ makes $Q(\boldsymbol{\eta})$ a pure quadratic form, as can be seen by inspection. The roots λ of $\det(\mathbf{A}' - \lambda \mathbf{I}) = 0$ are the same as those of (3.72). (This is proved in the Appendix.) Equation (3.70) coupled with (3.71) tells us that the pure quadratic $Q(\boldsymbol{\eta})$ is constructed in the same manner with respect to $\mathbf{A}'$ that $Q(\boldsymbol{\xi})$ is with respect to $\mathbf{A}$. Since Q is a scalar function of its respective vector it thus has a numerical value independent of the transformation. In particular, for a hyperbolic system, we have the characteristic condition that $Q(\boldsymbol{\xi}) = 0$ so that $Q(\boldsymbol{\eta}) = 0$.

A pure quadratic form $Q(\boldsymbol{\eta})$ can always be written in the following form:

$$Q(\eta) = \sum_{i=1}^{n} k_i \eta_i^2, \tag{3.73}$$

where the coefficients k_i can only have the values ± 1 or 0. The number of negative coefficients is called the *index of inertia* and the number of vanishing coefficients are invariant with respect to a linear or "affine" transformation.

The quadratic form as represented by (3.73) is now used to *classify* the second-order PDE, in particular, the principal part given by (3.66), into one of the three categories in a domain where the coefficients are approximately constants:

(1) *Elliptic* if each $k_i = 1$ or -1. Then, by a suitable linear transformation the differential equation can be written as

$$u_{x_1 x_1} + u_{x_2 x_2} + \cdots + u_{x_n x_n} + \cdots = 0, \tag{3.74}$$

where the dots represent, as usual, at most first-order derivatives. Equation (3.74) is Laplace's equation in n dimensions. $Q(\boldsymbol{\eta})$ is essentially positive definite so that none of the roots of $Q(\boldsymbol{\eta}) = 0$ are real, and therefore no real characteristic surfaces exist in n space. Physically, an elliptic equation means that a potential such as a gravitational potential exists, and no wave phenomena exists.

(2) *Parabolic* if one or more of the coefficients k_i in (3.73) vanish. For example, suppose $k_n = 0$. Then the PDE can be written as

$$u_{x_1 x_1} + u_{x_2 x_2} + \cdots + u_{x_{n-1} x_{n-1}} + \cdots = 0, \tag{3.75}$$

which may be interpreted as Fourier's unsteady heat conduction equation in $(n - 1)$ dimensions. There exists only a single parameter family of characteristic surfaces on which waves are propagated with infinite speed. Physically, this parabolic system implies a diffusion process.

(3) *Hyperbolic* if all the k_i's have the same sign, say positive, except one, k_n, for instance, which is negative. This is the case of interest to us, as it involves wave propagation phenomena. A suitable coordinate system puts the second-order PDE into the form

$$u_{x_1 x_1} + u_{x_2 x_2} + \cdots + u_{x_{n-1} x_{n-1}} - u_{x_n x_n} + \cdots = 0, \tag{3.76}$$

which is the wave equation in $(n - 1)$ dimensions, with unit wave speed, where the coordinate x_n plays the role of time. We shall discuss this equation in some detail below. It is sufficient to mention here that the three-dimensional wave equation with unit wave speed in (x, y, z) space is given by

$$u_{xx} + u_{yy} + u_{zz} - u_{tt} = 0. \tag{3.77}$$

3.7. Geometry of Characteristics for Second-Order Systems

We again consider the second-order quasilinear system described by (3.63), returning to the notation given in Section 3.6. To adapt (3.63) to the wave equation in n spatial dimensions, we introduce a slight change in notation for the independent variables by increasing the dimensions of the space to $(n + 1)$ in order to demark the time t. Specifically, let $\mathbf{x} = (x_0, x_1, x_2, \ldots, x_n)$ where $x_0 = t$. (Clearly, for (x, y, z) space, $n = 3$.)

We recall from Section 3.6 that the *Cauchy initial value problem* for our second-order system involves prescribing initial data for u and the u_i's $(i = 1, 2, \ldots, n)$ on an initial surface S defined by

$$\phi(s_0, s_1, s_2, \ldots, s_n) = 0, \qquad \text{grad } \phi \neq 0 \quad \text{on } S, \tag{3.78}$$

where $(s_1, s_2, \ldots, s_n)$ are the coordinates on the surface S, and s_0 is a parameter defining S, such that $s_0 = t_0 = 0$ on the initial surface. Actually, a specific surface is defined by a particular value of the parameter s_0: we may let S be a *characteristic surface* so that, as mentioned several times, $s_0 \neq 0$ on the characteristic surface, meaning that a characteristic surface can nowhere be tangent to an initial surface. In summary: $\mathbf{x} = \mathbf{s} = (s_0, s_1, \ldots, s_n)$ on S and $s_0 = 0$ if S is an initial surface. Courant and Hilbert [11, Vol. II] call $(s_0, s_1, s_2, \ldots, s_{n-1})$ "interior variables" since they are defined on S. They single out s_n by setting $s_n = \phi$, where $\phi = 0$ defines S. The pivotal point of this transformation from $\mathbf{x}$ to $\mathbf{s}$ space is to make full use of the initial data in the spirit of the Cauchy problem for our second-order system.

We have the functional relationships $u = u(\mathbf{x})$ and $\mathbf{x} = \mathbf{x}(\mathbf{s})$. Since u and the first derivatives are prescribed initial data on S, we know $u = u(0, s_1, s_2, \ldots, s_{n-1})$ and $u_i \equiv \partial u / \partial s_i$ $(i = 0, 1, 2, \ldots, n - 1)$ on S. Since $(0, s_1, \ldots, s_{n-1})$ are "interior variables" we can calculate all the second-order partial derivatives (and then the higher-order ones) on S, except $u_{\phi\phi}$. Now a key question arises: Can the PDE (3.63) and the initial data on the surface S allow us to determine $u_{\phi\phi}$ on S? To answer this question we transform (3.63) from $\mathbf{x}$ coordinates to $\mathbf{s}$ coordinates. The amazing fact is: In terms of the interior variables $s_0, s_1, \ldots, s_{n-1}$ and $s_n = \phi$ (3.63) is transformed (by using the chain rule) into

$$Q(\phi_0, \phi_1, \ldots, \phi_n)u_{\phi\phi} + \cdots = 0, \tag{3.79}$$

where the dots are expressions that only involve u and its first-order derivatives on S, which are known. Q is the characteristic determinant such that

$$Q(\phi_0, \phi_1, \ldots, \phi_n) = \sum_{i,j=0}^{n} a_{ij}\phi_i\phi_j, \qquad \phi_i \equiv \frac{\partial \phi}{\partial x_i}. \tag{3.80}$$

Recall that the ϕ_i's are the components of **grad** ϕ. The *principal part* of the PDE is

$$Q(\phi_0, \phi_1, \ldots, \phi_n)u_{\phi\phi}.$$

Equation (3.79) tells us that the principal part of the differential equation can be calculated in terms of the initial data, *as long as $Q \neq 0$*. We saw above that the *characteristic condition is $Q = 0$* on S, where S is a characteristic surface rather than an initial value surface. It is clear that if $Q = 0$, then the principal part is identically zero so that we cannot uniquely solve for $u_{\phi\phi}$ on the characteristic surface S. On the other hand, if $Q \neq 0$ then, as mentioned above, we can uniquely solve for the principal part in terms of the initial data on S so that S is indeed an initial value surface and not a characteristic surface. Therefore, the answer to the above question is that $u_{\phi\phi}$ can be calculated on S if and only if S is an initial value surface. We see that, if we can calculate $u_{\phi\phi}$ on S, then all the necessary partial derivatives can then be calculated on S so that the solution in a strip surrounding S can be constructed by a Taylor series expansion in terms of all the necessary derivatives on S. The solution for $u(\mathbf{x})$ can then be extended globally into the appropriate domain. Again, this can only be accomplished if S is an initial value surface and not a characteristic surface. If S is not a characteristic surface it is sometimes called a *free surface*. If S is not a free surface then it is a characteristic surface and the following *characteristic condition* is valid:

$$Q(\phi_0, \phi_1, \ldots, \phi_n) = \sum_{i,j=0}^{n} a_{ij}\phi_i\phi_j = 0 \quad \text{on } S. \tag{3.81}$$

Clearly, since $Q = 0$ on a characteristic surface S, and the principal part of the PDE Qu is identically zero, the remainder of the differential equation, which consists of the first-order derivatives (the dots in (3.79)), puts an *additional restriction on the data*. This again shows that initial data on a characteristic surface cannot be used to obtain the solution off the surface. We point out that characteristic surfaces can exist only if the characteristic condition (3.81) is satisfied by real functions ϕ (real characteristic surfaces). This means that the quadratic form $Q(\phi_0, \phi_1, \ldots, \phi_n)$ must be *indefinite*.

Note that the characteristic condition (3.81) is a first-order second-degree PDE in the $(n + 1)$ derivatives ϕ_i $(i = 0, 1, \ldots, n)$ whose solution yields a two-parameter family of characteristic surfaces $\phi = $ const. Since there are $(n + 1)$ variables $(x_0, x_1, \ldots, x_n)$, (3.81) need not satisfy the PDE identically, but only on the characteristic surfaces. However, Courant and Hilbert [11, Vol. II] point out that if the characteristic surface S is given in the form

$$x_0 = \psi(x_1, x_2, \ldots, x_n), \tag{3.82}$$

then (3.81) does represent the following first-order second-degree PDE for $\psi(x_1, \ldots, x_n)$ (which has, of course, only n variables):

$$\sum_{i,j=1}^{n} a_{ij}\psi_i\psi_j - 2 \sum_{i=1}^{n} a_{i0}\psi_i + a_{00} = 0, \qquad \psi_i \equiv \frac{\partial\psi}{\partial x_i}. \tag{3.83}$$

(Note that the coefficients a_{ij} are symmetric.) For the quasilinear case these coefficients are known functions of $(x_0, x_1, \ldots, x_n)$, u, and the u_i's. The expressions in $x_1, \ldots, x_n$ must be substituted for x_0 in these coefficients.

We know that $x_0 = t$. If we set $a_{00} = -1$ and $a_{i0} = 0$ for $i = 1, 2, \ldots, n$, and if a_{ij} is independent of t then the PDE (3.63) becomes

$$u_{tt} - \sum_{i,j=1}^{n} a_{ij} u_{ij} + \cdots = 0. \tag{3.84}$$

The characteristic PDE for ψ becomes

$$\sum_{i,j=1}^{n} a_{ij} \psi_i \psi_j = 1. \tag{3.85}$$

The system (3.84) and (3.85) often occurs in electromagnetic wave propagation. In fact, using the similarity transformation discussed above, we can transform the quadratic form $Q(\psi_i)$, given by the left-hand side of (3.85), into a pure quadratic form in terms of the transformed coordinates, so that (3.84) is transformed into its canonical form which is the wave equation in $(n - 1)$ dimensions given by (3.76) (with an obvious change in notation).

We continue with the geometry of the second-order PDE in n variables by discussing rays vis-à-vis bicharacteristics and wave fronts.

Bicharacteristics or Rays

We recall that (3.81), which is essentially $Q(\mathbf{grad}\ \phi) = 0$, is the characteristic condition for the PDE for u given by (3.63). Equation (3.81) is a fully nonlinear first-order second-degree PDE in the $(n + 1)$ components of $\mathbf{grad}\ \phi$. As such it falls into the category of PDEs discussed in Section 3.2. The same is true for the characteristic condition on $\mathbf{grad}\ \psi$ which is a quadratic form in the n components ψ_i of $\mathbf{grad}\ \psi$ set equal to zero, given by (3.83). It can be written as

$$Q(\mathbf{grad}\ \psi) = \sum_{i,j=1}^{n} a_{ij} \psi_i \psi_j - 2 \sum_{i=1}^{n} a_{i0} \psi_i + a_{00} = 0. \tag{3.83a}$$

For the generic form of the fully nonlinear PDE (3.7), given in Section 3.2, we identify F with Q, u with ϕ or ψ, and the p_i's with the ϕ_i's or ψ_i's. Recall that the characteristic ODE associated with (3.7) is given by the system (3.8). The correspondence between the first set of equations of the system (3.8) and the characteristic ODEs associated with (3.81) is

$$\frac{dx_i}{d\tau} = \frac{1}{2} \frac{\partial Q}{\partial \phi_i} = \frac{1}{2} Q_{\phi_i} = \sum_{j=0}^{n} a_{ij} \phi_j, \qquad i = 0, 1, \ldots, n, \tag{3.86}$$

where τ is the parameter that traces the trajectories which are the integral curves or solutions of (3.86). The system (3.86) is the set of characteristic ODEs associated with the PDE (3.81) for $\mathbf{grad}\ \phi$. Since (3.81) is the characteristic condition associated with (3.63) (the PDE for u) and (3.86) is the set of characteristic conditions for the characteristics of (3.63), we call the system

(3.86) the *bicharacteristics* associated with (3.63), the original differential equation for *u*. (The system (3.86) is the "characteristics of the characteristics".) Note that the one-dimensional wave equation (or any PDE in two independent variables) has no bicharacteristics, since the characteristics of the PDE are themselves ODEs. For the linear case, the coefficients a_{ij} are independent of *u* and the u_i's. This is not so for the quasilinear case. However, for this case, we think of a fixed solution for *u* at the field point x and insert that value of *u* and the u_i's into the coefficients.

Coming back to the characteristic system given by (3.8), the second equation yields

$$\frac{d\phi}{d\tau} = \sum_{i,j} a_{ij}\phi_i\phi_j = 0, \tag{3.87}$$

upon using (3.81). This tells us that $\phi = $ const. on a characteristic surface. Since Q does not depend on ϕ and the x_i's, the third equation of (3.8) yields

$$\frac{d\phi_i}{d\tau} = 0, \qquad i = 0, 1, 2, \ldots, n. \tag{3.87a}$$

As implied above, the solutions of the characteristic ODEs (3.86) yield a one-parameter family of integral curves $x_i = x_i(\tau)$, since (3.86) is a first-order system for the x_i's. This family of integral curves is called *the bicharacteristics or characteristic rays of the given second-order PDE* (3.63). These integral curves generate all the characteristic surfaces $\phi = $ const. associated with (3.63). Clearly, the conditions (3.87) and (3.87a) hold on each characteristic surface.

We make the following observation: If two distinct characteristic surfaces are tangent to each other at time $t = 0$, then at every time $t > 0$ they maintain that point of common tangency. Since, as will be described below, the characteristic surfaces are wave fronts, this means that the point of common tangency moves along a bicharacteristic or ray common to those two wave fronts.

We may go through the same theory to obtain the bicharacteristics or rays associated with the quadratic form in **grad** ψ given by (3.83a) which involves *n* components of **grad** ψ. However, we reserve this treatment for the subsection on wave fronts.

Wave Fronts

We recall, from previous discussions about wave propagation, that a wave front is a surface normal to the direction of wave propagation which travels with the wave speed or phase speed. For a plane wave the wave front is a planar surface. We also know that electromagnetic waves are transverse waves in the sense that both the electric vector **E** and the magnetic vector **H** oscillate normal to the direction of wave propagation, and also that **E** and **H** are normal to each other and lie in the plane of the wave front. For the one-dimensional wave equation a wave front is a characteristic curve in the (x, t) plane across which we can allow discontinuities in the second derivatives of $u(x, t)$, since we cannot use information on the characteristic curve to solve for the second derivatives—they are ambiguous. The direction of wave propagation is normal to the characteristics. If the wave speed is constant then the two-parameter family of characteristics are straight lines, so that the waves propagate in straight lines, the progressing waves being normal to the C_+ characteristics, and the regressing waves normal to the C_- characteristics.

When we consider the $(n + 1)$-dimensional hyperbolic PDE for $n > 2$ we know that the characteristic curves are now characteristic surfaces. We recall that a characteristic surface is not a free surface like an initial value surface. This means, as we saw above, that the second derivatives such as $u_{\phi\phi}$ are not uniquely determined if we were given the Cauchy initial data on the characteristic surface. We again emphasize that only on a free or initial value surface can the Cauchy initial data be used to calculate $u_{\phi\phi}$. There is an ambiguity in the second derivatives of $u(t, x_1, \ldots, x_n)$ so that these second derivatives might have discontinuities across the characteristic surface; meaning that there can be two different values on either side at a given point on a characteristic surface. Therefore a characteristic surface is one that can propagate discontinuities in second derivatives. But this is not the complete definition of a characteristic surface. In general, a characteristic surface is one that separates a disturbed region from an undisturbed region. We know that a propagating wave is caused by a disturbance in pressure, density, etc., at a source which propagates the wave with a characteristic speed (the wave speed). A characteristic surface is then a wave front in the sense that it separates the disturbed region, which involves times previous to where the wave front is, to the undisturbed region, involving times later than the position of the wave front. For example, in the phenomenon of sound wave propagation in a three-dimensional medium, a wave front or characteristic surface is a spherical surface across which there is an infinitesimal jump in pressure. If there is a finite jump in pressure, entropy, etc., across the wave front then a spherical shock wave is produced, so that a sound wave may be considered as a shock wave of infinitesmal strength. (The strength of a shock wave is defined as the ratio of the difference between the disturbed and the undisturbed pressure across the wave front to the undisturbed pressure.) Another more complicated example of a wave front or characteristic surface is the propagation of shock waves in a magnetohydrodynamic medium. The laws of fluid dynamics in an electromagnetically conducting medium must be satisfied in the disturbed region. These are the three conservation laws of physics. The electromagnetic material inside the shock wave (spherical shock in a three-dimensional medium) is disturbed in the sense that the pressure, density, etc., are in an unsteady state and can, in principle, be solved for from the field equations which stem from the conservation laws. Outside the wave front, the material is in an undisturbed or quiescent state. Actually, the surface forming the wave front for a shock wave is not a characteristic surface, since a characteristic surface travels with the speed of sound in the medium and a shock wave travels supersonically. But the shock front is approximately a characteristic surface for a very weak shock wave, since such a wave travels with approximately the speed of sound.

It is important to stress that wave fronts occur as "frontier surfaces" beyond which there is no excitation of the medium: quiescent state. The solutions of the field equations giving the field quantities such as pressure therefore yield unsteady solutions inside the wave front, and zero disturbances in the quiescent region.

We now consider $\psi(x_1, x_2, \ldots, x_n)$, as defined by (3.82), as a characteristic surface in n dimensions where $x_0 = t$ is time. We are then concerned with solutions $u = u(x_1, \ldots, x_n, t)$ of the PDE in the form given by (3.84). We interpret solutions u as being a function of n-dimensional or R_n space, with time t as a parameter, involving a characteristic surface defined by (3.82). For simplicity, we assume (3.84) to be a linear PDE with constant coefficients. For convenience, we rewrite (3.82)

$$\psi(x_1, x_2, \ldots, x_n) = t, \tag{3.82}$$

where the characteristic surface moves through R_n space with t as the parameter which generates the surface. This means all the x_i's are functions of t so that the parameter $\tau = t$. The characteristic PDE for ψ is given by (3.83). The system of characteristic equations associated with (3.83) is

$$\frac{dx_i}{dt} = \sum_{j=1}^{n} a_{ij}\psi_j, \qquad i = 1, 2, \ldots, n. \tag{3.88}$$

The system (3.88) consists of the *bicharacteristic equations* or the rays associated with the PDE (3.83).

We know that the wave front is given by the equation $\psi = t$. If we differentiate this equation with respect to t, use the chain rule and (3.85), we obtain

$$\sum_{i=1}^{n} \psi_i \dot{x}_i = \sum_{i,j=1}^{n} a_{ij}\psi_i\psi_j = 1, \quad \text{where } \dot{x}_i = \frac{dx_i}{dt} = v_i. \tag{3.89}$$

Let the vector $\dot{\mathbf{x}} = \mathbf{v} = (v_1, v_2, \ldots, v_n)$. Then (3.89) can be written in vector form as

$$\mathbf{grad}\, \psi \cdot \mathbf{v} = \sum_{i,j} a_{ij}\psi_i\psi_j = Q(\mathbf{grad}\, \psi) = 1. \tag{3.89a}$$

$\mathbf{v}$ is called the *ray velocity vector*. Since the PDE for u is hyperbolic, each of the n rowed matrices (a_{ik}) is positive definite. We may then consider the components of $\mathbf{grad}\, \psi$ as the coordinates of the ellipsoid given by (3.89), which tells us that the ray direction $\mathbf{v}$ is in the tangent plane at each point on the wave front $\psi - t = \text{const.}$ and also in the tangent plane of the above ellipsoid.

In addition to the ray velocity we must consider the *wave velocity* or velocity of the wave front. Let $\mathbf{c} = (c_1, c_2, c_3)$ be the wave velocity. $\mathbf{c}$ is normal to the wave front. It may therefore be determined by following a point on an orthogonal trajectory of a given wave front $\psi = t = \text{const.}$ as that point moves in time. This is another way of saying that $\mathbf{c}$ is normal to the surface $\psi = \text{const.}$

Clearly, this means that the components of $\mathbf{c}$ are proportional to the corresponding component of $\mathbf{grad}\ \psi$. This is given by

$$c_i = \frac{|c|\psi_i}{(\mathbf{grad}\ \psi)^2},\tag{3.90}$$

where $|c|$ is the magnitude of $\mathbf{c}$. Inserting (3.90) into (3.88) yields the following relation between the v_i's and c_i's:

$$v_i = \frac{(\mathbf{grad}\ \psi)^2}{|c|} \sum_{j=1}^{n} a_{ij}c_j.\tag{3.91}$$

3.8. Ray Cone, Normal Cone, Duality

We consider the quasilinear PDE (3.63) for $u(x, t)$ in R_n space, in the form given by (3.84), which is appropriate for wave propagation problems. We know that associated with (3.84) is the characteristic PDE $Q(\mathbf{grad}\ \phi) = 0$ given by (3.81) or $Q(\mathbf{grad}\ \psi) = 0$ given by (3.85). The integral surfaces are $\phi(t, x_1, x_2, x_n) = 0$ or $\psi(x_1 x_2, \ldots, x_n) = t$. The relationship between ϕ and ψ is given by

$$\phi(t, x_1, x_2, \ldots, x_n) = \psi(x_1, x_2, \ldots, x_n) - t = 0.\tag{3.92}$$

We now consider the characteristic condition $Q(\mathbf{grad}\ \phi) = 0$ given by (3.81). As we know, the characteristic surfaces $\phi(\mathbf{x}, t) = $ const. are determined by solving the bicharacteristic ODEs (3.86) for the rays, which are the trajectories $x_i(\tau)$. The directions of these rays through some field point P form a quadratic local ray cone or the *Monge cone* for the characteristic PDE (3.81). This cone was treated for the fully nonlinear PDE in two independent variables in Chapter 2. It is clear that (3.81) gives us a quadratic condition for the direction numbers ϕ_i which are components of $\mathbf{grad}\ \phi$ and are therefore normal to the characteristic surface elements. However, the rays which are in the direction given by the ray velocity v are in the tangent plane of the Monge cone.

Let each ϕ_i be represented by a vector ξ_i so that the set of vectors $\xi = (\xi_0, \xi_1, \xi_2, \ldots, \xi_n)$ emanate from the origin of the $(n + 1)$-dimensional ξ space (which, for convenience, we represent in the same coordinate system as $(t, \mathbf{x})$). Then the end points of these vectors lie on the cone

$$\sum_{i,j=0}^{n} a_{ij}\xi_i\xi_j = 0.\tag{3.93}$$

Equation (3.93) is sometimes called the *normal cone*, since it involves the normals to the quadratic surface (3.81).

The ray directions given by the components of the ray velocity $\dot{x}_i = v_i$ are given by

$$\dot{x}_i = v_i = \sum_{j=0}^{n} a_{ij}\xi_j,\tag{3.94}$$

which is seen from the bicharacteristic equations (3.86) (where the ϕ_i's are given by the ξ_i's). Clearly, the ξ_i's satisfy the characteristic condition (3.81). Since the matrix $\mathbf{A}$ is nonsingular it has an inverse $\mathbf{B}$, where $\mathbf{A}^{-1} = \mathbf{B}$. We can therefore solve (3.94) for the ξ_i's. We obtain

$$\xi_i = b_{ik}v_k \qquad \text{(summed over } k \text{ from 0 to } n\text{)}. \tag{3.95}$$

Substituting this equation for the ξ_i's into (3.81) we obtain

$$\sum_{i,j=0}^{n} b_{ij}v_iv_j = 0, \tag{3.96}$$

which is called the *ray cone*. If the v_i's satisfy the ray cone equation, then the ray velocity $\mathbf{v}$ is in the bicharacteristic direction. By definition, the ray cone is the envelope of all the characteristic *surface elements* passing through the field point P. We see that the normal cone, given by (3.93), is the envelope of all the planes *normal to the generating rays* passing through P. In this sense the two cones are *reciprocally related*. They are said to be *dual to each other*.

EXAMPLE. Consider the PDE

$$u_{tt} - u_{x_1 x_1} - k u_{x_2 x_2} = 0,$$

where k is a positive constant. Then the equation of the normal cone is

$$\xi_1^2 + k\xi_2^2 = \xi_0^2,$$

and the equation of the ray cone is

$$x_1^2 + \left(\frac{1}{k}\right)x_2^2 = t^2.$$

If $k = 1$ then we have the wave equation in R_2 space, and the two cones are identical.

We point out that the ray cone whose vertex is at a field point P is in two parts: the *forward* cone pointing toward increasing time or to the future, and the *backward* cone pointing into the past. If the coefficients a_{ij} of the PDE are not constant we merely construct a normal and local ray cone of the characteristic directions whose vertex is at each field point. The locus of all rays through a field point is sometimes called the *ray conoid* which is clearly a characteristic surface or wave front with the field point P as the "center of disturbance". This means that *a disturbance at P can only be propagated along the rays forming the characteristic surface*.

Finally, we remark that a similar analysis can be made where we use **grad** ψ instead of **grad** ϕ. Instead of considering a normal cone for the components of **grad** ϕ, we consider the ellipsoid (3.85) for the components of **grad** ψ. This is the *normal ellipsoid* for the ψ_i's. The ellipsoid dual to the normal ellipsoid is obtained by solving for the ψ_i's in the bicharacteristic ODEs (3.88). Again, letting b_{ij} be the (ij)th element of $\mathbf{A}^{-1}$, we solve (3.88) for the ψ_i's and obtain

$$\psi_i = \sum_{k=1}^{n} b_{ik} v_k, \qquad i = 1, 2, \ldots, n. \tag{3.97}$$

Inserting this expression for the ψ_i's into (3.85) yields

$$\sum_{i,j=0}^{n} b_{ij} v_i v_j = 1. \tag{3.98}$$

The ellipsoidal surface given by (3.98) is called the *ray ellipsoid*. We reiterate that the ellipsoid given by (3.85) is the normal ellipsoid, which is a quadratic in the components of **grad** ψ; these are the components of the normal to an element of the surface (3.85). On the other hand, the ellipsoid given by (3.98) is a quadratic in the components of $\dot{\mathbf{x}} = \mathbf{v}$, which are the direction numbers to an element of the tangent plane of the ellipsoidal surface (3.98). In this sense, these ellipsoids are reciporcally related. One is the dual of the other. Thus, we have a complete correspondence between the normal and ray cones with respect to the ϕ_i's, and the normal and ray ellipsoids with respect to the ψ_i's.

3.9. Wave Equation in n Dimensions

We shall consider the wave equation in R_n space in the form

$$u_{tt} - c^2(u_{x_1 x_1} + u_{x_2 x_2} + \cdots + u_{x_n x_n}) = 0, \tag{3.99}$$

where c is the wave speed which we assume to be constant.

The last example in Section 3.8 showed us that the wave equation in R_2 space has its normal cone identical to its dual or ray cone. We point out that this fact is also true for the wave equation in n dimensions (and, in particular, for $n = 3$).

The characteristic PDE for ψ, given by (3.85), becomes

$$\psi_1^2 + \psi_2^2 + \cdots + \psi_n^2 = 1. \tag{3.100}$$

Equation (3.100) is easily solved for ψ in terms of $x_1, x_2, \ldots, x_n$. We assume that we can separate variables in an additive way by letting $\psi = \sum_{i=1}^{n} f_i(x_i)$. We easily solve for the f_i's by inserting this expression into (3.100) and making use of the fact that the x_i's are independent. We get the result that ψ is a linear combination of the x_i's. Thus

$$\psi = \sum_{i=1}^{n} \alpha_i x_i, \qquad \text{where } \sum_i \alpha_i^2 = 1. \tag{3.101}$$

A wave front at a particular time is given by $\psi = ct = $ const., which is essentially (3.82), taking into account the wave speed c in the wave equation. Using (3.101) the wave front becomes

$$\sum_i \alpha_i x_i = ct, \tag{3.102}$$

where the components of **grad** ψ are the constants α_i. Equation (3.102), in the form $\sum \alpha_i x_i - ct = $ const., is a family of *planar surfaces* where the time t is the parameter that generates the wave front. (The reader is referred to Chapter 2 for a discussion of (3.100) in (x, y) and (x, y, z) space in connection with an example in *geometric optics*.)

The bicharacteristic or ray equations (3.88) become

$$\frac{dx_i}{dt} = v_i = c^2 \psi_i.\dagger \tag{3.103}$$

Thus the ray speed v is constant for a constant wave speed c. The bicharacteristic equations (3.103) can be integrated immediately from a given point P_0 whose coordinates are $\mathbf{x}^0 = (x_1^0, x_2^0, \ldots, x_n^0)$ to any point P: $\mathbf{x}$, where P_0 is a fixed point and P is a variable on the family of wave fronts. The solution of (3.103) then becomes

$$x_i - x_i^0 = v_i(t - t_0), \qquad i = 1, 2, \ldots, n, \tag{3.104}$$

where t_0 is the time corresponding to the point P_0 on the wave front. Upon squaring and summing over i (from 1 to n) in (3.104), and using the condition $\sum_{i=1}^{n} (v_i)^2 = c^2$, we obtain

$$\sum_{i=1}^{n} (x_i - x_i)^2 = c^2(t - t_0)^2. \tag{3.105}$$

Equation (3.105) is called the *characteristic cone* (whose vertex is P_0) corresponding to the wave equation (3.99) in R_n space.

$n = 3$. For the three-dimensional wave equation (R_3 space) we consider (x, y, z, t) space. The wave equation (3.100) becomes

$$u_{tt} - c^2(u_{xx} + u_{yy} + u_{zz}) = 0. \tag{3.106}$$

The characteristic cone (3.105) becomes

$$(x - x_0)^2 + (y - y_0)^2 + (z - z_0)^2 = c^2(t - t_0)^2. \tag{3.107}$$

The characteristic equation (3.100) or ellipsoid for **grad** ψ becomes

$$(\mathbf{grad}\ \psi)^2 = (\psi_x)^2 + (\psi_y)^2 + (\psi_z)^2 = 1. \tag{3.108}$$

Equation (3.108) is called the *eiconal$\ddagger$ equation*. It is a pivotal equation in wave propagation.

Let n be the *index of refraction* of an electromagnetic medium. By definition

$$n = \frac{c_0}{c},$$

$\dagger$ Note that $\sum_{i=1}^{n} v_i^2 = c^2$ so that $|\mathbf{v}| = |\mathbf{c}|$, which is correct since the ray speed is actually the wave speed.

$\ddagger$ Sometimes spelled "eikonal."

where c_0 is the wave speed in a vacuum and c is the wave speed in the medium. For a medium of variable index of refraction n is a known function of $\mathbf{x}$. In any case, the eiconal equation† takes the form

$$(\mathbf{grad}\ \psi)^2 = n^2, \tag{3.108a}$$

which is essentially (3.108).

The characteristic equation (3.81) or cone for $\mathbf{grad}\ \phi$ becomes

$$c^2[(\phi_x)^2 + (\phi_y)^2 + (\phi_z)^2] = (\phi_t)^2. \tag{3.109}$$

The equation (3.102) for the planar wave front becomes

$$\psi_x x + \psi_y y + \psi_z z - ct = \text{const.}, \tag{3.110}$$

where the components of $\mathbf{grad}\ \psi$ are constants.

Plane Waves

As pointed out above, if c and the ψ_i's are constants then the ray speed v is constant, so that the wave front is a planar surface traveling with the constant wave speed $v = c$ normal to the wave front. Propagating waves that have the property of their wave fronts being planar are called *plane waves*. In Chapter 1 we treated the subject of plane wave propagation in electromagnetic media, both nonconducting and conducting. We were concerned with time-harmonic wave propagation. For example, for the three-dimensional formulation, a harmonic progressing electric or $\mathbf{E}$ and magnetic or $\mathbf{B}$ wave is given by (1.52). This means that a time-harmonic solution of (3.106) for a progressing wave is a vector $\mathbf{u}(\mathbf{x}, t)$ having the form

$$\mathbf{u} = \mathbf{e}u \exp[i(\mathbf{k} \cdot \mathbf{x} - \omega t)], \tag{3.111}$$

where $\mathbf{e}$ is the unit vector in the direction of $\mathbf{u}$, u is the complex magnitude of $\mathbf{u}$, $\mathbf{k} = (k_x, k_y, k_z)$ is the wave number whose components in the (x, y, z) directions are shown, and ω is the frequency. $\mathbf{k}$ specifies the direction of wave propagation. If u represents either $\mathbf{E}$ or $\mathbf{B}$ or $\mathbf{H}$, then $\mathbf{e}$ is normal to the direction of wave propagation $\mathbf{k}$, since $\mathbf{u}$ represents an electromagnetic wave which, as was showed in Chapter 1, is transverse. The exponent in (3.111) becomes

$$\mathbf{k} \cdot \mathbf{x} - \omega t = k_x x + k_y y + k_z z - \omega t = \mathbf{k} \cdot (\mathbf{x} - \mathbf{c}t). \tag{3.112}$$

Now consider the left-hand side of (3.110), which is $\psi_x x + \psi_y y + \psi_z z - ct = \mathbf{grad}\ \psi \cdot \mathbf{x} - ct = \psi - ct$, upon using the definition of ψ for a plane wave given by (3.101). We set

$$\mathbf{u} = \mathbf{e}u \exp[i(\mathbf{grad}\ \psi \cdot \mathbf{x} - ct)]. \tag{3.113}$$

Inserting this solution for $\mathbf{u}$ into (3.106) and factoring out the exponential

---

† The importance of the eiconal equation will be seen in the subsector below on plane waves.

yields the eiconal equation (3.108) for ψ. Thus we see that ψ for a plane wave satisfies the characteristic or eiconal equation. Now the correspondence between $\mathbf{grad}\,\psi \cdot \mathbf{x}$, as given in the exponent of (3.113), and $\mathbf{k} \cdot \mathbf{x}$ as given in the exponent of (3.111) is obvious, since we may write this as $i(\mathbf{k} \cdot \mathbf{x} - \omega t) = i\mathbf{k} \cdot (\mathbf{x} - \mathbf{c}t)$, where $\mathbf{c} = \omega/\mathbf{k}$.

Note that a wave front is given by the equation

$$\mathbf{grad}\,\psi \cdot \mathbf{x} - ct = \text{const.},$$

where the constant defines the wave front. As time increases we see that a planar wave front propagates in R_3 space with a wave speed c. From (3.113) it is clear that $\mathbf{u}$ is constant on a given wave front. This means that $\mathbf{u}$ is propagated *undistorted* on a given wave front.

Phase for Plane Waves

Let

$$\mathbf{k} \cdot \mathbf{x} - \omega t = \mathbf{grad}\,\psi \cdot \mathbf{x} - \omega t = \phi(\mathbf{x}, t). \tag{3.114}$$

$\phi(\mathbf{x}, t)$ occurs in the exponent of (3.111); it is called the *phase* corresponding to a *progressing wave* which is a time-harmonic wave. If the phase is $\phi(\mathbf{x}, t) = \mathbf{k} \cdot \mathbf{x} + \omega t = \mathbf{grad}\,\psi \cdot \mathbf{x} + \omega t$ then we have a *regressing wave*. As mentioned, (3.111) is the representation for a time-harmonic wave. Note that not all waves need be time-harmonic in form. The only requirement is that the general plane wave solution of (3.106) be of the form

$$u(\mathbf{x}, t) = F(\mathbf{k} \cdot \mathbf{x} - \omega t) + G(\mathbf{k} \cdot \mathbf{x} + \omega t)$$

$$= F(\mathbf{grad}\,\psi \cdot \mathbf{x} - \omega t) + G(\mathbf{grad}\,\psi \cdot \mathbf{x} + \omega t), \tag{3.115}$$

where F and G are arbitrary functions of their arguments, and can be calculated knowing the two initial and two boundary conditions for a specific wave propagation problem. Suppose $G = 0$ so that $u = F$. If the phase $\phi(\mathbf{x}, t)$ is constant then F is constant so that the progressing wave is constant on the surface $\phi(\mathbf{x}, t) = \text{const}$. This can only mean that the wave is *undistorted* so that there are no dispersion or dissipation effects. This means that the wave travels with unchanging wave form. This can only be the case if $u = F(\phi(\mathbf{x}, t))$, and the wave number $\mathbf{k}$ is real. If $\mathbf{k}$ is complex then the imaginary part of $\mathbf{k}$ gives a dissipation effect or an exponential attenuation of the wave. Clearly, a similar argument is valid for a regressing wave ($F = 0$), where, which is constant on the surface $\phi(\mathbf{x}, t) = \text{const.}$; in this case $\phi(\mathbf{x}, t) = \mathbf{k} \cdot \mathbf{x} + \omega t = \mathbf{grad}\,\psi \cdot \mathbf{x} + \omega t$.

We make a more general statement, which is quite important in wave propagation phenomena: If the hyperbolic PDE for $u(\mathbf{x}, t)$ contains only the principal part, then not only exponential (time-harmonic) functions are solutions, but quite generally all functions of the form

$$u(\mathbf{x}, t) = f(\phi(\mathbf{x}, t)) \tag{3.115}$$

are solutions. $\phi(\mathbf{x}, t)$ is the general phase meaning that it stands for both a progressing and a regressing wave (with the obvious sign prevailing). The reason why (3.115) is valid for any hyperbolic equation containing only the principal part is that, as we saw, the wave equation can be obtained by a transformation of coordinates, thus transforming (3.84) (for $n = 3$) to the normal or canonical form (the three-dimensional wave equation (3.106)). We reiterate that the waves $u = f(\phi(\mathbf{x}, t)$ propagate signals in an undistorted or unchanging wave form along surfaces of constant phase with a constant wave speed c; the direction of wave propagation is given by the direction of the wave number $\mathbf{k}$.

Dispersion and Dissipation

We first take up the case of *dispersion*. The phenomenon of dispersion was discussed in Chapter 1 in connection with electromagnetic wave propagation. This was discussed in Section 1.4, the subsection on group velocity and dispersion. There it was shown that, if the wave speed c is a function of frequency (for a polychromatic wave), then the wave form is dispersed in the sense that different frequencies travel with different velocities. Clearly, this means that we no longer have an undistorted wave form as we did if c is constant (independent of frequency). A group velocity was introduced which is the velocity of the wave packet. This was given by (1.47). (The analysis was given for the one-dimensional wave equation, but can be extended to n dimensions.)

The phenemomen of *dissipation* was discussed in Section 1.5 of Chapter 1 which treated plane electromagnetic waves in a conducting medium. We saw that the generic PDE for wave propagation in a conducting medium was given by (1.73), which is the homogeneous wave equation with a damping term that is proportional to the velocity term $\partial \mathbf{V}/\partial t$ (where $\mathbf{V}$ stands for $\mathbf{E}$ or $\mathbf{H}$), the proportionality constant involving the conductivity σ. It was shown that the wave number k is complex, and was given by (1.79) and (1.80). It was also shown that both the real and imaginary parts of k contain the conductivity parameter; but for a nonconducting medium the imaginary part vanishes. For a conducting medium the imaginary part represents the pure dissipation or exponential decay of the wave form, as mentioned above.

Telegraph Equation

The telegraph equation is an important equation that is satisfied by the voltage or current in a cable or *transmission line*. It is given by

$$u_{tt} - c^2 u_{xx} + (a + b)u_t + abu = 0, \tag{3.116}$$

where u is a generic variable representing the current or voltage in a cable as a function of position x along the cable and time t, and a and b are positive constants which we give physical meaning to below. If $a = b = 0$, (3.116) reduces to the one-dimensional homogeneous wave equation yielding a dispersionless wave form.

The derivation of (3.116) on physical grounds is as follows: Consider a *transmission line* or *coaxial cable* composed of two concentric conductors which are thin hollow conducting cylinders. To simplify the model, we have essentially two parallel current-carrying wires of constant capacitance, inductance, and resistance. Let i be the current through each wire; v, the voltage between the two wires at some point x from the beginning of the transmission line at time t; L, the inductance of each wire; R, the resistance; and C, the capacitance. The current through the line is oscillating or is a.c. This means that at two neighboring points x and $x + \Delta x$ a long the line the voltage drop Δv is proportional to the rate of decrease of current (the inductive part); it is also proportional to the current (the resistive part). This gives

$$\Delta v = v(x + \Delta x, t) - v(x, t) = -\left[\frac{L\partial i(x, t)}{\partial t} + Ri(x, t)\right]\Delta x.$$

In the limit we have

$$v_x = -Li_t - Ri. \tag{3.117}$$

Since the current is oscillating, the voltage at x is also oscillating; this supplies some charge to the capacity in the element $(x, x + \Delta x)$ of the line, which is $\Delta q = -Cv\Delta x$. The capacitative part of the current becomes $q(x, t)_t = -Cv_t\Delta x$. The current also has a resistive part that is proportional to the voltage. Since $i = q_t(x, t)$, the current in the differential element $(x, x + dx)$ becomes

$$i = -Cv_t\Delta x - Gv\Delta x.$$

Taking the limit, we get

$$i_x = -Cv_t - Gv, \tag{3.118}$$

where the shunt conductance $G = 1/R$. By taking appropriate cross derivatives of (3.117) and (3.118) and substitutions we may either eliminate v and get the telegraph equation (3.116) for i or, conversely, eliminate i and get the telegraph equation for v. We have therefore derived (3.116), where u stands for either the voltage or current in the transmission line. The parameters are

$$c^2 = \frac{1}{LC}, \qquad a = \frac{1}{RC}, \qquad b = \frac{R}{L}. \tag{3.119}$$

If we set

$$v = \exp[\tfrac{1}{2}(a + b)t]u(x, t), \tag{3.120}$$

and substitute into (3.116), we obtain the following PDE for $v(x, t)$, where the v_t term is missing:

$$v_{tt} - c^2 v_{xx} - \left[\frac{a - b}{2}\right]^2 v = 0. \tag{3.121}$$

Note that this equation for $v(x, t)$ is also hyperbolic because of the nature of its principal part.

Equation (3.121) is dispersionless if and only if

$$a = b \qquad \text{or} \qquad c = \frac{1}{RC}. \tag{3.122}$$

If condition (3.122) holds then the general solution for v consists of progressing and regressing undistorted waves, so that the solution of (3.116) becomes

$$u(x, t) = \exp\left(-\frac{t}{RC}\right)[F(x - ct) + G(x + ct)], \qquad \text{where} \quad RC = \frac{L}{R}. \tag{3.123}$$

Equation (3.123) tells us that, if (3.122) is imposed, then the solution of the telegraph equation has wave forms which consist of linear combinations of "relatively" undistorted progressing and regressing waves (for $F(x - ct)$ and $G(x + ct)$, respectively). This means that there is no dispersion, only dissipation if the wave speed is constant (independent of frequency). Physically, this solution is very important in transmission line theory; for it tells us that, if the circuit parameters are adjusted so that $c^2 = 1/LC = \text{const.}$, then signals are propagated along the cable in a relatively undistorted form—there is no dispersion, only exponential dissipation or attenuation of energy, as previously mentioned.

We now consider the three-dimensional hyperbolic equation

$$u_{tt} - c^2(u_{xx} + u_{yy} + u_{zz}) + bu = 0, \tag{3.124}$$

where b is a constant. Clearly, if $b = 0$, then (3.124) reduces to (3.106). Coming back to the general case ($b \neq 0$), let us assume plane wave solutions of (3.124) of the form given by (3.115) where the phase $\phi(\mathbf{x}, t)$ is given by (3.114). Inserting $u = f(\phi)$ into (3.124) and using (3.114) yields the following equation for $f(\phi)$:

$$[\omega^2 - c^2(k_x^2 + k_y^2 + k_z^2)]f''(\phi) + bf(\phi) = 0 \qquad \text{or} \qquad f'' + rf = 0, \tag{3.125}$$

which is an ODE with constant coefficients where

$$r = \frac{b}{[\omega^2 - c^2|k|^2]}. \tag{3.126}$$

There are three cases for r:

(1) $r = 0$, yielding $b = 0$, which gives the wave equation (which is seen from (3.124)).
(2) $r < 0$, which means $\omega - c|k| < 0$ (since $b > 0$). This yields exponential solutions of (3.125). We rule out this case since it does not yield waves and one part of the solution becomes unbounded for large $\mathbf{x}$.
(3) $r > 0$, which means $\omega - c|k| > 0$, so that the velocity of the wave or phase velocity $c < \omega/|k|$. The solution of (3.125) is in the form of sinusoidal functions of ϕ and hence yields traveling waves. There is no attenuation or dissipation of energy if the frequency is real, since there are no exponential decay terms.

Limiting Wave Speed. Lastly, we consider the case where $r \to \infty$. This yields $c = w/|k|$. This is the upper limit on the wave speed. From (3.125) we obtain $f = 0$, since $b \neq 0$. This tells us that the wave speed must be less than $\omega/|k|$. If $b = 0$ (which is the three-dimensional wave equation case), then $c = \omega/|k|$. It follows that, if $b > 0$, then $c < \omega/|k|$.

Since the waves for the case $b > 0$ do not propagate with the same velocity as that for the wave equation, we have the phenomenon of dispersion; meaning of course, that the wave form is dispersed or changes form. Case 3 therefore demonstrates that time-harmonic solutions of (3.124) exist for $v(\mathbf{x}, t)$ which show dispersion. By extending (3.120) to three-dimensional space and solving for $u(\mathbf{x}, t)$ we obtain solutions for the three-dimensional extension of the telegraph equation (3.116), which is

$$u_{tt} - c^2(u_{xx} + u_{yy} + u_{zz}) + abu = 0. \tag{3.116a}$$

Invoking the condition $a = b$ yields relatively undistorted three-dimensional wave forms for (3.116a) which travel with a wave speed different from the wave speed $c_0 = \omega/|k|$.

A generalization of the time-harmonic solution of the telegraph equation, for a mononchromatic wave form to encompass wave forms having a continuous frequency distribution, can be obtained by recognizing that the phase velocity c is a given function of frequency. We multiply each frequency component by an amplitude that depends on that frequency and integrate over the frequency spectrum. This is the same as performing a Fourier transform analysis. It suffices to mention that, for a discrete frequency spectrum, for example, in a numerical analysis situation, the Fourier transform analysis is replaced by a Fourier series analysis. These topics will not be pursued here.

Since the telegraph equation has the same principle part as the wave equation, as mentioned above, the same results of the theory of characteristics for characteristic curves (in one spatial dimension) and surfaces (in two and three dimensions) prevail as for the corresponding wave equation.

APPENDIX

SIMILARITY TRANSFORMATIONS AND CANONICAL FORMS†

Introduction

In this chapter we discussed the *similarity transformation* on the symmetric matrix $\mathbf{A}$ to the diagonal matrix $\mathbf{A}'$ which was involved in transforming the quadratic form $Q(\xi) = \sum_{i,j} a_{ij}\xi_i\xi_j$ to the canonical form $Q(\eta) = \sum_i k_i \eta_i^2$, where k_i is the ith diagonal element of $\mathbf{A}'$. We saw that an appropriate orthogonal transformation, which mapped one space into another (in the same dimen-

† See, for example [4], [11, Vol. I], [13], [15], and [34].

sion), was involved in obtaining $\mathbf{A}'$. It was shown that $Q(\mathbf{\eta})$ was useful in enabling us to classify second-order PDEs into elliptic, parabolic, or hyperbolic types in their canonical forms. In particular, the characteristic condition $Q(\mathbf{\eta}) = 0$ guarantees the hyperbolicity of the differential equation. Note that the three types of differential equations are invariant with respect to a transformation of the quadratic form from one coordinate system to another. Indeed, these quadratic forms are intimately related to the principal parts which determine the character of these differential equations.

At the crux of a similarity transformation lies the transformation of a nonsingular† symmetric matrix into a diagonal matrix which is involved in a variety of physical situations. The concept of a similarity transformation has far-reaching applications, not only to electromagnetic theory, but also to many other aspects of mathematical physics. Some physical examples are: in electromagnetic theory and mechanics, electrically and mechanically coupled vibratiang systems—a similarity transformation uncouples the system putting it into its canonical form in terms of normal coordinates, in stress analysis—transformation of strain and stress axes into principal axes, in mechanics—rigid body motion, canonical forms for potential, and kinetic energy, etc. In all these examples the essence of obtaining a similarity transformation is to construct the proper orthogonal transformation from one space to another space, making use of eigenfunction theory.

In this chapter only a sketch of the similarity transformation was given with particular reference to the classification of the PDEs. In this appendix we round out the picture by giving a more detailed treatment of the similarity transformation in a general setting that applies to a variety of situations in mathematical physics.

3A.1 Geometric Considerations

Perhaps the simplest example of a similarity transformation is the rotation of an ellipse from tilted axes to those coinciding with the (x, y) axes, which puts the ellipse into normal or canonical form. The ellipse is an example of a conic section. It is represented by a quadratic form $Q(\mathbf{x}) = ax^2 + 2bxy + cy^2$ (where $\mathbf{x} = (x, y)$). $Q(\mathbf{x})$ can easily be put into canonical form by the elementary process of "completing the square".

$$ax^2 + 2bxy + cy^2 = a\left[x + \left(\frac{b}{a}\right)y\right]^2 + \left(c - \frac{b^2}{a}\right)y^2.$$

The form of the right-hand side suggests we introduce the following

---

† A nonsingular matrix is one that has a unique inverse so that it must be a square matrix having a non-zero determinant.

"transformed" variables:

$$x' = x + \left(\frac{b}{a}\right)y, \qquad y' = y.$$

This is a linear transformation from (x, y) to (x', y') coordinates. In the latter coordinate system the quadratic form becomes

$$Q(x') = ax'^2 + \left(c - \frac{b^2}{a}\right)y'^2,$$

which is the canonical form corresponding to $Q(x)$. This analysis requires $a \neq 0$. If $a = 0$, but $c \neq 0$, a similar transformation works. Finally, if $a = c = 0$, we get $Q(x) = 2bxy$ and $2bxy = 0$ represents an equilateral hyperbola. In this case, the transformation $x = x' + y', y = x' - y'$, reduces the form to $Q(x') = 2b(x'^2 - y'^2)$. $Q(x') = 0$ is the canonical form for the hyperbola. It is easily seen that this is obtained from $2bxy = 0$ by a pure rotation of the (x, y) axes through $45°$ to the $(x'y')$ axes.

Similarly, the technique of completing the square can be performed on the quadratic form $Q(x) = \sum_{i,j=1}^{n} a_{ij}x_i x_j$ in n space. We assume $a_{11} \neq 0$ and write $Q(x) = a_{11}(\sum_{i,j=1}^{n} b_{ij}x_i x_j)$, where $b_{ij} = a_{ij}/a_{11}$ and $b_{11} = 1$. Because of the symmetry of the matrix $\mathbf{A}$, the terms which actually involve x_1 become

$$x_1^2 + 2\sum_{j=2}^{n} b_{1j}x_1 x_j = \left(x_1 + \sum_{j=2}^{n} b_{1j}x_j\right)^2 - \left(\sum_{j=2}^{n} b_{1j}x_j\right)^2.$$

This leads to the linear transformation

$$y_1 = x_1 + \sum_{j=2}^{n} b_{1j}x_j, \qquad y_i = x_i \quad \text{for } i = 2, 3, \ldots, n.$$

This puts the quadratic in the form

$$a_{11}y_1^2 + \sum_{j,k=2}^{n} c_{jk}y_j y_k,$$

which is a pure quadratic in y_1 with a "residual" part which is a quadratic form in $(n - 1)$ variables. The same process is repeated on this residual part which picks out the y_2^2 term, leaving a residual in $(n - 2)$ variables. By induction we can repeat this process a finite number of times until the new coefficients in the last residual term are zero. This tells us that

$$Q(\mathbf{x}) = \sum_{i,j=1}^{n} a_{ij}x_i x_j \tag{3A.1}$$

(which is essentially the same as (3.68)), can be transformed into the canonical form

$$Q(\mathbf{y}) = \sum_{i=1}^{n} d_i y_i^2 \tag{3A.2}$$

(essentially (3.73)), providing the matrix $\mathbf{A}$ is nonsingular and symmetric.

To describe conic sections in Euclidian geometry we need quadratic forms whose coefficients lie in the field of real numbers (which we have been assuming all along). Therefore each diagonal element d_i of (3A.2) is a real number. We may thus simplify $Q(\mathbf{y})$ by letting $z_i = d_i^{1/2} y_i$, $i = 1, 2, \ldots, n$. Therefore any quadratic form may be put into the form

$$Q(\mathbf{z}) = z_1^2 + \cdots + z_p^2 - z_{p+1}^2 - \cdots - z_n^2. \tag{3A.3}$$

(The same as (3.73).)

A comment on this section: The technique of completing the square for the two-dimensional case and the more general n-dimensional case led to an "affine" or linear transformation from $\mathbf{x}$ space to $\mathbf{x}'$ space (having a unique inverse). The transformation is not necessarily orthogonal. Geometrically, this means that for the two-dimensional case we have a rotation of the (x, y) into the (x', y') coordinate system, the primed coordinate system being oblique if the transformation is not orthogonal. For the general case of n dimensions the transformation from $\mathbf{x}$ to $\mathbf{x}'$ space yields an oblique primed coordinate system where the n-dimensional conic section (ellipsoid, paraboloid, or hyperboloid) is in canonical although oblique coordinates (the axes of the conic sections coincide with the $\mathbf{x}'$ axis system).

We state in passing that, given an $(n \times n)$ nonsingular matrix that is not symmetric, it is possible to find a transformation (not necessarily orthogonal) that will transform the matrix to a triangular matrix (either upper or lower) where all the elements below or above the diagonal, respectively, are zero. This means that a similarity transformation on a nonsingular nonsymmetric matrix yields a triangular matrix. There are a variety of problems in mathematic physics that deal with this type of transformation. However, we do not take up this method here.

The following treatment specializes to orthogonal transformations which means that the transformed coordinate system (as well as the original) is composed of orthogonal axes. This has the advantage of allowing us to use an analysis which makes use of eigenvectors and eigenvalues in a rather elegant way. The minor disadvantage is that we cannot treat oblique coordinate systems by this method. We recall that the matrix $\mathbf{A}$ associated with the quadratic form $Q(\mathbf{x})$ is symmetric. This is a necessary and sufficient condition for an appropriate orthogonal transformation to yield the diagonal matrix $\mathbf{A}'$.

3A.2. Orthogonal Transformations and Eigenvectors in Relation to Similarity Transformations

In Section 3.6, the subsection entitled "Relation to Quadratic Forms", the similarity transformation was discussed in relation to the classification of second-order PDEs, as mentioned above. The similarity transformation given by (3.71) is related to the orthogonal transformation given by the orthonormal matrix **R**. We shall now examine these concepts in more detail, in a more general setting.

There is no loss in generality if we restrict ourselves to a three-dimensional $\mathbf{x} = (x, y, z)$ space. The interested reader can easily generalize to n dimensions. The quadratic form becomes (in extended form)

$$Q(\mathbf{x}) = \mathbf{x}^*\mathbf{A}\mathbf{x} = a_{11}x^2 + 2a_{12}xy + 2a_{13}xz + a_{22}y^2 + 2a_{23}yz + a_{33}z^2,$$

where $\mathbf{A}: a_{ij}$ is a (3×3) nonsingular, symmetric matrix.

Eigenvectors and Eigenvalues of the Linear Vector Function A

For simplicity, in this treatment we continue to deal with three-dimensional space. We consider the matrix **A** as a *linear vector function*. This means that **A** operates on a vector **v** producing another vector, say **w**, or $\mathbf{A}\mathbf{v} = \mathbf{w}^*$, where **v** is a column vector and $\mathbf{w}^*$ is a row vector (the transpose of the column vector **w**). We now raise the question: Given the linear vector function **A**, can we find a vector **v** such that the vector **w** is in the same direction as **v**? This means that **w** is proportional to **v**, or there exists a scalar λ such that $\mathbf{w} = \lambda\mathbf{v}$

$$\mathbf{A}\mathbf{v} = \lambda\mathbf{v} \tag{3A.4}$$

If it is possible to find a **v** that has the property given by (3A.4) then we must find at least one λ which satisfies this equation. Let **I** be the three-dimensional *identity matrix* meaning that the ijth element $I_{ij} = \delta_{ij}$. Then (3A.4) can be put in the matrix form

$$(\mathbf{A} - \lambda\mathbf{I})\mathbf{v} = 0. \tag{3A.5}$$

An elementary theorem in algebra tells us that, in order to obtain nontrivial solutions for **v**, we must have

$$\det(\mathbf{A} - \lambda\mathbf{I}) = 0. \tag{3A.6}$$

Equation (3A.6) is called the *characteristic equation* or sometimes the *secular equation* if the λ's are associated with frequencies in vibration problems. (Note that the term "characteristic equation" is used for a variety of equations in physics.) This equation showed up in (3.72) for the n-dimensional case where we saw that the λ's are the eigenvalues). Equation (3A.6) is a third-degree polynomial for the λ's, which may be written as

$$(\lambda - \lambda_1)(\lambda - \lambda_2)(\lambda - \lambda_3) = 0, \tag{3A.7}$$

where the λ_i's are the root of (3A.6). They are called the *eigenvalues* corresponding to the *eigenvectors* $\mathbf{v}_i$ for $i = 1, 2, 3$. Therefore, if we can find the roots λ_i of (3A.6), then the answer to the above question is "yes". **w** is in the same direction as **v** if we can find these roots. If the λ_i's are real and distinct (they do not have to be, for example, there may be one real and two complex conjugate roots) then there are three real directions for **w** given by $\lambda_i\mathbf{v}$ for $i = 1, 2, 3$.

(Note that, for the n-dimensional case, (3A.6) is an nth-degree polynomial in λ yielding n roots for the λ_i's.) Expanding (3A.6), for the three-dimensional case, we get

$$
\det(\mathbf{A} - \lambda \mathbf{I}) = \begin{vmatrix} a_{11} - \lambda & a_{12} & a_{13} \\ a_{21} & a_{22} - \lambda & a_{23} \\ a_{31} & a_{32} & a_{33} - \lambda \end{vmatrix}
$$

$$
= \lambda^3 - I_1 \lambda^2 + I_2 \lambda - I_3 = 0. \tag{3A.8}
$$

The coefficients I_1, I_2, I_3 are given by

$$
I_1 = a_{11} + a_{22} + a_{33},
$$

$$
I_2 = \begin{vmatrix} a_{22} & a_{23} \\ a_{23} & a_{33} \end{vmatrix} + \begin{vmatrix} a_{33} & a_{13} \\ a_{13} & a_{11} \end{vmatrix} + \begin{vmatrix} a_{11} & a_{12} \\ a_{12} & a_{22} \end{vmatrix} = 0,
$$

$$
I_3 = \begin{vmatrix} a_{11} & a_{12} & a_{13} \\ a_{21} & a_{22} & a_{23} \\ a_{31} & a_{32} & a_{33} \end{vmatrix}. \tag{3A.9}
$$

I_1 is called the *trace* or the sum of the diagonal elements of $\mathbf{A}$, I_2 is called the sum of the diagonal minors of $\mathbf{A}$, and I_3 is called the determinant of $\mathbf{A}$.

I_1, I_2, and I_3 are scalars and are therefore invariant with respect to a transformation of the coordinate system. The eigenvalues are also independent of the coordinate system. At this point it is useful to supply the proof (which is straightforward) of these statements for a rotation of the (x, y, z) orthogonal coordinate system to the (x', y', z') orthogonal coordinate system. (The proof may be extended to any linear or affine transformation of coordinates.) We reintroduce the rotation matrix $\mathbf{R}$, this time in three dimensions which we use to rotate the $\mathbf{x}$ into the $\mathbf{x}'$ coordinate system. We have

$$
\mathbf{x} = \mathbf{R}\mathbf{x}'. \tag{3A.10}
$$

We now apply the matrix $\mathbf{A}$ acting as a linear vector function on the vector $\mathbf{x}$ in the same sense as (3A.4), obtaining

$$
\mathbf{A}\mathbf{x} = \lambda \mathbf{x}. \tag{3A.11}
$$

Inserting (3A.10) into (3A.11), premultiplying by $\mathbf{R}^*$, and making use of the fact that $\mathbf{R}^{-1}\mathbf{R} = \mathbf{R}^*\mathbf{R} = \mathbf{I}$, we obtain

$$
\mathbf{R}^*\mathbf{A}\mathbf{R}\mathbf{x}' = \mathbf{A}'\mathbf{x}' = \lambda \mathbf{x}', \tag{3A.12}
$$

where

$$
\mathbf{A}' = \mathbf{R}^*\mathbf{A}\mathbf{R},
$$

which is the similarity transformation from $\mathbf{A}$ to $\mathbf{A}'$ given by (3.71). Comparing (3A.11) and (3A.12) we observe that the linear vector function $\mathbf{A}'$ in the transformed or primed coordinate system operates on $\mathbf{x}'$, producing the same multiple of $\mathbf{x}'$ that $\mathbf{A}$ does on $\mathbf{x}$ in the original or unprimed coordinate system. This should be sufficient to demonstrate that the eigenvalues are unchanged with respect to a rotation of coordinate systems (since we have done nothing in the analysis).

However, we now supply a direct proof that the eigenvalues, as well as the I_i's are invariant with respect to a rotation of the coordinate systems. The proof involves relating the characteristic equation (3A.6) for $\mathbf{A}$ with the corresponding equation for $\mathbf{A}'$ and making use of the similarity transformation. The characteristic equation for $\mathbf{A}'$ is obtained from (3A.12) (assuming λ' corresponding to $\mathbf{x}'$). We obtain

$$\det(\mathbf{A}' - \lambda'I) = \lambda'^3 - I_1'\lambda'^2 + I_2'\lambda' - I_3' = \det(\mathbf{R}^*\mathbf{A}\mathbf{R} - \lambda'\mathbf{I})$$

$$= \det[\mathbf{R}^*(\mathbf{A} - \lambda'\mathbf{I})\mathbf{R}] = \det(\mathbf{A} - \lambda I)$$

$$= \lambda^3 - I_1\lambda^2 + I_2\lambda - I_3 = 0, \qquad (3A.13)$$

where we have used the properties of determinants and the fact that

$$\det(\mathbf{R}^*)\det(\mathbf{R}) = \det(\mathbf{R}^*\mathbf{R}) = \det(\mathbf{I}) = 1.$$

since $\mathbf{R}$ is an orthonormal matrix.

Let $\mathbf{u}$ be a normalized eigenvector, meaning that $\mathbf{u}\cdot\mathbf{u} = 1$ or, in matrix notation, $\mathbf{u}^*\mathbf{u} = 1$. The quadratic form with respect to $\mathbf{A}$ is

$$Q(\mathbf{u}) = \mathbf{u}^*\mathbf{A}\mathbf{u} = \mathbf{u}^*\lambda\mathbf{u} = \lambda\mathbf{u}^*\mathbf{u} = \lambda. \qquad (3A.14)$$

Thus each eigenvalue λ of a linear vector function $\mathbf{A}$ is the value of the quadratic form $\mathbf{u}^*\mathbf{A}\mathbf{u}$, where $\mathbf{u}$ is a normalized eigenvector associated with the eigenvalue λ. Clearly, for any eigenvector $\mathbf{v}$, we must divide the quadratic form $\mathbf{v}^*\mathbf{A}\mathbf{v}$ by $\mathbf{v}^*\mathbf{v}$.

3A.3. Diagonalization of $\mathbf{A}'$

We are now in a position to construct a rotation matrix $\mathbf{R}$ which makes $\mathbf{A}'$ a diagonal matrix. It turns out that the diagonal elements of $\mathbf{A}'$ are the eigenvalues. The crux of the method is to let each column vector of $\mathbf{R}$ be an eigenvector. If $\mathbf{r}_i$ is the ith column vector of $\mathbf{R}$, we have

$$\mathbf{A}\mathbf{r}_i = \lambda_i\mathbf{r}_i, \qquad i = 1, 2, 3. \qquad (3A.15)$$

The similarity transformation can be written as

$$\mathbf{A}' = \mathbf{R}^*\mathbf{B},$$

where

$$\mathbf{B} = \mathbf{A}\mathbf{R}. \qquad (3A.16)$$

The ith column of $\mathbf{B}$ is the scalar product of the linear vector function $\mathbf{A}$ and the ith column of $\mathbf{r}$, in other words the ith column of $\mathbf{B}$, is $\lambda_i\mathbf{r}_i$, upon using

(3A.15). We may therefore write **B** in terms of these three column vectors, as

$$\mathbf{B} = (\lambda_1 \mathbf{r}_1, \lambda_2 \mathbf{r}_2, \lambda_3 \mathbf{r}_3).$$

To construct **A'** we must premultiply **B** by **R*** which are the row vectors of **R**. Using the orthonormality of **R** gives

$$\mathbf{A'} = \begin{pmatrix} \lambda_1 & 0 & 0 \\ 0 & \lambda_2 & \\ 0 & 0 & \lambda_3 \end{pmatrix}. \tag{3A.17}$$

This tells us that **A'** is a diagonal matrix whose ith diagonal element is the ith eigenvalue λ_i.

The matrix **A'** in the transformed coordinate system **x'** is therefore symmetric, since it is diagonal. We have $\mathbf{A'} = \mathbf{R^*AR}$. Taking the transpose of this equation yields

$$(\mathbf{A'})^* = (\mathbf{AR})^*\mathbf{R} = \mathbf{R^*A^*R} = \mathbf{A'} = \mathbf{R^*AR},$$

where we used the fact that, for any two matrices **A** and **B** of the same dimension, if $\mathbf{C} = \mathbf{AB}$ then $\mathbf{C^*} = \mathbf{B^*A^*}$. It follows that **A'** is symmetric. (Note that we used the symmetry property of A in constructing the quadratic form $Q(\mathbf{x})$.)

We now put the above material in extended form, for ease of comprehension. As before, the ijth element of **A** is a_{ij}. Let r_{ij} be the ijth element of **R**. The three rows of **R** consist of a set of orthonormal row vectors; and similarly, the columns are composed of orthonormal column vectors. Let the column vectors be eigenvectors with respect to the eigenvalues λ. Then we have

$$\sum_{k=1}^{3} a_{ik} r_{kj} = \lambda_l r_{ij}, \qquad i, j, l = 1, 2, 3. \tag{3A.18}$$

The system (3A.18) consists of nine equations, three for each of the three eigenvalues λ_l. The ith equation is a linear combination of the components of the ith column vector of **R** for the lth eigenvalue. When written out in detail, this system is the extended form of (3A.15). Using (3A.18), (3A.16) becomes

$$\mathbf{B} = \begin{pmatrix} \lambda_1 r_{11} & \lambda_2 r_{12} & \lambda_3 r_{13} \\ \lambda_1 r_{21} & \lambda_2 r_{22} & \lambda_3 r_{23} \\ \lambda_1 r_{31} & \lambda_2 r_{32} & \lambda_3 r_{33} \end{pmatrix}. \tag{3A.19}$$

Premultiplying (3A.19) by **R*** yields (3A.17), where we have made use of the orthonormal properties of **R** which are

$$(r_{11})^2 + (r_{21})^2 + (r_{31})^2 = 1,$$

$$(r_{11})(r_{12}) + (r_{21})(r_{22}) + (r_{31})(r_{32}) = 0, \quad \text{etc.}$$

The reader is referred to [14, Chaps. 1 and 2] for worked-out examples and applications to obtaining principal axes.

CHAPTER 4

Variational Methods

Introduction

This chapter is a modification and extension, of some of the material in [15, Chap. 9], to our specific needs. The emphasis will be on the Lagrange and Hamilton canonical equations of motion, with applications to wave propagation in electromagnetic media. For the convenience of the reader, some of the essential features of the *calculus of variations*, as well as D'Alembert's principle, Hamilton's principle and other variational principles, will be reviewed in the context of phase space. This is the setting for a proper understanding of the Hamilton–Jacobi theory which gives us a deep insight into the partial differential equations (PDEs) of wave propagation. For a more thorough treatment, the reader is referred to the above reference, and to the standard works on the calculus of variations such as [5], for a more refined mathematical treatment of this subject which involves existence theorems, and so forth.

Up to now, the approach used in developing the mathematical apparatus for electromagnetic wave propagation has been to obtain the field equations, which are the hyperbolic PDEs that arise from the conservation laws of physics, namely, mass or continuity, momentum, and energy. This approach is essentially a *local* one in the sense that the conservation laws are applied locally to a volume element surrounding a field point in an appropriate space. Then the appropriate initial and boundary conditions are applied to these field equations to obtain solutions to specific problems. On the other hand, the *global* approach used by the calculus of variations, in the application of variational methods to wave propagation in electromagnetic media, allows us to start with the solution domain and by variational procedures minimize (or more generally extremize) certain *functionals* from which the conservation laws are then derived. This will now be made clear.

We shall describe the essence of the calculus of variations by first appealing to an elementary situation in the differential calculus where we have a differentiable function $y = f(x)$ which we want to either maximize or minimize. (We say we want to "extremize" the function when we do not care whether it is a maximum or a minimum.) In any case, we know that we set the first derivative

$dy/dx = 0$. This is a necessary condition for an extremum but does not ensure it, since we may have a horizontal inflection point, which is neither a maximum nor a minimum. We also know that a sufficient condition that ensures a minimum is that $d^2y/dx^2 > 0$, and that $d^2y/dx^2 < 0$ ensures a maximum.

In the calculus of variations this concept of extremizing a function of a single independent variable is generalized to that of extremizing a functional, which is roughly defined as "a function of a function". Specifically, a functional is an integral whose integrand involves a class of functions, each function defining a particular value of the integral. Of this class of functions we desire to find that function which makes the integral an extremum. In other words, *the functional is an integral that involves a class of functions, one of which extremizes the functional.* We assume constant end points of the integral, for simplicity.

4.1. Principle of Least Time

It is best to give an insight into the meaning of a functional by an example. To this end, we now give an important example in optics which illustrates the concept of a functional; this is the famous *principle of least time* or *Fermat's principle*, since it was first enunciated by the French mathematician Pierre Fermat (1601–1665). The statement of Fermat's principle is: Of all the possible paths taken by a light ray in going from point A to point B (A and B are fixed in space), the actual path or trajectory taken by the beam is the one that *takes the least time.* Clearly, for a medium of constant index of refraction, the path is a straight line from A to B, since the wave speed c is constant. But, for a medium of variable index of refraction, the answer is not so simple (since c depends on the path and the field point in the medium), and we need to apply the methods of the calculus of variations. For simplicity, we consider two fixed points A and B in the (x, y) plane in a medium of variable index of refraction. According to Fermat's principle, the actual path is the particular curve in the (x, y) plane given by the function $y = y(x)$ that minimizes the time it takes the light beam to go from A to B. This means that, of all the possible trajectories given by the class of functions $y = y(x)$, we want that trajectory which minimizes the time t from A to B, which is given by

$$I(y) = \int_{t_A}^{t_B} dt = \int_A^B \frac{dt}{ds} ds = \int_A^B \frac{\sqrt{1 + y'^2}\, dx}{c} = t_B - t_A, \qquad y' \equiv \frac{dy}{dx}, \qquad (4.1)$$

where ds is the element of arc length along the trajectory $y = y(x)$, $c = c(x, y(x))$ is the wave speed or phase of the light beam, and the integral $I(y)$ is the required functional. $I(y)$ represents the time it takes for the light beam to go from A to B, so that this functional depends on the class of functions $y = y(x)$; I clearly depends on the trajectory given by the curve $y = y(x)$. How does $I(y)$ depend on y? We note that the integrand of (4.1) involves the derivative y' as well as $c(x, y(x))$. We would expect that the method of the calculus of variations would allow us to derive a differential equation whose integral would yield the required curve $y = y(x)$ that minimizes the functional $I(y)$; this trajectory would, of course, depend on the known function $c(x, y(x)) = ds/dt$, the velocity of light in the medium. (In this case the extremum of the functional is a minimum.)

How do we minimize $I(y)$ to get the required trajectory? The answer is given in the next section where we take up the calculus of variations in one dimension.

4.2. One-Dimensional Calculus of Variations, Euler's Equation

The integrand in (4.1) corresponding to the functional $I(y)$ is of the form $F(y')$. More generally, the integral is in the form $F(x, y, y')$ so that we can generalize Fermat's principle as follows:

Of all the admissible functions y in the (x, y) plane passing through the fixed end points A: (x_A, y_A), B: (x_B, y_B) find that function y = y(x) which minimizes the integral

$$I(y) = \int_A^B F(x, y, y') \, dx. \tag{4.2}$$

This statement is the basic formulation of the calculus of variations in one dimension. The assumption is that a minimum (or extremum) exists. Actually, we settle for a weaker condition, namely, that an extremum exists. This is also called a *stationary value* (it allows for a maximum as well, which may occur in some problems). There are problems in physics and geometry where we cannot find a curve which gives a stationary value for the functional.

We now present a simple example in optics which shows that no minimum of a functional exists. This example also illustrates the use of Fermat's principle. Consider Fig. 4.1, in which we have the (x, y) plane, which is optically transparent, embedded in a medium of constant index of refraction. Consider the fixed point A on the x axis which is a distance L from the origin (the length of the line segment $OA = L$). Suppose a light beam starts at point O, travels along the y axis a variable distance to the point P_i, then goes along the straight line from P_i to A. Each trajectory is given by the curve $y = y_i(x)$ for $i = 1, 2, \ldots$. This means that every admissible path is composed of the broken line segments $y_i = OP_i + P_iA$ for $i = 1, 2, \ldots$. According to Fermat's principle, the time taken to go along each path $OP_i + P_iA$ is a minimum for that path. Therefore (for each y_i) the functional $I(y_i)$ represents the time taken to traverse the path $OP_i + P_iA$. Clearly, the length of this path is greater than L by the amount OP_i. In the limit $P_i \to P_x = 0$. The minimum distance is the line segment OA of length L. But this path OA is *not an admissible path* since the light beam must start at O in the direction of the y axis and then travel on a line of sight to A. As P_i tends to O the length of each admissible path gets

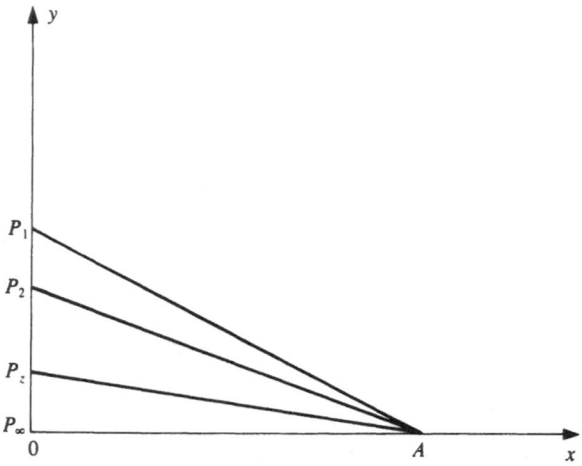

Fig. 4.1. Optical example of no extremum.

closer to L; but there is *no minimal path* since we obtain an infinite number of admissible paths as $P_i \to 0$. We have thus demonstrated that *an extremum or stationary value for the functional does not exist.*

We return to the functional given by (4.2). Let y be the set of admissible curves and let $y = \bar{y}(x)$ be any curve in this set; $I(\bar{y})$ is not necessarily an extremum. Finally, let $y = y(x)$ be the minimizing curve (we assume here that an extremum is a minimum). Clearly $I(y)$ is a minimum. Then any $\bar{y}$ in the neighborhood of the minimizing function y can be represented in the form

$$\delta y(x) = \bar{y}(x) - y(x) = \varepsilon\eta(x), \tag{4.3}$$

where $\delta y(x)$ is called the *variation of* $y(x)$, ε is a small number or parameter which determines the amount of the variation, and $\eta(x)$ is an arbitrary function of x. Since all curves pass through the end points, we must have the homogeneous boundary conditions

$$\eta(x_A) \equiv \eta_A = \eta(x_B) \equiv \eta_B = 0. \tag{4.4}$$

We assume $I(y)$ is never negative, so that

$$I(y + \varepsilon\eta(x)) > I(y).$$

The left-hand side of this inequality is a differentiable function of ε, so that the necessary condition for y to minimize $I(y)$ is

$$\left.\frac{dI(y + \varepsilon\eta)}{d\varepsilon}\right|_{\varepsilon=0} = 0.$$

Using (4.2) we obtain

$$\left.\frac{dI(y + \varepsilon\eta)}{d\varepsilon}\right|_{\varepsilon=0} \equiv (I_\varepsilon)_0 = \int_{x_A}^{x_B} (\eta F_y + F_{y'}\eta') \, dx = 0, \qquad y' \equiv \frac{dy}{dx}, \quad \eta' \equiv \frac{d\eta}{dx},$$

by the usual method of differentiating under the integral sign, where we also used (4.3). Upon integrating the second term in the integral by parts and using the homogeneous boundary conditions, we obtain

$$I_\varepsilon = \int_{x_A}^{x_B} \left(F_y - \frac{dF_{y'}}{dx} \right) y_\varepsilon \, dx, \qquad \eta = y_\varepsilon.$$

To obtain the extremum condition we multiply through by a differential $d\varepsilon$ and evaluate the derivatives at $\varepsilon = 0$. This results in

$$(I_\varepsilon)_0 \, d\varepsilon = \delta I = \int_{x_A}^{x_B} \left(F_y - \frac{dF_{y'}}{dx} \right) (y_\varepsilon)_0 \, d\varepsilon \, dx = 0, \qquad (4.5)$$

where δI is the variation of I. Since δy represents some arbitrary variation of y, (4.5) can only be valid if

$$F_y - \frac{dF_{y'}}{dx} = 0. \qquad (4.6)$$

Equation (4.6) is most important in the calculus of variations. It is called *Euler's equation* for the one-dimensional case. It is a necessary condition that $I(y)$ have an extremum. The curve $y = y(x)$ that minimizes I is determined by solving Euler's equation for a given $F(x, y, y')$. Equation (4.6) tells us that the rate of change of momentum $dF_{y'}/dx$ equals the external force F_y, which is Newton's equation of motion.

We digress a moment to give a physical interpretation of the variations δy and δI. Figure 4.2 shows a minimizing curve $y = y(x)$ in the (x, y) plane and a neighboring curve $y = \bar{y}(x)$ (for some nonzero value of ε) passing through the same end points A and B. δy is the change of y for a fixed x from the minimizing curve ($\varepsilon = 0$) to the curve $\bar{y}$. This is an important point; for, when we come to the principle of virtual work y plays the role of displacement and

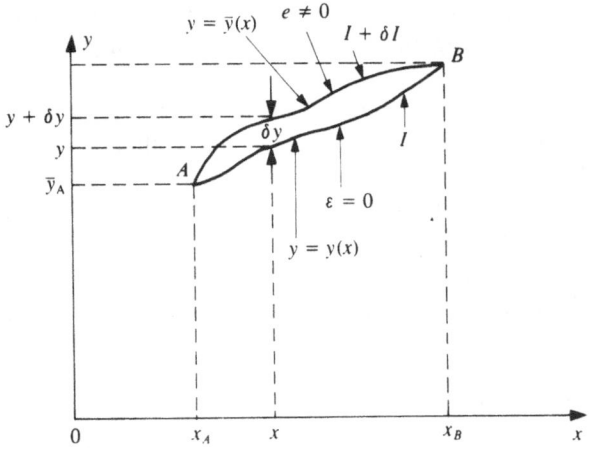

Fig. 4.2. Minimizing curve y and neighboring curve $\bar{y}$.

x represents time, so that a virtual displacement δy occurs with no change in time. Clearly, δI is the variation of the functional from the minimizing to the neighboring curve.

We now give several examples in various branches of physics to illustrate the use of Euler's equation.

EXAMPLE 1. We first investigate a simple example in electricity by considering a series alternating current circuit consisting of a coil of inductance $\mathscr{L}$ in series with a condenser of capacitance C. Let y stand for the charge and x for time. We know that, for a series circuit the total voltage is equal to the sum of the voltages across the condenser and coil, and the current in each of the two elements is the same. Therefore, the circuit equation for the charge becomes $\mathscr{L} \, d^2y/dx^2 + (1/C)y = v(x)$, where v is the externally applied voltage or forcing function. If we switch off the voltage, $v = 0$ and we have a simple undamped harmonic oscillator expressed by

$$\mathscr{L}\ddot{y} + \left(\frac{1}{C}\right)y = 0, \qquad \frac{d^2y}{dx^2} \equiv \ddot{y}.$$

If we let $T = (\mathscr{L}/2)\dot{y}^2$ and $V = (1/2C)y^2$, we interpret T as the kinetic energy and V as the potential energy. We let the function $L = T - V$. L is called the *Lagrangian* and is important in variational methods. We shall have much more to say about L below, in connection with Hamilton's principle and Lagrange's equations of motion. We now identify L with F in Euler's equation (4.6). Thus

$$F = L = T - V = \tfrac{1}{2}\mathscr{L}\dot{y}^2 - \left(\frac{1}{2C}\right)y^2.$$

Substituting this expressing for F into (4.6) yields the simple harmonic oscillator or circuit equation for the series $\mathscr{L} - C$ circuit with no external voltage. Note that the term $F_{y'} = \mathscr{L}\dot{y}$ plays the role of momentum and $F_y = -(1/C)y$ represents the external force derivable from the potential energy $(1/C)y^2$. (This is analogous to the spring force in a simple mass–spring harmonic oscillator.)

EXAMPLE 2. We now give a simple example in optics. Consider a light beam traveling from points A to B in the (x, y) plane with a constant speed c_0 (the speed of light in a vacuum) which we normalize by setting $c_0 = 1$. From Fermat's principle we obtain the trivial result that the beam travels in a straight line from A to B with unit velocity, since the medium has a constant index of refraction. We shall "use an elephant gun to kill a fly" by proving this result analytically from Euler's equation. The integrand in (4.1) is

$$F = \sqrt{1 + y'^2}.$$

Substituting into (4.6) with

$$F_y = 0, \qquad F_{y'} = \frac{y'}{\sqrt{1 + y'^2}},$$

yields

$$\frac{d}{dx} \frac{y'}{\sqrt{1 + y'^2}} = 0,$$

which, upon integration, gives

$$\frac{y'}{\sqrt{1 + y'^2}} = k,$$

where k is some constant; which means $y' = m$, where m is the constant slope of the trajectory. Therefore, the solution of this first-order differential equation for $y(x)$ is the straight line going from A to B with the velocity $c_0 = 1$.

$$y(x) = \left(\frac{y_B - y_A}{x_B - x_A}\right) x.$$

EXAMPLE 3 (The Brachistochrone Problem). This is a well-known example in mechanics. The attention of the scientific world was first drawn, in 1696, to problems of this type by the great Swiss mathematical physicist John Bernoulli. Its solution led to the foundations of the calculus of variations. The statement of the problem is: Find the curve (in the (x, y) plane, for simplicity) joining two fixed points A and B along which a particle slides without friction under the influence of gravity from the higher point A to the lower point in the least time. Note that there is no real analogy between this problem and Fermat's principle for the actual path of a light beam. Fermat's principle states that the path of any light beam is such that the time taken to go between two points is a minimum; while the brachistochrone problem asks for a particular path such that the time taken to slide down it is a minimum (clearly not all particles slide down paths having such a property).

Let v be the speed of the particle along the required curve. Then the time taken for the particle to travel a distance ds is $dt = ds/v$. If A is the upper point and B the lower point, then the time of flight from A to B is given by

$$t_B - t_A = \int_{t_A}^{t_B} dt = \int_A^B \frac{ds}{v}.$$

We now want to obtain v as a function of y, which we define as the vertical distance measured downward from the initial point A. We need this in order to obtain the functional $\int_{t_A}^{t_B} dt = I(y)$. This is easily accomplished by appealing to the law of conservation of energy for the particle in a conservative force field, which is $T = V$ or $\frac{1}{2}mv^2 = mgv$, where we used the initial condition that the particle falls from rest. This gives

$$v = \sqrt{2gy}.$$

Then the expression for the functional becomes

$$t_B - t_A = I(y) = \int_{t_A}^{t_B} \frac{\sqrt{1 + y'^2}}{\sqrt{2gy}} \, dx,$$

and F is identified as

$$F = \frac{\sqrt{1 + y'^2}}{\sqrt{2gy}}.$$

We use this value of F in Euler's equation (4.6). However, we observe that F does not depend explicitly on x, or $F_x = 0$. This means that

$$\frac{dF'}{dx} = y''F_{y'} + y'F_y,$$

and, upon using (4.6), we obtain the following identity

$$\frac{d}{dx}(y'F_{y'} - F) = 0,$$

which is valid only if F is explicitly independent of x. From this equation we get

$$y'F_{y'} - F = \text{const.} = k.$$

Upon using the above expression for F and after some simplification we obtain the first-order second-degree differential equation for $y(x)$

$$(y')^2 = \frac{a^2}{y} - 1,$$

where

$$a^2 = \tfrac{1}{2}k^2 g.$$

Integrating this differential equation yields

$$x = \int_{y_A}^{y_B} \frac{\sqrt{y}}{\sqrt{a^2 - y}} \, dy.$$

Setting $y = a^2 \sin^2(\theta/2)$ and simplifying, we obtain

$$x = \int a^2 \sin^2\left(\frac{\theta}{2}\right) d\theta = \left(\frac{a^2}{2}\right)[\theta - \sin \theta] + C,$$

where C is the constant of integration. From this, we obtain

$$x(\theta) = \left(\frac{a^2}{2}\right)[\theta - \sin \theta] + C,$$

$$y(\theta) = \left(\frac{a^2}{2}\right)[1 - \cos \theta].$$

This pair of equations represents a cycloid in parametric form, which can be visualized geometrically as a curve swept out by a point on the circumference of a circle of radius $a/\sqrt{2}$, rolling without slipping along the x axis. By adjusting the constants a and C we can get the cycloid to pass through the points $A: (x_A, y_A)$, $B: (x_B, y_B)$, thus solving the brachistochrone problem.

EXAMPLE 4 (Theory of Surfaces). Our last example is one in the elementary theory of surfaces. We desire to find a minimum surface of revolution. Specifically, suppose we form a surface of revolution by taking some curve $y = y(x)$ which passes through the points A and B. The problem in the calculus of variations is to find that curve $y = y(x)$ which minimizes the surface area and which is given by the functional

$$I(y) = 2\pi \int_{x_A}^{x_B} x\sqrt{1 + y'^2} \, dx.$$

Again we appeal to (4.6) to get the extremum of $I(y)$, where

$$F = 2\pi x \sqrt{1 + y'^2},$$

so that (4.6) becomes

$$\frac{dF_{y'}}{dx} = \frac{2\pi d[xy'/\sqrt{1 + y'^2}]}{dx} = 0,$$

which yields

$$\frac{xy'}{1 + y'^2} = a,$$

where a is the integration constant. Squaring this equation, factoring terms, and solving for y', we obtain

$$\frac{dy}{dx} = \frac{a}{\sqrt{x^2 - a^2}},$$

where clearly $a^2 < x^2$ for a real minimizing curve. The general solution of this first-order differential equation is

$$y = a \int \frac{dx}{\sqrt{x^2 - a^2}} + b = a \text{ arc cosh}\left(\frac{x}{a}\right) + b$$

or

$$x = a \cosh\left(\frac{1}{a}\right)(y - b),$$

which is the equation of a catenary. Again the constants of integration (a, b) are determined by the requirement that the curve pass through the end points (A, B). This solves the problem of finding the minimizing surface for this surface of revolution.

We have given an example in each of the fields of electrical circuits, optics, mechanics, and the elementary theory of surfaces to illustrate the use of the one-dimensional Euler equation in extremizing a functional.

4.3. Generalization to Functionals with More Than One Dependent Variable

We extend our investigations in the calculus of variations to the problem of finding the extremum of a functional $I(\mathbf{u})$, where the integrand depends on n dependent scalar variables u_i $(i = 1, 2, \ldots, n)$ which are the components of a vector $\mathbf{u} = (u_1, u_2, \ldots, u_n) = \mathbf{u}(x)$. We formulate a typical problem of this type as follows

Let $F(x, \mathbf{u}, \mathbf{u}'(x))$ be a function of the $(2n + 1)$ arguments $(x, u_1, u_2, \ldots, u_n,$ $u_1', u_2', \ldots, u_n')$. Let F be a continuously differentiable function of x up to and including the second-order derivatives. The functional

$$I(\mathbf{u}) = \int_{x_A}^{x_B} F(x, \mathbf{u}, \mathbf{u}') \, dx, \qquad \mathbf{u}' = \frac{d\mathbf{u}}{dx}, \tag{4.7}$$

over a fixed interval $[x_A, x_B]$, has a definite value determined by the choice of $\mathbf{u}(x)$. In comparison, we regard all functions $\mathbf{u}(x)$ admissible which satisfy the above continuity conditions and for which the boundary values $\mathbf{u}(x_1)$ and $\mathbf{u}(x_2)$ have prescribed values. This means that we consider curves $u_i(x)$ $(i = 1, 2, \ldots, n)$ joining the two fixed end points A and B in the $(n + 1)$-dimensional space in which the coordinates are $(x, u_1, u_2, \ldots, u_n)$. The variational problem now requires us to find, among all sets of admissible functions $\mathbf{u}(x)$, the optimizing one which is the set that extremizes the functional $I(\mathbf{u})$ (makes the functional stationary). We do not discuss here the actual nature of the stationary value, which brings up the subtle question of the existence of solutions, etc. See, for example, [5] for answers to such questions. We confine ourselves here to inquiring: For what set of $\mathbf{u}(x)$ is the functional stationary?

The concept of a stationary value of $I(\mathbf{u})$ is defined in the same way as in the one-dimensional case. We include the set of functions $\mathbf{u}(x)$ in a one-parameter family of functions which depends on the small parameter ε in the following manner: Let $\boldsymbol{\eta} = \boldsymbol{\eta}(\eta_1, \eta_2, \ldots, \eta_n)$, where the components consist of a set of n arbitrarily chosen functions of x which possess continuous second derivatives in the interval $[x_A, x_B]$, and which satisfy the homogeneous boundary conditions

$$\boldsymbol{\eta}(x_A) = \boldsymbol{\eta}_A = 0, \qquad \boldsymbol{\eta}(x_B) = \boldsymbol{\eta}_B = 0.$$

We now consider the family of functions

$$\bar{\mathbf{u}}(x) = \mathbf{u}(x) + \varepsilon\boldsymbol{\eta}(x) \quad \text{or} \quad \bar{u}_i = u_i + \varepsilon\eta_i, \qquad i = 1, 2, \ldots, n, \tag{4.8}$$

where $\bar{\mathbf{u}}(x)$ is any admissible member of the family and $\mathbf{u}(x)$ is the optimizing member. The first variation of $\mathbf{u}$ is defined as $\delta\mathbf{u}(x)$, where

$$\delta\mathbf{u}(x) = \bar{\mathbf{u}}(x) - \mathbf{u}(x) = \varepsilon\boldsymbol{\eta}(x). \tag{4.9}$$

Note that the variation $\delta\mathbf{u}(x)$ has the same significance as for the one-dimensional case, namely, the variation of $\mathbf{u}(x)$ from any admissible neighboring curve $\bar{\mathbf{u}}(x)$ to the extremizing or optimizing curve $\mathbf{u}(x)$.

Replacing $\mathbf{u}$ by $\mathbf{u} + \varepsilon\boldsymbol{\eta}$ in (4.7) gives I in terms of the parameter ε.

$$I(\mathbf{u} + \varepsilon\boldsymbol{\eta}) = F(x, \mathbf{u} + \varepsilon\boldsymbol{\eta}, \mathbf{u}' + \varepsilon\boldsymbol{\eta}') \, dx = I(\varepsilon), \tag{4.10}$$

where we consider I as a differentiable function of ε for fixed x, $\mathbf{u}$, and $\mathbf{u}'$. A necessary condition that we have a stationary value for I is that the variation $\delta I = 0$ at $\varepsilon = 0$. This gives

$$\delta I(0) = I_\varepsilon(0) \, d\varepsilon = 0. \tag{4.11}$$

To find the stationary value of $I(\mathbf{u})$ we proceed as follows: Pick a particular $\mathbf{u}(x)$ and let $u_j(x) = 0$ for all the $(n - 1)$ i's such that $i \neq j$. Then u_j is the only component of $\mathbf{u}$ that varies, so that the condition (4.11), upon using the result of the one-dimensional case, is equivalent to

$$F_{u_j} - \frac{dF_{u_j'}}{dx} = 0. \tag{4.12}$$

Since $u_j(x)$ is arbitrarily chosen, (4.12) is valid for $j = 1, 2, \ldots, n$. Equation (4.12) is the generalization of (4.6) for functionals of n dependent variables. This gives the following theorem:

A necessary condition that the functional $I(\mathbf{u})$ be stationary is that the n functions u_j shall satisfy the system of n Euler equations given by (4.12).

The Euler equations (4.12) consist of a system of n second-order differential equations for the optimizing function $\mathbf{u}(x)$. They are the equations of motion for a dynamical system. Therefore, the problem of finding the stationary value of the functional is reduced to that of solving these Euler equations. The solution of these equations of motion yields an optimizing curve (for a given set of initial conditions), which may be considered as a trajectory in n-dimensionl or $\mathbf{u}$ space, generated by varying x (time) as a parameter. This expresses a solution for a dynamical system, and will be discussed more fully in the succeeding sections.

4.4. Special Case

We consider the special case where $F = F(\mathbf{u}, \mathbf{u}')$ so that x does not appear explicitly in the integrand of the functional. We easily show that this special case implies that the expression

$$F(\mathbf{u}, \mathbf{u}') - \sum_{i=1}^{n} u_i' F_{u_i} = \text{const.} = E \tag{4.13}$$

is an integral of Euler's equations (4.12). We set the constant of integration to E for the total energy of a conservative system. The truth of (4.13) follows immediately if we differentiate (4.13) with respect to x and use (4.12) for the n u_i's. The pivotal importance of this first integral (4.13) of Euler's equations will be seen when we investigate Lagrange's equations of motion. It will be shown that F is identified with the Lagrangian L so that $F = L = T - V$, where T is the kinetic energy and V the potential energy of a conservative system, and **u** are the generalized coordinates and **u'**, the generalized velocities. These important concepts will be fully discussed below in the appropriate places.

4.5. Hamilton's Variational Principle and Configuration Space

Having sketched the main ideas of the calculus of variations, we now have a reliable guide for formulating and treating wave propagation problems in electromagnetic media. Wave phenomena are best formulated in terms of *Hamilton's variational principle* which will be described in this section. This principle leads to the Lagrange equations of motion—expressed as a second-order system in terms of generalized coordinates and velocities. Hamilton's variational principle is one which considers the *entire motion* of a dynamical system between times t_A and t_B, and small variations of this motion from the actual motion. This is in the spirit of the calculus of variations, and is called an *integral principle*.

Although wave propagation problems involve continuous media (which is composed of a continuous distribution of particles), for the sake of clarity we shall at first consider *discrete systems*, which are dynamical systems composed of a finite number of particles. It is easier to visualize terms like "generalized coordinates", "configuration space", etc., for a discrete system. Therefore, when we discuss Hamilton's integral principle, we shall be concerned with the entire motion of a "discrete ensemble" or a dynamical system consisting of a finite number of particles.

Configuration Space

We now put the meaning of the phrase "entire motion of a dynamical system between times t_A and t_B" in more precise language. Specifically, we consider the motion of a dynamical system in *configuration space*. To define this space we first introduce the concept of *number of degrees of freedom* of our system; this is defined as the number of *independent* coordinates necessary to describe the location of all the particles of a system. These independent coordinates are called *generalized coordinates*; they are characterized by the vector $\mathbf{q} = (q_1, q_2, \ldots, q_n)$ having n components corresponding to an n degree of freedom system. A system of N particles undergoing translational motion with no constraints has $3N$ degrees of freedom, since each particle needs three independent coordinates to describe its position in space. For this system we have $n = 3N$. An example from mechanics: A rigid body under no constraints has three degrees of freedom in translational motion (the motion of its center of gravity in space), and three degrees of freedom in rotation (corresponding to the three "Euler angles" which define an arbitrary rotation) giving six degrees of freedom. The concept of "constraints" will be discussed below.

With this background we now define *configuration space* as an n-dimensional hyperspace whose coordinates are given by the n components of $\mathbf{q}$. Each point in configuration space is defined by a particular value of $\mathbf{q}$ (having n coordinates). Since a solution of the equations of motion in configuration space for each particle involves knowing its initial position and velocity, the position and velocity of the particle at each time completely describes its motion in configuration space. Therefore, associated with each generalized coordinate vector $\mathbf{q}$ is a *generalized velocity* vector $\dot{\mathbf{q}} \equiv d\mathbf{q}/dt = (\dot{q}_1, \dot{q}_2, \ldots, \dot{q}_n)$. For a given set of initial conditions if we know $\mathbf{q}(t)$ and $\dot{\mathbf{q}}(t)$, that is, the generalized coordinates and velocity at each point in configuration space as functions of time, then we have a complete dynamical description of our system of particles. Looking at the situation geometrically: For a given set of initial conditions, a path in configuration space is generated with t as the parameter (increasing with time), each trajectory corresponding to a given set of initial conditions. We emphasize that configuration space has no necessary connection with three-dimensional space, just as the generalized coordinates are not necessarily position coordinates. We point out again that generalized coordinates consist of any set of independent variables that completely describe the position of a system of particles in configuration space (they need not have the dimension of length). A trajectory in configuration space will not have any necessary resemblance to the path in space of any actual particle. At each point on the trajectory, $\mathbf{q}(t)$ and $\dot{\mathbf{q}}(t)$ represent the generalized coordinates and velocities *of all the particles* at time t; this is what is meant by saying that each point on the trajectory represents the *entire system* at some t. Solving the equations of motion for $\mathbf{q}(t)$ and $\dot{\mathbf{q}}(t)$ (for a given set of initial conditions) gives a complete dynamical description of the system of particles, thus generating a path in configuration space.

Hamilton's Principle

We shall take as our model a finite system of particles in a conservative force field so that $T + V = E$, meaning that the total energy of the system is conserved—there are no dissipative forces. (Later on we shall extend the investigations to include dissipative forces.) Therefore the dynamical properties of our system are determined by the kinetic energy T and potential energy V. In general, we assume $T = T(\mathbf{q}, \dot{\mathbf{q}})$ and $V = V(\mathbf{q})$. Specifically, we take T to be a quadratic function of $\dot{\mathbf{q}}$ of the form

$$T = \sum_{i,j=1}^{n} P_{ij}(q_1, q_2, \ldots, q_n)\dot{q}_i\dot{q}_j,\dagger \tag{4.14}$$

---

† For the ith particle, q_i is a generalized coordinate vector having three components (in three space) and (similarly for the generalized velocity $\dot{q}_i$.

where the P_{ij}'s are prescribed functions of q. Equation (4.14) allows T to be a quadratic in $\dot{\mathbf{q}}$, depending arbitrarily on q. We also assume that V is a quadratic function of $\mathbf{q}$ of the form

$$V = \sum_{i,j} b_{ij} q_i q_j, \tag{4.15}$$

where the b_{ij}'s are given constants.

Using these definitions, Hamilton's variational principle states:

Between any two instants of time, t_A and t_B, the actual motion of a system of particles, defined by their trajectories $\mathbf{q} = \mathbf{q}(t)$ in configuration space, proceeds in such a way was to make the functional

$$I(\mathbf{q}) = \int_{t_A}^{t_B} (T - V)\, dt = \int_{t_A}^{t_B} L(\mathbf{q}(t)\dot{\mathbf{q}}(t))\, dt, \tag{4.16}$$

an extremum (stationary), with respect to the neighboring trajectories in configuration space given by $\mathbf{q} = \bar{\mathbf{q}}(t)$, where $\mathbf{q} = \mathbf{q}(t)$ is the optimizing path. This means that we want to find that path $\mathbf{q} = \mathbf{q}(t)$ in configuration space that makes the first variation

$$\delta I(\mathbf{q}) = \delta \int_{t_A}^{t_B} L(\mathbf{q}, \dot{\mathbf{q}})\, dt = 0. \tag{4.17}$$

All the trajectories are subject to the boundary conditions

$$\bar{\mathbf{q}}(t_A) = \mathbf{q}(t_A), \qquad \bar{\mathbf{q}}(t_B) = \mathbf{q}(t_B). \tag{4.18}$$

Equation (4.17) is the expression of Hamilton's principle, subject to (4.18).

We point out again that $T - V = L$ the *Lagrangian*. It is clear that L depends on $\mathbf{q}$ and $\dot{\mathbf{q}}$, so that (4.17) is a generalization to n dependent variables given by $\mathbf{q}$ of the functional I described in the section on the calculus of variations of several dependent variables, where L corresponds to F, $\mathbf{q}$ to $\mathbf{u}$, $\dot{\mathbf{q}}$ to $\mathbf{u}'$, and t to x, in (4.7). Note that L does not depend on t so that the special case given by (4.13) is valid.

4.6. Lagrange's Equations of Motion

We recall that a necessary condition for Hamilton's variational principle (4.17) to be valid is that the Euler equations (4.12) must hold. Using the above correspondence we obtain

$$\frac{d}{dt}\frac{\partial L}{\partial \dot{q}_i} - \frac{\partial L}{\partial q_i} = 0, \qquad i = 1, 2, \ldots, n. \tag{4.19}$$

The system (4.19) of n equations is called *Lagrange's equations of motion*. They arise as a necessary consequence of (4.17). Equations (4.19) are the same as

(4.12). It is for this reason that (4.12) is sometimes called the "Euler–Lagrange" equations. Lagrange's equations of motion (4.19) consist of a set of n PDEs, so that (4.19) is an alternative way of representing Newton's law of motion for an n degree of freedom system. Note that each degree of freedom is described by a single equation of motion, and the system (4.19) is uncoupled since $\mathbf{q}$ is a set of independent coordinates.

The reason why (4.19) represents Newton's law is: $\partial L/\partial \dot{q}_i$ represents the linear momentum of the ith particle, and $\partial L/\partial q_i$ represents the "generalized force" derivable from the potential V (we shall have more to say about generalized forces below). Then (4.19) tells us that the rate of change of the momentum $(d/dt)(\partial L/\partial \dot{q}_i)$ is equal to the generalized force $\partial L/\partial q_i$ acting on the ith particle due to V. Let $\mathbf{p} = (p_1, p_2, \ldots, p_n)$ be the linear momentum vector so that $\mathbf{p} = m\mathbf{q}$ where m is the mass of a particle. Then Newton's law of motion in vector notation is

$$\frac{d\mathbf{p}}{dt} \equiv \dot{\mathbf{p}} = \mathbf{F}, \qquad (4.20)$$

where $\mathbf{F} = (F_1, F_2, \ldots, F_n)$ is the resultant of the external forces acting on the system. Clearly, the correspondence with Lagrange's equations of motion is

$$p_i = \frac{\partial L}{\partial \dot{q}_i}, \qquad \dot{p}_i = \frac{d}{dt}\frac{\partial L}{\partial \dot{q}_i}, \qquad F_i = \frac{\partial L}{\partial q_i}, \qquad \dot{p}_i = F_i, \qquad i = 1, 2, \ldots, n. \quad (4.20')$$

The way to use (4.19) is as follows: For a given system of n particles we construct T, V, and then L, insert L into (4.19), and solve the resulting system for the trajectories $q_i = q_i(t)$, $i = 1, 2, \ldots, n$. These trajectories turn out to be integrals of the equations of motion which are second-order ODEs.

EXAMPLE 1. This is easily illustrated by the trivial example of a simple harmonic oscillator of mass m and spring constant k having a single degree of freedom given by q. This means the mass–spring system represents a single particle system oscillating in the one-dimensional direction q. The kinetic energy $T = \frac{1}{2}mq^2$ and the potential energy $V = \frac{1}{2}kq^2$, so that the Lagrangian becomes

$$L = T - V = \tfrac{1}{2}m\dot{q}^2 - \tfrac{1}{2}kq^2.$$

Lagrange's equations (4.19) reduce to a single equation ($i = 1$). The momentum is $\partial L/\partial \dot{q} = m\dot{q} = p$, the external force is $\partial L/\partial q = -kq$, so that (9.19) becomes

$$\frac{dp}{dt} = m\frac{d^2q}{dt^2} = -kq,$$

which is the differential equation of a simple harmonic oscillator with no external force other than the spring force which is derivable from V.

EXAMPLE 2. We now consider a less trivial example: a system composed of n uncoupled simple harmonic oscillators in three-dimensional space.

The ith particle has mass m_i and all the particles have the spring constant k. Since the particles are uncoupled (noninteracting), there are no constraints acting on the system so that we may use the Cartesian reference frame for the generalized coordinates. Therefore $\mathbf{q} = (x_1, y_1, z_1, \ldots, x_n, y_n, z_n)$ and $\dot{\mathbf{q}} = (u_1, v_1, w_1, \ldots, u_n, v_n, w_n)$, where $\dot{x}_i = u_i$, $\dot{y}_i = v_i$, $\dot{z}_i = w_i$, $i = 1, 2, \ldots, n$. Then the Lagrangian becomes

$$L = T - V = \tfrac{1}{2} \sum_{i=1}^{n} m_i[(u_i)^2 + (v_i)^2 + (w_i)^2] - \tfrac{1}{2}k \sum_{i=1}^{n} [(x_i)^2 + (y_i)^2 + (z_i)^2].$$

Differentiating L with respect to u_i, v_i, and w_i yields the x, y, and z components of the momentum for the ith particle.

$$\frac{\partial L}{\partial u_i} = m_i u_i = (p_i)_x,$$

$$\frac{\partial L}{\partial v_i} = m_i v_i = (p_i)_y,$$

$$\frac{\partial L}{\partial w_i} = m_i w_i = (p_i)_z, \qquad i = 1, 2, \ldots, n.$$

The spring force becomes

$$\frac{\partial L}{\partial x_i} = -kx_i, \qquad \frac{\partial L}{\partial y_i} = -ky_i, \qquad \frac{\partial L}{\partial z_i} = -kz_i, \qquad i = 1, 2, \ldots, n.$$

Using these expressions in Lagrange's equations (4.19) yields

$$m_i\ddot{x}_i = -kx_i, \qquad m_i\ddot{y}_i = -ky_i, \qquad m_i\ddot{z}_i = -kz_i, \qquad i = 1, 2, \ldots, n,$$

which are the differential equations of motion for our system of n uncoupled simple harmonic oscillators in three-space.

What we have accomplished up to now is to state Hamilton's variational principle, and from this integral principle use the Euler's equations which become the Lagrange's equations of motion. This means that we have derived Lagrange's equations from Hamilton's principle. As pointed out above, this "integral approach" uses a variation technique to compare admissible paths with the actual path in configuration space; the actual path is the one that makes the functional $I = \int_{t_A}^{t_B} L \, dt$ stationary, which is Hamilton's principle.

There is another approach to deriving Lagrange's equations of motion, which is called a "differential approach". In this method we start from a consideration of the *instantaneous state* of the system and superimpose small *virtual displacements* about this state. From this concept is developed the *principle of virtual work*. Making use of the constraints acting on the system, this principle is used to derive Lagrange's equations. This approach was suggested by the Swiss mathematician James Bernoulli and developed by the French mathematician D'Alembert; it is aptly called *D'Alembert's principle*. This method will now be treated.

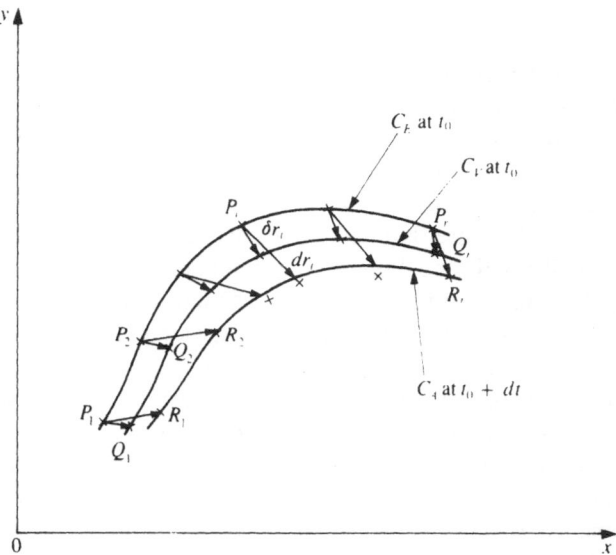

Fig. 4.3. Virtual and actual displacements of n particles in the (x, y) plane.

4.7. D'Alembert's Principle, Constraints, and Lagrange's Equations†

We first discuss the concept of *virtual displacement*. A virtual displacement represents an arbitrary infinitesmal change or displacement of the configuration of the system from its original state, consistent with the forces and constraints imposed on the system at a given instant. The important point is: *time does not change during a virtual displacement*, this is in contrast to an actual infinitesmal displacement where time changes by an amount dt. Specifically, let $\mathbf{r}_i$ be the radius vector of the ith particle in a three-dimensional Cartesian coordinate system. Let $\delta\mathbf{r}_i$ be an arbitrary virtual displacement of the ith particle consistent with the constraints on the system at a given time t. $d\mathbf{r}_i$ is the actual displacement of the particle. Figure 4.3 shows an example in the (x, y) plane of a system of n particles in equilibrium at time t_0, their virtual displacements and their actual displacements. The particles in the equilibrium state are on the (fictitious) curve C_{E}. These particles are transformed to the curve C_{V} under a virtual displacement, and to the curve C_{A} under an actual displacement. The position of the ith particle in the equilibrium state is given by the point P_i. Upon undergoing a virtual displacement $\delta\mathbf{r}_i$ it is transformed to the point Q_i. If the particle at P_i undergoes an actual displacement $d\mathbf{r}_i$ it is transformed to the point R_i. These transformations hold for all the n particles. The particles on C_{V} are at time t_0 and are consistent with the constraints acting on them. The particles on C_{A} are at $t_0 + dt$.

---

† See, for example, [22].

Constraints

We observe that the radius vector $\mathbf{r}_i$ for the ith particle is not necessarily the generalized coordinate vector $\mathbf{q}_i$, since the $\mathbf{q}_i$'s are independent coordinates and the $\mathbf{r}_i$'s involve any constraints imposed on the particle. We therefore discuss the nature of constraints acting on our system. Suppose a system of n particles is subject to k constraints. Clearly $0 < k < n$; for if $k = 0$, then the system is "free" in the sense that no constraints act on it, and if $k = n$ then the system is static. Roughly speaking, a constraint is a restriction on the motion of a particle. It is important to understand the nature of the constraints on a system in order to properly interpret Lagrange's equations of motion with constraints. Constraints will now be classified and examples will be give.

We first consider the simple case of a single particle in three-space. If it is free to move anywhere on a surface but cannot leave it, then there is a single constraint since the equation defining the surface on which the particle moves ties up one of the three coordinates. Therefore $k = 1, n = 2$. This is an example of a *holonomic constraint*, which is defined as one expressed by an equation connecting the coordinates. Another example of a holonomic constraint involves a rigid body composed of n particles. A rigid body means there is no relative motion between any two neighboring particles, so that the condition of constraint is given by

$$(\mathbf{r}_i - \mathbf{r}_j)^2 = c_{ij}, \qquad i \neq j, \qquad i, j = 1, 2, \ldots, n,$$

where the c_{ij}'s are constants. For a system of n particles acted on by k holonomic constraints we have the k constraining equations which can be put in the form

$$f_i(\mathbf{r}_1, \mathbf{r}_2, \ldots, \mathbf{r}_n) = 0, \qquad i = 1, 2, \ldots, k.$$

Constraints that are not holonomic are called *nonholonomic*. An example of a nonholonomic constraint is a particle placed on the surface of a sphere of radius R and given an initial velocity. The particle is in a gravitational field so that it will slide down the surface and eventually fall off. This type of constraint is not governed by an equation, but by an *inequality*. In this case the inequality is

$$\mathbf{r} - R > 0,$$

where $\mathbf{r}$ is the radius vector of the particle measured from the center of the sphere.

Constraints are also classified as *scleronomic* (time-independent) and *rheonomic* (time-depending). The above examples involve scleronomic constraints since the surface is assumed to be stationary. A bead sliding on a stationary wire is another example of a scleronomic constraint. If the bead slides on a moving wire, then the constraint is rhenonomic. Both examples involve holonomic constraints, since the wire is governed by an equation.

Regardless of the type of constraint, if a system of n particles has k constraints expressed by k equations and/or inequalities, then the number of generalized coordinates is $n - k$. This means that the n Lagrange equations (4.19) are not independent but must incorporate the k constraint conditions. This was treated by Lagrange where he introduced *undetermined multipliers* to eliminate the constraints and derive the appropriate equations of motion. This will be discussed in the appropriate place.

Principle of Virtual Work

Having defined virtual displacements and constraints, we are now in a position to discuss the concept of *virtual work* which will be used to develop D'Alembert's principle. Virtual work is defined as the work done on a system due to the external forces acting on it, such that the system undergoes a virtual displacement compatible with the constraints. We again consider a system of n particles in an equilibrium configuration. Then the resultant external force on each particle vanishes so that $\mathbf{F}_i = 0$, $i = 1, 2, \ldots, n$. Let δW_i be the *virtual work* on the ith particle due to a virtual displacement $\delta \mathbf{r}_i$. Then $\delta W_i = \mathbf{F}_i \cdot \delta \mathbf{r}_i$, and the condition that the system be in equilibrium is that the virtual work due to the resultant force $\mathbf{F}$ shall vanish.

$$\sum_{i=1}^{n} \mathbf{F}_i \cdot \delta \mathbf{r}_i = 0. \tag{4.21}$$

Let the resultant of the external forces on the ith particle $\mathbf{F}_i$ be given by

$$\mathbf{F}_i = \mathbf{F}_i' + \mathbf{f}_i, \tag{4.22}$$

where $\mathbf{F}_i'$ is the applied force and $\mathbf{f}_i$ is the constraining force on the ith particle. If we restrict ourselves to a system where the virtual work done by the forces of constraint is zero,† then $\mathbf{F}_i$ is replaced by $\mathbf{F}_i'$ in (4.21). Then (4.21) states that the condition for a system of n particles to be in equilibrium is that the virtual work of the applied forces on the system must vanish.

We point out the important fact that the virtual work δW on the system is due to the virtual displacement $\delta \mathbf{r}$, where time does not change during the action of the forces causing δW. This must be distinguished from the *actual* work dW on the system where the forces act during the time dt.

We now have a problem: The principle of virtual work, as expressed by (4.21), is based on the condition that the system is in equilibrium, meaning that we are dealing with a problem in statics. The problem is: Since we desire to investigate dynamical systems, how do we generalize the principle of virtual work to include such systems? The answer was given by D'Alembert. He ingeniously wrote the equations of motion for the ith particle in the following form:

$$\mathbf{F}_i - \dot{\mathbf{p}}_i = 0, \qquad i = 1, 2, \ldots, n. \tag{4.23}$$

Clearly the inertial force on the ith particle is $d\mathbf{p}_i/dt \equiv \dot{\mathbf{p}}_i$. The right-hand side

---

† This will not be true for a system undergoing frictional forces.

of (4.23) is the zero vector resultant force. This seemingly innocuous device of putting all the forces on the left-hand side of the equation has the advantage of treating the inertial force $\dot{\mathbf{p}}_i$ in the same manner as the external force $\mathbf{F}_i$. Therefore (4.23) says the vector sum of the inertial and external forces on the ith particle vanishes, so that we have the same situation as a particle in equilibrium. This means that D'Alembert, by this simple algebraic trick, transformed the system of n particles from a problem in dynamics to an apparent problem in statics. Therefore, the principle of virtual work can be applied to a system, each particle of which is under the action of the $\mathbf{F}_i - \dot{\mathbf{p}}_i$ force system. Another way of stating the principle of virtual work applied to a dynamical system is: A system of n particles, in equilibrium under the action of the resultant of the external forces $\mathbf{F}$ and the "reversed inertial forces" $-\dot{\mathbf{p}}_i$, has zero virtual work acting on it. This is given by

$$\sum_{i=1}^{n} (\mathbf{F}_i - \dot{\mathbf{p}}_i) \cdot \delta \mathbf{r}_i = 0. \tag{4.24}$$

We again restrict ourselves to systems for which the virtual work of the forces of constraint vanishes so that $\mathbf{F}$ is interpreted as the resultant applied force on the system. Then (4.24), which arises from (4.23), is called *D'Alembert's principle*. As we saw, (4.24) is the principle of virtual work applied to a dynamical system under no constraints.

We would like to derive Lagrange's equations of motion from D'Alembert's principle. However, we cannot do so if this principle is in the form given by (4.24). The reason is that the $\mathbf{r}_i$'s do not form an independent coordinate system. We therefore need to transform (4.24) to the generalized coordinate system given by the $\mathbf{q}_i$'s. The reason for performing such a transformation is: Suppose $\mathbf{r}$ is transformed into $\mathbf{q}$, and (4.24) is transformed to the corresponding equation in the $\mathbf{q}$ coordinate system. Then the coefficient of each $\delta \mathbf{q}_i$ must vanish or $\mathbf{F}_i - \mathbf{p}_i = 0$ for each i, since the $\delta \mathbf{q}_i$'s are independent. We thus obtain the equation of motion for each particle.

Transformation to Generalized Coordinates

We now carry out our program of transforming the coordinate system $\mathbf{r}$ to the generalized coordinate system $\mathbf{q}$. We seek a coordinate transformation

$$\mathbf{r}_i = \mathbf{r}_i(q_1, q_2, \ldots, q_n), \qquad i = 1, 2, \ldots, n, \tag{4.25}$$

such that, as mentioned, the virtual work on the dynamical system of n particles expressed by (4.24) will be transformed into the corresponding system, where each $\delta \mathbf{q}_i$ is independent so that the equation of motion $\mathbf{F}_i - \dot{\mathbf{p}}_i = 0$ is valid for each particle. In this case $\mathbf{p}_i$ is called the *generalized momentum* corresponding to the generalized coordinate $\mathbf{q}_i$ of the ith particle. $\mathbf{p}_i$ is sometimes called the momentum "conjugate to" $\mathbf{q}_i$ since they both refer to the same particle. Knowing $\mathbf{q}_i$ and $\mathbf{p}_i$ as functions of time (from the solution of the equations of motion) gives the complete time history of the dynamical system.

The velocity of the ith particle is $\mathbf{v}_i = d\mathbf{r}_i/dt$. We now express $\mathbf{v}_i$ as a linear combination of the generalized velocity components $dq_i/dt = \dot{q}_i$ by using the chain rule of differentiation. We obtain

$$\mathbf{v}_i = \frac{d\mathbf{q}_i}{dt} \equiv \dot{\mathbf{q}}_i = \sum_{j=1}^{n} \frac{\partial \mathbf{r}_i}{\partial q_j} \dot{q}_j + \frac{\partial \mathbf{r}_i}{\partial t}. \tag{4.26}$$

Note that $\mathbf{v}_i$ is a vector while $\dot{q}_j$ is the jth component of the generalized velocity of a particular particle and is thus a scalar. (The notation will be made more precise when we specialize below to a single particle.) In a similar manner, the virtual displacement of the ith particle $\delta \mathbf{r}_i$ is given as a linear combination of the virtual generalized displacements δq_i by the expansion

$$\delta \mathbf{r}_i = \sum_{j=1}^{n} \frac{\partial \mathbf{r}_i}{\partial q_j} \delta q_j. \tag{4.27}$$

(Note that time is constant for a virtual displacement so that $\partial \mathbf{r}_i/\partial t = 0$.) In terms of the $\delta \mathbf{q}_i$'s the virtual work δW of the system due to the $\mathbf{F}_i$'s becomes

$$\delta W = \sum_{i=1}^{n} \mathbf{F}_i \cdot \delta \mathbf{r}_i = \sum_{i,j} \mathbf{F}_i \cdot \frac{\partial \mathbf{r}_i}{\partial q_j} \delta q_j = \sum_{j} Q_j \delta q_j, \tag{4.28}$$

where Q_j is the *generalized force* on the jth particle, and is defined by

$$Q_j = \sum_{i} \mathbf{F}_i \cdot \frac{\partial \mathbf{r}_i}{\partial q_j}. \tag{4.29}$$

We now apply D'Alembert's principle (4.24). Using the fact that $\mathbf{p}_i = m_i \mathbf{r}_i$, we have

$$\sum_{i} \dot{\mathbf{p}}_i \cdot \delta \mathbf{r}_i = \sum_{i} m_i \ddot{\mathbf{r}}_i \cdot \delta \mathbf{r}_i = \sum_{i} m_i \ddot{\mathbf{r}}_i \cdot \frac{\partial \mathbf{r}_i}{\partial q_j} \delta q_j.$$

We express the right-hand side of this equation in the form

$$\sum_{i} m_i \ddot{\mathbf{r}}_i \cdot \frac{\partial \mathbf{r}_i}{\partial q_j} = \sum_{i} \left[\frac{d}{dt} (m_i \dot{\mathbf{r}}_i) \cdot \frac{\partial \mathbf{r}_i}{\partial q_j} - m_i \dot{\mathbf{r}}_i \cdot \frac{d}{dt} \left(\frac{\partial \mathbf{r}_i}{\partial q_j} \right) \right].$$

In the last term of the right-hand side we exchange the operators d/dt and $\partial/\partial q_j$ and obtain

$$\frac{d}{dt} \left(\frac{\partial \mathbf{r}_i}{\partial q_j} \right) = \sum_{k} \left(\frac{\partial^2 \mathbf{r}_i}{\partial q_j \partial q_k} \right) \dot{q}_k + \frac{\partial^2 \mathbf{r}_i}{\partial q_j \partial t} = \frac{\partial \mathbf{v}_i}{\partial q_j},$$

using (4.26). From these expressions we get

$$\sum_{i} m_i \ddot{\mathbf{r}}_i \cdot \frac{\partial \mathbf{r}_i}{\partial q_j} = \sum_{i} \left[\frac{d}{dt} \left(m_i \mathbf{v}_i \cdot \frac{\partial \mathbf{v}_i}{\partial q_j} \right) - m_i \mathbf{v}_i \cdot \frac{\partial \mathbf{v}_i}{\partial q_j} \right].$$

This gives

$$\sum_{i} \dot{\mathbf{p}}_i \cdot \delta \mathbf{r}_i = \sum_{j} \left[\frac{d}{dt} \left(\frac{\partial}{\partial \dot{q}_j} \right) \left(\sum_{i} \tfrac{1}{2} m_i v_i^2 \right) - \frac{\partial}{\partial q_j} \sum_{i} \tfrac{1}{2} m_i v_i^2 \right] \delta q_j.$$

$$= \sum_{j} \left[\frac{d}{dt} \left(\frac{\partial T}{\partial \dot{q}_j} \right) - \frac{\partial T}{\partial q_j} \right] \delta q_j, \tag{4.30}$$

using the definition of the kinetic energy T of the system. Using (4.28) and (4.30) in D'Alembert's principle (4.24) we get

$$\sum_j \left[\frac{d}{dt}\left(\frac{\partial T}{\partial \dot{q}_j}\right) - \frac{\partial T}{\partial q_j} - Q_j \right]\delta q_j = 0.$$

Since the virtual displacements δq_j are independent, the coefficients of δq_j must vanish, yielding

$$\frac{d}{dt}\left(\frac{\partial T}{\partial \dot{q}_j}\right) - \frac{\partial T}{\partial q_j} = Q_j, \qquad j = 1, 2, \dots, n. \tag{4.31}$$

The n equations (4.31) are the Lagrange equations for the kinetic energy T and the generalized forces Q_j in terms of the generalized coordinates and velocities. We have thus derived Lagrange's equations in terms of the kinetic energy and generalized forces from the principle of virtual work. So far we have not necessarily imposed a conservative force system on the Q_j's. We do so now.

Conservative System. For a conservative system we know that the potential energy V is a quadratic function of $\mathbf{q}$. Therefore, the force $\mathbf{F}_i$ on ith particle is derivable from V in the sense that

$$\mathbf{F}_i = -\mathbf{grad}_i\, V,$$

so that the generalized force on the jth particle becomes

$$Q_j = \sum_i \mathbf{F}_i \cdot \frac{\partial \mathbf{r}_i}{\partial q_j} = -\sum_i \mathbf{grad}_i\, V \cdot \frac{\partial \mathbf{r}_i}{\partial q_j} = -\frac{\partial V}{\partial q_j}. \tag{4.32}$$

Introducing the Lagrangian $L = T - V$ and recalling that T is a function of $\mathbf{q}$ and $\dot{\mathbf{q}}$, and that V is a function of $\mathbf{q}$, (4.31) becomes (4.19), which are the Lagrange equations of motion for a system of n particles in a conservative force field. We have thus derived Lagrange's equations of motion for a conservative force field from D'Alembert's principle. Since (4.19) was previously derived from Hamilton's principle, the analysis presented here shows Hamilton's principle in the light of the principle of virtual work.

A Single Particle. It is instructive to derive Lagrange's equations *ab initio* from D'Alembert's principle for a single particle ($n = 1$) in three-space where $\mathbf{r} = (x, y, z)$. The generalized coordinates and velocities are $\mathbf{q} = (q_1, q_2, q_3)$, $\dot{\mathbf{q}} = (\dot{q}_1, \dot{q}_2, \dot{q}_3)$. The transformation (4.25) becomes

$$x = x(q_1, q_2, q_3), \qquad y = y(q_1, q_2, q_3), \qquad z = z(q_1, q_2, q_3). \tag{4.33}$$

From this transformation we calculate the components of the velocity $\mathbf{v} = (\dot{x}, \dot{y}, \dot{z})$ by expanding the operator d/dt, obtaining

$$\dot{x} = \left(\frac{\partial x}{\partial q_1}\right)\dot{q}_1 + \left(\frac{\partial x}{\partial q_2}\right)\dot{q}_2 + \left(\frac{\partial x}{\partial q_3}\right)\dot{q}_3, \tag{4.34}$$

and similar expansions for $\dot{y}$ and $\dot{z}$. It is clear from (4.33) that the coefficients of the $\dot{q}_i$'s in these expansions are functions of the q_i's. This tells us that $\mathbf{v} = \mathbf{v}(q_1, \ldots, \dot{q}_3)$. From the above expressions we obtain

$$\frac{\partial \dot{x}}{\partial \dot{q}_i} = \frac{\partial x}{\partial q_i}, \qquad \frac{\partial \dot{y}}{\partial \dot{q}_i} = \frac{\partial y}{\partial q_i}, \qquad \frac{\partial \dot{z}}{\partial \dot{q}_i} = \frac{\partial z}{\partial q_i}, \qquad i = 1, 2, 3. \qquad (4.35)$$

The partial derivatives $\partial x/\partial q_i$, etc., are functions of the q_i's.

We now follow the motion of the particle; this means that the term $\partial/\partial t$ in the operator d/dt vanishes. We also have

$$\frac{d}{dt}\left(\frac{\partial x}{\partial q_i}\right) = \frac{\partial \dot{x}}{\partial q_i},$$

$$\frac{d}{dt}\left(\frac{\partial y}{\partial q_i}\right) = \frac{\partial \dot{y}}{\partial q_i}, \qquad (4.36)$$

$$\frac{d}{dt}\left(\frac{\partial z}{\partial q_i}\right) = \frac{\partial \dot{z}}{\partial q_i}.$$

The kinetic energy for the particle is

$$T = \left(\frac{m}{2}\right)(\dot{x}^2 + \dot{y}^2 + \dot{z}^2)$$

$$= \tfrac{1}{2}(a_{11}(\dot{q}_1)^2 + a_{12}\dot{q}_1\dot{q}_2 + a_{13}\dot{q}_1\dot{q}_3 + a_{22}(\dot{q}_2)^2 + a_{23}\dot{q}_2\dot{q}_3 + a_{33}(\dot{q}_3)^2) \qquad (4.37)$$

Since $(\dot{x}, \dot{y}, \dot{z})$ are functions of the $\dot{q}_i$'s we wrote T as a quadratic form in the $\dot{q}_i$'s where the coefficients a_{ij} are functions of the q_i's and $a_{ij} = a_{ji}$. From (4.37) and (4.35) we obtain

$$\frac{\partial T}{\partial \dot{q}_i} = \left(\frac{\partial T}{\partial \dot{x}}\right)\frac{\partial \dot{x}}{\partial \dot{q}_i} + \left(\frac{\partial T}{\partial \dot{y}}\right)\frac{\partial \dot{y}}{\partial \dot{q}_i} + \left(\frac{\partial T}{\partial \dot{z}}\right)\frac{\partial \dot{z}}{\partial \dot{q}_i}$$

$$= m\dot{x}\frac{\partial x}{\partial q_i} + m\dot{y}\frac{\partial y}{\partial q_i} + m\dot{z}\frac{\partial z}{\partial q_i}, \qquad i = 1, 2, 3. \qquad (4.38)$$

Differentiating (4.38) yields

$$\frac{d}{dt}\left(\frac{\partial T}{\partial \dot{q}_i}\right) = m\ddot{x}\frac{\partial x}{\partial q_i} + m\ddot{y}\frac{\partial y}{\partial q_i} + m\ddot{z}\frac{\partial z}{\partial q_i}$$

$$+ m\dot{x}\frac{\partial \dot{x}}{\partial q_i} + m\dot{y}\frac{\partial \dot{y}}{\partial q_i} + m\dot{z}\frac{\partial \dot{z}}{\partial q_i}, \qquad i = 1, 2, 3. \qquad (4.39)$$

From (4.37) we obtain

$$\frac{\partial T}{\partial q_i} = m\dot{x}\frac{\partial \dot{x}}{\partial q_i} + m\dot{y}\frac{\partial \dot{y}}{\partial q_i} + m\dot{z}\frac{\partial \dot{z}}{\partial q_i}. \qquad (4.40)$$

The components of the equation of motion $m\ddot{\mathbf{r}} = \mathbf{F}$ are

$$m\ddot{x} = X, \qquad m\ddot{y} = Y, \qquad m\ddot{z} = Z. \qquad (4.41)$$

Inserting (4.41) into (4.39) and subtracting (4.40) from (4.39) yields

$$\frac{d}{dt}\left(\frac{\partial T}{\partial \dot{q}_i}\right) - \frac{\partial T}{\partial q_i} = X\frac{\partial x}{\partial q_i} + Y\frac{\partial y}{\partial q_i} + Z\frac{\partial z}{\partial q_i}, \qquad i = 1, 2, 3. \qquad (4.42)$$

We now introduce the virtual work δW done by the force $\mathbf{F} = (X, Y, Z)$ due to the virtual displacements $(\delta x, \delta y, \delta z)$, and get

$$\delta W = X\delta x + Y\delta y + Z\delta z = Q_1\delta q_1 + Q_2\delta q_2 + Q_3\delta q_3, \qquad (4.43)$$

where $\mathbf{Q} = (Q_1, Q_2, Q_3)$ is the *generalized force* whose components are given by

$$Q_i = X\frac{\partial x}{\partial q_i} + Y\frac{\partial y}{\partial q_i} + Z\frac{\partial z}{\partial q_i}, \qquad i = 1, 2, 3. \qquad (4.44)$$

Upon using (4.44), (4.42) becomes

$$\frac{d}{dt}\left(\frac{\partial T}{\partial \dot{q}_i}\right) - \frac{\partial T}{\partial q_i} = Q_i, \qquad i = 1, 2, 3.\dagger \qquad (4.45)$$

These are the three Lagrange equations of motion for a particle in terms of T and $\mathbf{Q}$.

If the particle is under the action of a *conservative force*, then there exists a potential function $V(\mathbf{q})$ such that the components of the generalized force are given by

$$Q_i = -\frac{\partial V}{\partial q_i} \qquad \text{or} \qquad \mathbf{Q} = -\mathbf{grad\ q}. \qquad (4.46)$$

Introducing the Lagrangian $L = T - V$, (4.45) becomes

$$\frac{d}{dt}\left(\frac{\partial L}{\partial \dot{q}_i}\right) - \frac{\partial L}{\partial q_i} = 0, \qquad i = 1, 2, 3, \qquad (4.47)$$

which are the three Lagrange equations of motion for a particle acted upon by a conservative force (a force derivable from a potential).

4.8. Nonconservative Force Field, Velocity-Dependent Potential

We now consider an n degree of freedom system (of N particles, where $n = 3N$) acted upon by a *nonconservative force field*. This means that $\mathbf{F}$ is not derivable from a potential V that depends only on $\mathbf{q}$, so that the generalized force $\mathbf{Q}$ is not given by (4.46) since the potential depends on $\dot{\mathbf{q}}$ as well as $\mathbf{q}$. An example is the case of a frictional or viscous force acting on the system. Such a force dissipates energy in the form of heat, so that energy is not conserved. However, we may still use Lagrange's equations (4.47), but in a generalized sense.

This is done by replacing the potential V by a *velocity-dependent potential*

---

† Clearly (4.31) is the generalization of (4.45) for an n degree of freedom system.

$U = U(\dot{\mathbf{q}}, \mathbf{q})$. U is sometimes called a *generalized potential*. (Note that we distinguish U from V which only depends on $\mathbf{q}$ and is used for a conservative force field.) The Lagrangian is now given by

$$L = T - U. \tag{4.48}$$

We now extend Lagrange's equations (4.45) or (4.47) to an n degree of freedom system in a nonconservative force field. (Equation (4.47) becomes (4.49).) In order to do this we set

$$T(\dot{\mathbf{q}}, \mathbf{q}) = L(\dot{\mathbf{q}}, \mathbf{q}) + U(\dot{\mathbf{q}}, \mathbf{q})$$

in (4.45) and obtain

$$\frac{d}{dt}\left(\frac{\partial L}{\partial \dot{q}_i}\right) - \frac{\partial L}{\partial q_i} = Q_i + \frac{\partial U}{\partial q_i} - \frac{d}{dt}\left(\frac{\partial U}{\partial \dot{q}_i}\right), \qquad i = 1, 2, \ldots, n. \tag{4.49}$$

If we set the right-hand side of (4.49) equal to zero we obtain Lagrange's equations for an n degree of freedom system, where L is given by (4.48). Then the generalized force Q_i for a nonconservative system becomes

$$Q_i = -\frac{\partial U}{\partial q_i} + \frac{d}{dt}\left(\frac{\partial U}{\partial \dot{q}_i}\right). \tag{4.50}$$

For the special case of a conservative force system (4.50) reduces to (4.46).

Application to Electromagnetic Fields

At this stage it is instructive to show how we can construct the generalized potential U for a special type of force field, namely, the electromagnetic forces acting on moving charges. It will be shown that U is a linear combination of a scalar potential of $\mathbf{q}$ and a vector potential of $\dot{\mathbf{q}}$. Then U will be used to construct the Lagrangian for a charged particle in an electromagnetic field.

We start be considering Maxwell's equations which are given in Chapter 1 by (1.1) $\cdots$ (1.4) in Gaussian units. The force $\mathbf{F}$ on a particle of charge q is called the *Lorentz force* and is given by (1.7). The electric field $\mathbf{E}$ is not the gradient of a scalar potential since $\nabla \times \mathbf{E} \neq 0$, as seen from (1.2). In Section 1.3 we introduced the *vector potential* $\mathbf{A}$ related to $\mathbf{B}$ by (1.22). Some authors call $\mathbf{A}$ the *magnetic vector potential*. We then defined the scalar potential by (1.24). In terms of the potentials Φ and $\mathbf{A}$ the Lorentz force becomes

$$\mathbf{F} = q\left[-\mathbf{grad}\ \Phi - \left(\frac{1}{c}\right)\frac{\partial \mathbf{A}}{\partial t} + \left(\frac{1}{c}\right)(\mathbf{v} \times \nabla \times \mathbf{A})\right]. \tag{4.51}$$

We use the following vector identity for the last term on the right-hand side of (4.51):

$$\mathbf{v} \times \nabla \times \mathbf{A} = \mathbf{grad}(\mathbf{A} \cdot \mathbf{v}) - \mathbf{A} \cdot \mathbf{grad}\ \mathbf{v} - \mathbf{v} \cdot \mathbf{grad}\ \mathbf{A}$$

$$= \mathbf{grad}(\mathbf{A} \cdot \mathbf{v}) - \mathbf{A} \cdot \mathbf{grad}\ \mathbf{v} - \frac{d}{dt}[\mathbf{grad}_v(\mathbf{A} \cdot \mathbf{v})] + \frac{\partial \mathbf{A}}{\partial t}, \tag{4.52}$$

where we used the fact that

$$\mathbf{v} \cdot \mathbf{grad\ A} = \frac{d\mathbf{A}}{dt} = \frac{d}{dt}[\mathbf{grad}_v(\mathbf{A} \cdot \mathbf{v})] + \frac{\partial \mathbf{A}}{\partial t}.$$

Equation (4.52) can be justified by expanding the x component of $\mathbf{v} \times \nabla \times \mathbf{A}$). Upon setting $\mathbf{v} = (u, v, w)$ we have

$$(\mathbf{v} \times \nabla \times \mathbf{A})_x = v\left(\frac{\partial A_y}{\partial x} - \frac{\partial A_x}{\partial y}\right) - w\left(\frac{\partial A_x}{\partial z} - \frac{\partial A_z}{\partial x}\right)$$

$$= v\frac{\partial A_y}{\partial x} + w\frac{\partial A_z}{\partial x} + u\frac{\partial A_x}{\partial x} - v\frac{\partial A_x}{\partial y} - w\frac{\partial A_z}{\partial z} - u\frac{\partial A_x}{\partial x},$$

$$\left(\text{upon adding and subtracting } u\frac{\partial A_x}{\partial x}\right)$$

$$= \left(\frac{\partial}{\partial x}\right)(\mathbf{v} \cdot \mathbf{A}) - \mathbf{A} \cdot \frac{\partial \mathbf{v}}{\partial x} - \frac{dA_x}{dt} + \frac{\partial A_x}{\partial t}.$$

Using (4.52), (4.51) becomes

$$\mathbf{F} = q\left[\mathbf{grad}\left(-\Phi + \left(\frac{1}{c}\right)\mathbf{A} \cdot \mathbf{v}\right) - \left(\frac{1}{c}\right)\mathbf{A} \cdot \mathbf{grad\ v} - \left(\frac{1}{c}\right)\frac{d}{dt}[(\mathbf{grad}_v(\mathbf{A} \cdot \mathbf{v})].\right.$$

If we define the generalized potential U by

$$U = q\Phi - \left(\frac{q}{c}\right)\mathbf{A} \cdot \mathbf{v} \tag{4.53}$$

the Lorentz force becomes

$$\mathbf{F} = -\mathbf{grad\ } U + \left(\frac{d}{dt}\right)[\mathbf{grad}_v(\mathbf{A} \cdot \mathbf{v})] - \left(\frac{q}{c}\right)\mathbf{A} \cdot \mathbf{grad\ v}. \tag{4.54}$$

These results tell us that $\mathbf{F}$ is given in terms of the scalar potential Φ and vector potential $\mathbf{A}$.

Rayleigh's Dissipation Function

We again consider an n degree of freedom system of particles acted on by a frictional or viscous force. If the motion is small enough, the viscous force is proportional to the particle velocity. (As an aside, for aircraft or rockets that fly at greater velocities than those that occur in the materials sciences, the viscous drag is proportional to the square of the velocity.) We therefore restrict ourselves to a physical situation where the particles (charged or otherwise) move with a velocity such that the following law applies (relating the frictional force to the particle velocity):

$$\mathbf{f}_i = -k\mathbf{v}_i, \tag{4.55}$$

where $\mathbf{f}_i$ is the frictional force and $\mathbf{v}_i = (u_i, v_i, w_i)$ is the velocity of the ith particle. The friction coefficient $\mathbf{k} = (k_x, k_y, k_z) > 0$ is assumed to be the same on each particle. Lord Rayleigh [37, Vol. I, Chap. IV], introduced the scalar quadratic function of the velocity called the *Rayleigh dissipation function†* $\mathscr{F}$, which is defined by

$$\mathscr{F} = \frac{1}{2} \sum_{i=1}^{n} [k_x(u_i)^2 + k_y(v_i)^2 + k_z(w_i)^2]. \qquad (4.56)$$

From this definition we obtain the following expression for the frictional force $\mathbf{f}$ on the system:

$$\mathbf{f} = -\mathbf{grad}_v. \qquad (4.57)$$

We now give a physical interpretation of $\mathscr{F}$ by considering a differential amount of work dW_f done by the system against the frictional force $\mathbf{f}$.

$$dW_f = -\mathbf{f} \cdot d\mathbf{r} = -\mathbf{f} \cdot \mathbf{v}\, dt = -(k_x u^2 + k_y v^2 + k_z w^2), \qquad \sum_{i=1}^{n} (u_i)^2 = u^2, \quad \text{etc.}$$

Relating this expression to (4.56) we see that $\mathscr{F}$ is half the rate of energy dissipated by the frictional force $\mathbf{f}$. Let Q_j be the jth component of the generalized force due to the frictional force. Then

$$Q_j = \sum_i \mathbf{f}_i \cdot \frac{\partial \mathbf{r}_i}{\partial q_j} = -\sum_i \mathbf{grad}_v\, \mathscr{F} \cdot \frac{\partial \dot{\mathbf{r}}_i}{\partial \dot{q}_j} = -\frac{\partial \mathscr{F}}{\partial \dot{q}_j}. \qquad (4.58)$$

Lagrange's equations now become

$$\frac{d}{dt}\left(\frac{\partial L}{\partial \dot{q}_j}\right) - \frac{\partial L}{\partial q_j} = -\frac{\partial \mathscr{F}}{\partial \dot{q}_j}, \qquad j = 1, 2, \ldots, n. \qquad (4.59)$$

EXAMPLE. As an example of the use of (4.59) we consider an n degree of freedom system acted upon by a potential $V(\mathbf{q})$ and the Rayleigh dissipation function $\mathscr{F}(\dot{\mathbf{q}})$ given by (4.56). Let m be the mass of each particle, and let K be the spring constant. The system is assumed to be uncoupled (no constraints) so that we can use the generalized coordinates and velocities $(\mathbf{q}, \dot{\mathbf{q}})$ for the Lagrangian, which is

$$L = \frac{m}{2} \sum_i (\dot{q}_i)^2 - \frac{K}{2} \sum_i (q_i)^2.$$

It is clear, for our n degree of freedom system in three-dimensional space, that (as mentioned previously) the generalized coordinate and velocity (q_j and $\dot{q}_j$) are actually vectors for the jth particle and are represented in Cartesian three-space by (x_j, y_j, z_j) and $(u_j = \dot{x}_j, v_j = \dot{y}_j, w_j = \dot{z}_j)$, respectively. The gen-

---

† In the above-cited reference, Lord Rayleigh mentions that this function first appeared in an article written by him in a paper entitled, "General Theorems Relating to Vibrations", published in the *Proceedings of the London Mathematical Society*, June, 1873.

eralized momentum for the jth particle becomes

$$p_j = \frac{\partial L}{\partial \dot{q}_j} = m\dot{q}_j$$

$$= (mu_j, mv_j, mw_j).$$

The generalized force for the jth particle becomes, upon inserting (4.56) into (4.58)

$$Q_j = -\mathbf{k}\dot{q}_j$$

$$= (-k_x u_j, -k_y v_j, -k_z w_j).$$

Using these expressions for L and $\mathscr{F}$, Lagrange's equations (4.59) become

$$m\ddot{q}_j = -Kq_j - \mathbf{k}\dot{q}_j.$$

This is the vector equation of the jth particle. It tells us that the rate of change of momentum of the jth particle is equal to the vector sum of the spring restoring force and the viscous damping force which is proportional to the particle velocity. When written out in component form it becomes

$$m\ddot{x}_j = -Kx_j - k_x \dot{x}_j,$$

$$m\ddot{y}_j = -Ky_j - k_y \dot{y}_j,$$

$$m\ddot{z}_j = -Kz_j - k_z \dot{z}_j,$$

which are the second-order equations of motion for a damped harmonic oscillator in three dimensions.

Generalization of Hamilton's Principle

In our investigations of a nonconservative force field a generalized potential U was introduced which is velocity-dependent, and the generalized force Q_i was given by (4.50) which shows that $\mathbf{Q}$ is derivable from U. Also, for the special case where the nonconservative force was a viscous force proportional to the particle velocity, the Rayleigh dissipation function serves the purpose of the velocity potential.

Now suppose we have a more general type of nonconservative force field that is not derivable from any velocity potential. Do Lagrange's equations of motion hold for such a system? The answer is that these equations are still given by (4.31) in terms of T, where the generalized force Q_i is not derivable from any potential. We reiterate an important point concerning (4.31), which is the basis for generalizing Hamilton's principle: As mentioned previously, (4.31) are Lagrange's equations for T in terms of an arbitrary force field given by the generalized force Q_j. They were derived from D'Alembert's principle and do not involve L explicitly. Therefore, we are free to give a more general definition of L, which we now do.

Generalized Lagrangian. Recall that Hamilton's principle was given by (4.17) where $L = T - V$ for a conservative system, or where $L = T - U$ for a non-conservative force derivable from the generalized or velocity potential U. For a more general nonconservative force field (which is not derivable from either V or U) we replace these expressions for the Lagrangian by a more general definition, a *generalized Lagrangian*, given by

$$L = T + W. \tag{4.60}$$

Equation (4.60) tells us that the generalized Lagrangian equals the sum of the kinetic energy and the work done by the external forces on the system. This is a natural definition for the Lagrangian; for it reduces to $L = T - V$ for a conservative system. This is seen as follows: For such a system the external force $\mathbf{F} = -\mathbf{grad}\ V$, so that Q_j is given by (4.32). However, from $\mathbf{F}$ derivable from the potential V and (4.32) we obtain (4.19), Lagrange's equations for a conservative field, by setting $T = L + V$ in (4.31). By setting $T = L + U$ and by inserting the generalized force given by (4.50) into (4.31) we obtain Lagrange's equations (4.19).

In order to generalize Hamilton's principle for a nonconservative force not derivable from a potential, we insert (4.60) into the functional in $\delta I = \delta \int_{t_1}^{t_2} L\, dt$ given by (4.17); still keeping the end points t_1, t_2 fixed. Then the generalized form of Hamilton's principle becomes

$$\delta I = \delta \int_{t_1}^{t_2} (T + W)\, dt = 0. \tag{4.61}$$

The virtual work δW is given by (4.28), where now the $\mathbf{F}_i$'s are nonconservative forces not derivable from a potential. However, the Q_j's in (4.28) are still given by (4.29). (For a conservative force field $\mathbf{F}_i = -\mathbf{grad}\ V$, so that W is replaced by $-V$ and $T + W$ is replaced by $T - V = L$, and (4.61) reduces to (4.17) for a conservative system. For a nonconservative force field derivable from a generalized potential $T + W$ is replaced by $T - U = L$.)

The virtual work δW has the following important physical significance: We recall that the virtual displacement δq_j of each generalized coordinate q_j in (4.28), as already mentioned, does not involve a change in time. We now study our n degree of freedom dynamical system in configuration space. This is shown in Fig. 4.4 where C represents the actual path of the system between the fixed end points at time t_1, t_2, respectively, and C' is a neighboring path between the same end points. At an interior point P on C the forces given by the $\mathbf{F}_i$'s do work on the system given by

$$W_P = \sum_{i=1}^{n} \mathbf{F}_i \cdot \mathbf{r}_i.$$

By a virtual displacement $\delta \mathbf{r}$ we displace the system from P to a corresponding point P' on the path C'. If t_P is the time at P then the time at P' is the same $(t_{P'} = t_P)$. It is clear that the virtual displacement $\delta \mathbf{r}_i$ on the ith particle is

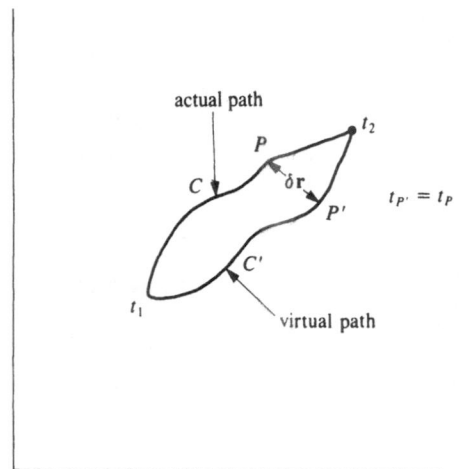

Fig. 4.4. Construction of a varied path in configuration space by a virtual displacement.

transformed by (4.25) to the generalized coordinates **q** so that each $\delta \mathbf{r}_i$ is related to the δq_j's by (4.27). Therefore, we use (4.28) in the form

$$\delta W = \sum_{j=1}^{n} Q_j \delta q_j. \tag{4.62}$$

We now apply Hamilton's principle in the form given by (4.61) to Fig. 4.3. In (4.61) the operator δ can be put inside the integral sign since the end points are fixed. Because of this fact (referring to the figure), we do not integrate $T + W$ with respect to time from t_1 to t_2 along C, do the same along C', and subtract the integrals to get the variation. Instead, we do the following: We form $T_{P'} - T_P = \delta T$, which we may call the "virtual kinetic energy" plus the virtual work $W_{P'} - W_P = \delta W$. We then set the integral of this result with respect to time from t_1 to t_2 equal to zero. This is written as

$$\delta \int_{t_1}^{t_2} (T + W)\, dt = \int_{t_1}^{t_2} (\delta T + \delta W)\, dt = \int_{t_1}^{t_2} \delta T\, dt + \int_{t_1}^{t_2} \sum_j Q_j \delta q_j\, dt = 0, \tag{4.63}$$

upon using (4.62). Since T is a function of the q_j's and $\dot{q}_j$'s, the first integral of the right-hand side of (4.63) becomes, upon integrating by parts,

$$\int_{t_1}^{t_2} \delta T\, dt = \int_{t_1}^{t_2} \sum_j \left[\frac{\partial T}{\partial q_j} - \frac{d}{dt}\left(\frac{\partial T}{\partial \dot{q}_j}\right) \right] \delta q_j\, dt.$$

Combining the two integrals, (4.63) becomes

$$\int_{t_1}^{t_2} \sum_j \left[\frac{\partial T}{\partial \dot{q}_j} - \frac{d}{dt}\left(\frac{\partial T}{\partial q_j}\right) + Q_j \right] \delta q_j\, dt = 0. \tag{4.64}$$

Equation (4.64) is Hamilton's principle for the general case where Q_j is not necessarily derivable from a potential. We assume holonomic constraints so that the δq_j's are independent. Therefore the integral can vanish if and only if the separate coefficients of the δq_j's vanish, yielding (4.31). We have thus shown that (4.61) represents the proper generalization of Hamilton's principle for a general nonconservative force system.

4.9. Constraints Revisited, Undetermined Multipliers

Up to now we have only considered holonomic constraints so that the set of generalized coordinates $\mathbf{q}$ and their conjugate velocities $\dot{\mathbf{q}}$ are independent (involve no constraints). Suppose we have a system of N particles involving $3N$ coordinates with m holonomic constraints (each of which is represented by an equation connecting the coordinates of the particles); then, as we know, the number of degrees of freedom $n = 3N - m$. This means that there are n uncoupled Lagrange equations of motion.

Now suppose that the constraints are nonholonomic. We can still treat the system by Lagrange's method, providing we doctor it up to account for these constraints. Lagrange figured out how to do this by introducing certain parameters (one for each constraint) which he called *undetermined multipliers*. His idea was to fix up the Lagrangian to account for these constraint vis-à-vis these multipliers. We now describe this method.

If the system of N particles involves k nonholonomic constraints that are rheonomic, then there are m equations relating the differentials dq_j and dt. We represent these constraints in the form

$$\sum_{j=1}^{n} a_{ij}\, dq_j + a_{i,n+1}\, dt = 0, \qquad i = 1, 2, \ldots, m.\dagger \qquad (4.65)$$

In (4.65) the subscript "i" represents the ith constraint. If these constraints are scleronomic then the coefficients $a_{i,n+1}$ are identically zero. Clearly, the differential displacements in (4.65) are not virtual. Now the variational process in Hamilton's principle involves virtual displacements so that $dt = 0$ for all these displacements. Hence the virtual displacements occurring in Hamilton's principle must satisfy equations of constraint of the form

$$\sum_{j=1}^{n} a_{ij}\delta q_j = 0, \qquad i = 1, 2, \ldots, m. \qquad (4.66)$$

Lagrange associated a parameter (the undetermined multiplier) λ_i with each of the constraints given by (4.66). We therefore have

$$\lambda_i \sum_j a_{ij}\delta q_j = 0. \qquad (4.67)$$

---

† This system does not include constraints in the form of inequalities.

We now introduce an "augmented" Lagrangian L' defined by

$$L' = L - \sum_{i,j=1,1}^{m,n} \lambda_i a_{ij} \delta q_j. \tag{4.68}$$

Note that $L' = L$ identically, since (4.67) must hold. However, the multipliers λ_i, and the coefficients a_{ij} defining the constraints, enter Lagrange's equations vis-à-vis L'. Let us consider a conservative system. Then Lagrange's equations for an n degree of freedom system is given by (4.19). Solving (4.68) for L and substituting the result into (4.19) yields

$$\frac{d}{dt}\left(\frac{\partial L'}{\partial \dot{q}_j}\right) - \frac{\partial L'}{\partial q_j} = \sum_{i=1}^{m} \lambda_i a_{ij}, \qquad j = 1, 2, \ldots, n. \tag{4.69}$$

These are the n Lagrange equations in terms of L'. Note that m of these equations are coupled by the constraints whereas, as previously mentioned, (4.19) are the uncoupled Lagrange equations in terms of L for an n degree of freedom unconstrained system. It is curious that mathematically $L' = L$, but *conceptually L' is different from L* in that it involves the m constraints. Since the δq_j's are connected by the m equations of constraint (4.66) we may choose the first $(n - m)$ equations of (4.69) independently, the last m are fixed by the constraining equations (4.66). That is the reason for the given range of j. We have the n unknown q_j's to be determined from the n equations (4.69) in terms of the m λ_i's. However, these m unknowns can be solved from the constraint equations (4.65) (in terms of the differentials, rather than the virtual displacements). Therefore, the two systems (4.65) and (4.69) represent $(m + n)$ equations for the $(m + n)$ unknowns.

A question now arises: What is the significance of the m λ_i's? Suppose we removed all the constraints on the system and replaced them by applied external forces Q_j' in such a manner as to keep the motion the same as it was when the constraints were applied. Clearly these extra applied forces must equal the forces of constraint, since they are the forces applied to satisfy the conditions of constraint. Lagrange's equations of motion under the influence of these forces Q_j' are

$$\frac{d}{dt}\left(\frac{\partial L}{\partial \dot{q}_j}\right) - \frac{\partial L}{\partial q_j} = Q_j'. \tag{4.70}$$

Relating (4.70) with (4.69) we obtain the following expression for Q_j':

$$Q_j' = \sum_{i=1}^{n} \lambda_i a_{ij}, \qquad j = 1, 2, \ldots, m. \tag{4.71}$$

In this type of problem, where we deal with nonholonomic constraints, we do not eliminate the forces of constraint, but use the m Q_j''s as the generalized forces of constraint as part of the answer.

We now turn to an alternative formulation of the equations of motion developed by Hamilton.

4.10. Hamilton's Equations of Motion

Thus far we have been investigating dynamical systems from the point of view of Lagrange's equations of motion. We recall that for an n degree of freedom system these equations involve n generalized coordinates $\mathbf{q}$ and their conjugate velocities $\dot{\mathbf{q}}$, so that there are n second-order PDEs for $L(\mathbf{q}, \dot{\mathbf{q}})$. Solutions of these equations yield families of trajectories defined by $\mathbf{q} = \mathbf{q}(t)$ and $\dot{\mathbf{q}} = \dot{\mathbf{q}}(t)$. A particular trajectory requires that $\mathbf{q}(0)$ and $\dot{\mathbf{q}}(0)$ be prescribed, so that we must know $2n$ initial conditions, the q_i's and $\dot{q}_i$'s at $t = 0$. These are the same requirements on the initial conditions demanded by Newton's equations.

Hamilton developed an alternative formulation of the equations of motion. He still used the n generalized coordinates $\mathbf{q}$, but instead of using $\dot{\mathbf{q}}$ he substituted the *generalized momentum vector* $\mathbf{p} = (p_1, p_2, \ldots, p_n)$. $\mathbf{q}$ are called the *canonical coordinates*. Associated with each q_i is a *canonical momentum* or *conjugate momentum* p_i which is defined by (4.20'), namely, $p_i = \partial L/\partial \dot{q}_i$.

We shall now be concerned with conservative holonomic, scleronomic dynamical systems. In order to motivate Hamilton's formulation, we develop the theorem of conservation of energy from Lagrange's formulation of the equations of motion. We know that for a conservative system the external force $\mathbf{F}$ is derivable from a potential V, or $\mathbf{F} = -\mathbf{grad}\, V(\mathbf{q})$. Since the constraints are scleronomic L is independent of t so that $L = L(\mathbf{q}, \dot{\mathbf{q}})$ or $\partial L/\partial t = 0$. Expanding the total derivative by the chain rule yields

$$\frac{dL}{dt} = \sum_j \left(\frac{\partial L}{\partial q_j}\right)\dot{q}_j + \sum_j \left(\frac{\partial L}{\partial \dot{q}_j}\right)\ddot{q}_j$$

$$= \sum_j \frac{d}{dt}\left(\frac{\partial L}{\partial \dot{q}_j}\right)\dot{q}_j + \sum_j \left(\frac{\partial L}{\partial \dot{q}_j}\right)\ddot{q}_j$$

$$= \sum_j \frac{d}{dt}\left(\dot{q}_j \frac{\partial L}{\partial \dot{q}_j}\right),$$

upon using Lagrange's equations (4.19). It therefore follows that

$$\frac{d}{dt}\left(\sum_j \dot{q}_j \frac{\partial L}{\partial \dot{q}_j} - L\right) = 0,$$

from which we obtain

$$\sum_j \dot{q}_j \frac{\partial L}{\partial \dot{q}_j} - L = \text{const.} = H.$$

The constant H is called the *Hamiltonian*. Using the fact that $\partial L/\partial \dot{q}_j = p_j$, we get

$$H = \sum_j \dot{q}_j p_j - L. \tag{4.72}$$

The right-hand side of (4.72) is thus an integral of motion obtained from Lagrange's equations (meaning that H is constant for a conservative system where L is independent of t). The important point is that Hamilton defined H

by (4.72) and visualized H as being a function of $\mathbf{q}$ and $\mathbf{p}$. From this hypothesis he derived the equations of motion in the form of a pair of first-order PDEs. But this is getting ahead of ourselves.

We first want to show that H is indeed the total energy of the conservative system or that $H = T + V$. This is another way of saying that the expression on the right-hand side of (4.72) (which, as mentioned, is an integral of motion) is actually the total energy of the conservative system. Since $L = T - V$, and V is independent of $\dot{\mathbf{q}}$, we obtain

$$p_j = \frac{\partial T}{\partial \dot{q}_j}.$$

Since T is a homogeneous quadratic function of the $\dot{q}_j$'s the first term on the right-hand side of (4.72) becomes

$$\sum_j \dot{q}_j p_j = \sum_j \dot{q}_j \frac{\partial T}{\partial \dot{q}_j} = 2T.$$

Inserting this expression into (4.72) yields

$$H = 2T - (T - V) = T + V,$$

which proves that H is the total energy.

We now derive the pair of first-order PDEs that was mentioned above. In this treatment we can be a little more general and allow H to depend on t thus giving $H = H(\mathbf{p}, \mathbf{q}, t)$. Using this functional relationship we first expand dH, obtaining

$$dH = \sum_j \frac{\partial H}{\partial q_j} dq_j + \sum_j \frac{\partial H}{\partial p_j} dp_j + \frac{\partial H}{\partial t}. \tag{4.73}$$

We now form dH from (4.72) allowing L to depend on t.

$$dH = \sum_j \dot{q}_j \, dp_j + \sum_j p_j \, d\dot{q}_j - \sum_j \frac{\partial L}{\partial q_j} dq_j - \sum_j \frac{\partial L}{\partial \dot{q}_j} d\dot{q}_j - \frac{\partial L}{\partial t} dt. \tag{4.74}$$

The coefficients of the $d\dot{q}_j$'s are zero because of the definition of p_i. We also have (since T is independent of $\mathbf{q}$)

$$\frac{\partial L}{\partial q_j} = \dot{p}_j.$$

Then (4.74) reduces to

$$dH = \sum_j \dot{q}_j \, dp_j - \sum_j \dot{p}_j \, dq_j - \frac{\partial L}{\partial t} dt. \tag{4.75}$$

Comparing (4.75) with (4.73) yields the following system of $(2n + 1)$ equations:

$$\dot{q}_j = \frac{\partial H}{\partial p_j}, \qquad \dot{p}_j = -\frac{\partial H}{\partial q_j}, \qquad j = 1, 2, \ldots, n, \tag{4.76}$$

$$\frac{\partial L}{\partial t} = -\frac{\partial H}{\partial t}. \tag{4.77}$$

The pair of $2n$ equations given by (4.76) are called *Hamilton's canonical equations of motion*. (This is what Hamilton set out to prove.) They constitute the system of $2n$ first-order PDEs for $H(\mathbf{p}, \mathbf{q}, t)$, which replace the Lagrange system of n second-order equations for $L(\mathbf{q}, \dot{\mathbf{q}}, t)$. The equation (4.77) merely expresses the relationship between L and H for a system where t is given explicitly.

As an *example* of the use of Hamilton's equations (4.76) we consider a system of n simple harmonic oscillators, where $\mathbf{q} = (q_1, q_2, \ldots, q_n)$ is interpreted as a set of normal coordinates so that each mass oscillates in an uncoupled manner and satisfies the equations of motion

$$\dot{p}_j = -kq_j, \qquad j = 1, 2, \ldots, n.$$

For an electrical circuit analogy we may consider a system consisting of n oscillating elements each of which is a series coupled condenser and coil. The normal coordinate q_j corresponds to the current divided by the inductance of the jth element.

Our task is now to derive these equations of motion by using Hamilton's equations. To do this we need to construct the Hamiltonian, which depends only on T and V. We therefore write the kinetic and potential energies of the system.

$$T = \frac{1}{2} \sum_j \frac{(p_j)^2}{m_j}, \qquad V = \frac{k}{2} \sum_j (q_j)^2,$$

where the spring constant k is assumed to be the same for each mass. The Hamiltonian becomes

$$H = T + V = \frac{1}{2} \sum_j \frac{(p_j)^2}{m_j} + \frac{k}{2} \sum_j (q_j)^2 = H(\mathbf{p}, \mathbf{q}).$$

From this we obtain

$$\frac{\partial H}{\partial q_j} = kq_j, \qquad \frac{\partial H}{\partial p_j} = m_j^{-1} p_j.$$

Inserting these expressions into Hamilton's equations, (4.76) yields

$$kq_j = -\dot{p}_j, \qquad \frac{p_j}{m_j} = \dot{q}_j.$$

The first equation is the equation of motion for the simple harmonic oscillator, for the jth normal coordinate. This is what we set out to demonstrate. The second equation is merely the definition of the momentum of the jth particle.

4.11. Cyclic Coordinates

If the Lagrangian of a system does not contain a given generalized coordinate q_i (although it may contain the corresponding velocity $\dot{q}_j$) then that q_j is called a *cyclic coordinate*. If q_j is a cyclic coordinate then $\partial L / \partial q_j = 0$, so that the

Lagrange equation for the jth particle becomes

$$\frac{d}{dt}\left(\frac{\partial L}{\partial \dot{q}_j}\right) = \dot{p}_j = 0, \tag{4.78}$$

yielding the important result that

$$p_j = \text{const.} \quad \text{for } q_j \text{ cyclic.} \tag{4.79}$$

This gives us the following conservation theorem:

The generalized momentum conjugate to a cyclic coordinate is conserved (meaning that it is constant).

Looking at the definition of H given by (4.72), we observe that if q_j is cyclic then $H = -L$ for that coordinate. This means that the above conservation theorem can readily be transformed to the Hamiltonian formulation. In addition, if p_j is conserved then the second of Hamilton's equations (4.76) tells us that H is independent of the cyclic coordinate q_j. We obtain the following theorem:

A cyclic coordinate q_j will be absent from the Hamiltonian H. Conversely, if a generalized coordinate q_j does not occur in H then the conjugate momentum p_j is conserved, and q_j is cyclic.

The converse is easily seen since, if H does not contain q_j then $\partial H/\partial q_j = 0$.

An important advantage that the Hamiltonian formulation has over the Lagrangian is as follows: Suppose we label the n generalized coordinates so that q_n is cyclic. Then we can write the Lagrangian as

$$L = L(q_1, q_2, \ldots, q_{n-1}, \dot{q}_1, \dot{q}_2, \ldots, \dot{q}_n, t).$$

Note that all the n generalized velocities occur in L so that there are still n Lagrange equations of motion. This means that we must still solve an n degree of freedom system in obtaining the trajectories for the n q_j's, even though q_n is cyclic. On the other hand, if we use the Hamiltonian formulation, the cyclic coordinate q_n truly deserves the alternative description *ignorable coordinate*, since H contains $(n-1)$ degrees of freedom. The conjugate momentum $p_n = \alpha = \text{const.}$ so that

$$H = H(q_1, q_2, \ldots, q_{n-1}, p_1, p_2, \ldots, p_{n-1}, \alpha, t).$$

This tells us that, in contrast to L, H actually describes an $(n-1)$ degree of freedom system if one of the coordinates is cyclic. This means that we need only solve $(n-1)$ trajectories from Hamilton's equations, which are thus valid for $j = 1, 2, \ldots, n-1$. We completely ignore q_n and recognize the conjugate $p_n = \alpha$ as an initial condition on the momentum, since it is a constant of integration arising from the law of conservation of momentum for α. The trajectory of the q_n can then be determined by integrating the corresponding equation of motion $\dot{q}_n = \partial H/\partial \alpha$.

We shall capitalize on the use of cyclic coordinates in Hamilton's formulation when we investigate the Hamilton–Jacobi theory of PDEs. We now turn to another important variational principle.

4.12. Principle of Least Action

At this stage we discuss another variational principle also associated with the Hamilton canonical equations. It is called the *principle of least action* and will be useful when we study the Hamilton–Jacobi theory in relation to wave propagation. It is also related to Fermat's principle. We recall that in Hamilton's principle a varied path was compared with the actual path of a dynamical system in configuration space, which is the one making the functional $\int_{t_1}^{t_2} L \, dt$ a stationary value. All admissible paths of the system take the same time to go from t_1 to t_2 (which are the fixed end points).

Up to now we have not pointed out the fact that the neighboring paths might not necessarily conserve the energy of the system. This means that we are assured that $T + V = E$ is constant only on the actual path. The purpose of the principle of least action is to rectify this weakness of Hamilton's principle. Like Hamilton's principle it compares neighboring paths with the actual path in configuration space. But it differs in that corresponding points on the actual path compared to a neighboring path are not reached at the same time, and therefore the varied path is not necessarily described in the same time as the dynamical path. Instead, the varied path is taken so that the total energy $T + V$ is the same on every path. It turns out that the functional that is made stationary is the *action integral* (described below). The method of making this functional stationary is a little different from that involved in Hamilton's principle. This will now be discussed.

In physics, especially in quantum mechanics, there is an integral called the *action integral* A defined by

$$A = \int_{t_1}^{t_2} \sum_j p_j \dot{q}_j \, dt. \tag{4.80}$$

From (4.72), we have

$$\sum_j p_j \dot{q}_j = H + L = 2T,$$

so that an alternative form for A is

$$A = \int_{t_1}^{t_2} 2T \, dt. \tag{4.80'}$$

The principle of least action states that in a conservative system we have the following condition imposed on the action integral;

$$\Delta A = \Delta \int_{t_1}^{t_2} \sum_j p_j \dot{q}_j \, dt = \Delta \int_{t_1}^{t_2} (H + L) \, dt = \Delta \int_{t_1}^{t_2} 2T \, dt = 0, \tag{4.81}$$

where the operator Δ represents a new type of variation of the path of a system of particles in configuration space from t_1 to t_2. It allows for a variation in time, even at the end points. We now give the meaning of Δ. First recall that in our discussion of virtual work the operator δ was shown to correspond to virtual displacements in which time was held fixed during the displacement. A virtual displacement does not always coincide with an actual physical displacement. Consequently, as mentioned above, H might not be conserved in a conservative system when we vary the path using the operator δ. In contrast, the operator Δ (which will be mathematically described below) is defined as dealing with displacements which do involve changes in time, so that the varied path is obtained by a succession of displacements each of which includes a differential change in time dt. The points in configuration space obtained by using the δ operator all travel with the same speed. However, it is clear that the times of transit of the points along the varying paths, by using the Δ operator, need no longer be constant (the speed may vary) in order to satisfy the requirement that E be conserved on all admissible paths. As a result the Δ operator allows for a variation in time even at the end points, although the variations of the q_j's at the end points still remain zero.

To describe the Δ variation process mathematically we start by tagging each curve (system path in configuration space) with a parameter α. Since the variation includes the time associated with each point, t must also be a function of α. It follows that, for the configuration or $\mathbf{q}$ space, each $q_j = q_j(\alpha, t)$. We now expand Δq_j in the limit

$$\Delta q_j \rightarrow d\alpha \left(\frac{dq_j}{d\alpha} \right) = d\alpha \left(\frac{\partial q_j}{\partial \alpha} \right) + d\alpha \left(\frac{dt}{d\alpha} \right) \frac{dq_j}{dt}. \tag{4.82}$$

The first term $d\alpha(\alpha q_j/\partial\alpha) = \delta q_j$, while the coefficient of dq_j/dt represents the change in t occurring as a result of the Δ variation and can therefore be designated as Δt. Consequently, (4.82) can be written as

$$\Delta q_j = \delta q_j + \dot{q}_j \Delta t. \tag{4.83}$$

The relation between δ and Δ, given by (4.83), holds for any function $f(\mathbf{q}, t)$ so that

$$\Delta f(\mathbf{q}, t) = \delta f + \dot{f}\Delta t = \sum_j \left(\frac{\partial f}{\partial q_j} \right) \delta q_j + \left[\sum \left(\frac{\partial f}{\partial \dot{q}_j} \right) \delta \dot{q}_j + \frac{\partial f}{\partial t} \right] \Delta t. \tag{4.84}$$

The first summation term represents the expansion of δf, while the terms in the brackets represent the total time derivative of f.

We now prove (4.81). Applying Δ to A gives

$$\Delta A = \Delta \int_{t_1}^{t_2} (L + H) \, dt = \Delta \int_{t_1}^{t_2} L \, dt + H(\Delta t_2 - \Delta t_1),$$

since H is constant. We now work on the functional $\int_t^t L \, dt = I$, making use of the fact that $\Delta f = \delta f + \dot{f}\Delta t$.

$$\Delta \int_{t_1}^{t_2} L\, dt = I(t_2) - I(t_1) = \delta I(t_2) - \delta I(t_1) + \dot{I}(t_2)\Delta t_2 - \dot{I}(t_1)\Delta t_1.$$

$$= \delta \int_{t_1}^{t_2} L\, dt + L\Delta t \Big|_{t_1}^{t_2}.$$

Note the important fact that $\int_{t_1}^{t_2} L\, dt$ does not obey Hamilton's principle and hence does not vanish, because in the Δ process used here the variations of the δq_j's do not vanish at the end points (the Δq_j's do however, but they are not used here). We now calculate $\delta \int_{t_1}^{t_2} L\, dt$ with the aid of Lagrange's equations

$$\delta \int_{t_1}^{t_2} L\, dt = \int \delta L\, dt = \sum_j \int \left[\left(\frac{\partial L}{\partial q_j}\right)\delta q_j + \left(\frac{\partial L}{\partial \dot{q}_j}\right)\delta \dot{q}_j \right] dt$$

$$= \sum_j \int \left[\frac{d}{dt}\left(\frac{\partial L}{\partial \dot{q}_j}\right)\delta q_j + \left(\frac{\partial L}{\partial \dot{q}_j}\right)\left(\frac{d}{dt}\right)\delta q_j \right] dt$$

$$= \sum_j \int \left[\frac{d}{dt}\left(\frac{\partial L}{\partial \dot{q}_j}\delta q_j\right) \right] dt$$

$$= \sum_j \int \frac{d}{dt}\left[\left(\frac{\partial L}{\partial \dot{q}_j}\right)\Delta q_j - \left(\frac{\partial L}{\partial \dot{q}_j}\right)\dot{q}_j\Delta t \right] dt \quad \text{(upon using (4.83))}$$

$$= \sum_j \left[\left(\frac{\partial L}{\partial \dot{q}_j}\right)\Delta q_j - \left(\frac{\partial L}{\partial \dot{q}_j}\right)\dot{q}_j\Delta t \right].$$

The Δq_j's vanish at the end points, but Δt does not vanish there because the time of transit is not constant. Consequently, upon using the fact that $\partial L/\partial \dot{q}_j = p_j$, the last equation becomes

$$\delta \int_{t_1}^{t_2} L\, dt = -\sum_j p_j \dot{q}_j \Delta t \Big|_{t_1}^{t_2}. \tag{4.84}$$

Combining the above results we obtain

$$\Delta A = \left[-\sum_j p_j q_j + L + H \right]\Delta t \Big|_{t_1}^{t_2} = 0.$$

This completes the proof of the principle of least action given by (4.81), which states that the Δ variation of the action integral shall vanish.

We first consider a single particle of mass m. The kinetic energy is

$$T = \left(\frac{m}{2}\right)\left(\frac{d\mathbf{r}}{dt}\right)\cdot\left(\frac{d\mathbf{r}}{dt}\right).$$

Setting $d\mathbf{r}\cdot d\mathbf{r} = ds^2$, where ds is the element of arc length, the principle of least action becomes

$$\Delta \int 2T\, dt = \Delta \int_{t_1}^{t_2} \sqrt{2mT}\, ds = \Delta \int_{t_1}^{t_2} \sqrt{2m(H-V)}\, ds = 0. \tag{4.85}$$

We now extend the principle of least action to an n degree of freedom system, the ith particle having mass m_i. We seek a more general expression for dt in terms of an arc length using generalized coordinates. Writing the kinetic energy as a quadratic function of the generalized velocities, we have

$$T = \frac{1}{2} \sum_{ij} m_{ij} \dot{q}_i \dot{q}_j = \frac{1}{2} \sum_{i,j} m_{ij} \frac{dq_i \, dq_j}{(dt)^2}. \tag{4.86}$$

We now define the differential $d\rho$ by

$$(d\rho)^2 = \sum_{i,j} m_{ij} \, dq_i \, dq_j, \tag{4.87}$$

This expression represents the most general form of the element of path length for a trajectory in configuration space $\mathbf{q}$. In the application given here the coefficients m_{ij} are the elements of the *metric tensor*. In a Cartesian coordinate system the elements of the metric tensor reduce to $m_{ij} = \delta_{ij}$, as can be seen by comparing the expression $(ds)^2 = (dx)^2 + (dy)^2 + (dz)^2$ with (4.87). Using this equation we may put the kinetic energy in the following form:

$$T = \frac{1}{2} \left(\frac{d\rho}{dt} \right)^2 \quad \text{or} \quad dt = \frac{d\rho}{\sqrt{2T}}. \tag{4.88}$$

Using this relation for dt, we put the principle of least action in the form

$$\Delta \int_{t_1}^{t_2} T \, dt = \Delta \int T \, d\rho = \Delta \int \sqrt{H - V(\mathbf{q})} \, d\rho = 0. \tag{4.89}$$

Equation (4.89) is sometimes called *Jacobi's form of the principle of least action*. It should be emphasized that this form of the principle of least action is concerned with the path of the system point (the n degree of freedom system) in configuration space, rather than its motion in time. In fact, the time nowhere appears in the integrand $\sqrt{H - V}$. We shall use this treatment of the principle of least action when we discuss the Hamilton–Jacobi theory in its relation to wave propagation.

We now have the mathematical apparatus and physical ideas necessary to discuss the relation between these variational principles and wave propagation phenomena. To this end, we turn our attention to a discussion of continuous media. We start by extending Lagrange's equations to a continuum, and then deal with a similar extension of Hamilton's equations.

4.13. Lagrange's Equations of Motion for a Continuum

For applications to electromagnetic phenomena we may consider the generic vector $\mathbf{u}(\mathbf{x}, t)$ to represent either the electric field $\mathbf{E}$ or the magnetic field $\mathbf{H}$, or also the displacement vector. However, for our purpose here (until further notice) we interpret $\mathbf{u}$ to be the displacement vector both in electromagnetic phenomena and in the dynamics of nonconducting continuous media. This means that **u** represents the displacement of a particle from its equilibrium configuration as a function of $(\mathbf{x}, t)$. **x** represents the Cartesian coordinates of three-space. We use a symmetric notation $\mathbf{u} = \mathbf{u}(u_1, u_2, u_3)$ and $\mathbf{x} = (x_1, x_2, x_3)$. The particle velocity is $\mathbf{v} = \mathbf{u}_t = (u_{1,t}, u_{2,t} u_{3,t})$, where $u_{i,t} \equiv \partial u_i/\partial t$, and **grad u** $= (u_{1,1}, u_{1,2}, \ldots, u_{3,3})$, where $u_{i,j} \equiv \partial u_i/\partial x_j$. Using this notation and the tensor summation convention (which we use from now on), the kinetic energy T and potential energy V become

$$T = \tfrac{1}{2}\rho u_{i,t}u_{i,t},$$

$$V = \tfrac{1}{2}\mu u_{i,j}u_{i,j}.$$

where ρ is the density of the medium and μ is the spring constant (the reciprocal of the capacitance for an electrical medium, or proportional to Young's modulus for an elastic medium). These equations for T and V tell us that T is a quadratic form in **v** and V is a quadratic form in **u**. (Those readers not too comfortable with the tensor notation should write out these expressions for T and V in extended form.) For a continuous medium (either conducting or nonconducting) **u** is analogous to **q** (the generalized coordinates) and **v** is analogous to **q̇** (the generalized velocities).

Recall that for an n degree of freedom discrete system we have n Lagrange equations of motion. For a continuous medium there is clearly a continuous distribution of degrees of freedom so that it would be absurd to use Lagrange's equations in the form given for a discrete system. An important fact to grasp is that, for a continuous medium, the position coordinates **x** are not generalized coordinates as they are in a discrete system. For a discrete system each value of i corresponds to a different one of the generalized coordinates q_i, which means **q** is, in a sense, a function of i. For a continuous system **x** merely serves as a continuous index replacing the discrete index i. Therefore, for a continuous system, **u** is a function of the "index" **x** so that **u** plays the role of generalized coordinates. More accurately $\mathbf{u} = \mathbf{u}(\mathbf{x}, t)$ indicating that **x**, like t, can be considered as continuously varying parameters entering into the Lagrangian. For a continuous distribution of particles the Lagrangian L is equal to the volume integral of $T - V$, or

$$L = \int (T - V)\,d\tau,$$

where $d\tau$ is the element of volume $dx_1\,dx_2\,dx_3$ and the integration is taken over the appropriate volume. Note that the integrand $T - V$ represents the Lagrangian at a point **x** at time t. This integrand has the dimensions of the Lagrangian divided by the volume; it is the density distribution of L. For this reason, we therefore introduce it as a separate variable called the *Lagrange density function* $\mathscr{L}$† defined by

$$\mathscr{L} = T - V = \tfrac{1}{2}\rho u_{i,t}u_{i,t} - \tfrac{1}{2}\mu u_{i,j}u_{i,j}. \tag{4.90}$$

---

† See, for example, [22].

Then the Lagrangian functional is of the form

$$L = \int \mathscr{L}(\mathbf{u}, \mathbf{u}_t)\, d\tau. \tag{4.91}$$

The generalization of Hamilton's principle for a three-dimensional continuum for the various modes of wave propagation described by $\mathbf{u}(\mathbf{x}, t)$ can be formulated as follows: Find that function $\mathbf{u} = \mathbf{u}(\mathbf{x}, t)$ which makes the functional $I(\mathbf{u})$ an extremum, or which makes the first variation $\delta I(\mathbf{u}) = 0$. $I(\mathbf{u})$ is defined by

$$I(\mathbf{u}) = \int_{t_1}^{t_2} L\, dt = \int_{t_1}^{t_2} \int_{\mathscr{R}} \mathscr{L}\, d\tau\, dt. \tag{4.92}$$

We introduce the virtual displacement vector $\delta \mathbf{u}$, which is the extension for a continuum of the generalized virtual displacement vector $\delta \mathbf{q}$. Consequently, Hamilton's principle takes on the form

$$\delta I(\mathbf{u}) = \delta \int_{t_1}^{t_2} \int_{\mathscr{R}} \mathscr{L}\, d\tau\, dt = \int_{t_1}^{t_2} \int_{\mathscr{R}} \delta \mathscr{L}\, d\tau\, dt = 0. \tag{4.93}$$

Note that we integrate over the region $\mathscr{R}$ in space and over t from t_1 to t_2. The operator δ operates on $\mathbf{u}$ and $\mathbf{v}$ but does not operate on $\mathbf{x}$ or t. Used here, the operator δ is a generalization of the same operator used in developing the principle of virtual work for a discrete system of particles where t was fixed while performing the variation δ. Since $\mathscr{L}$ is a function of $\mathbf{u}$ and $\mathbf{u}_t$, upon expanding $\mathscr{L}$ in terms of the operator δ, we obtain

$$\delta \mathscr{L} = \frac{\partial \mathscr{L}}{\partial u_i} \delta u_i + \frac{\partial \mathscr{L}}{\partial u_{i,t}} \delta u_{i,t} + \frac{\partial \mathscr{L}}{\partial u_{i,j}} \delta u_{i,j}. \tag{4.94}$$

Clearly, the reason why we have the $\partial \mathscr{L}/\partial \mathbf{u}_x$ terms in (4.94) is that V involves **grad u**. We now insert (4.94) into Hamilton's principle (4.93), and first integrate by parts with respect to t, obtaining

$$\int_{t_1}^{t_2} \left(\frac{\partial \mathscr{L}}{\partial u_{i,t}}\right) \delta u_{i,t}\, dt = -\int_{t_1}^{t_2} \frac{d}{dt}\left(\frac{\partial \mathscr{L}}{\partial u_{i,t}}\right) \delta u_i\, dt,$$

using the homogeneous boundary conditions $\delta \mathbf{u}(t_1) = \delta \mathbf{u}(t_2) = 0$. Next we work on the integrals involving the spatial derivatives of $\mathbf{u}$ by interchanging the $\partial/\partial \mathbf{x}$ and δ operators and then integrating by parts. We obtain

$$\int \left(\frac{\partial \mathscr{L}}{\partial u_{i,j}}\right) \delta u_{i,j}\, d\tau = \int \left(\frac{\partial \mathscr{L}}{\partial u_{i,j}}\right)\left(\frac{\partial \delta u_i}{\partial x_j}\right) d\tau$$

$$= -\int \frac{d}{dx_j}\left(\frac{\partial \mathscr{L}}{\partial u_{i,j}}\right) \delta u_i\, d\tau,$$

again using the fact that $\delta \mathbf{u}$ vanishes at the end points. Collecting these results yields the following form for Hamilton's principle:

$$\delta I(\mathbf{u}) = \int_{t_1}^{t_2} \int_{\mathcal{R}} \delta u_i \left[\frac{\partial \mathcal{L}}{\partial u_i} - \frac{d}{dt}\left(\frac{\partial \mathcal{L}}{\partial u_{i,t}} \right) - \frac{d}{dt}\left(\frac{\partial \mathcal{L}}{\partial u_{i,j}} \right) \right] d\tau \, dt = 0. \quad (4.95)$$

The four-dimensional integral can vanish only if the coefficients of each of the linearly independent virtual displacements δu_i vanish. This gives the three *Euler equations†* *for a continuum*

$$\frac{d}{dt}\left(\frac{\partial \mathcal{L}}{\partial u_{i,t}} \right) + \frac{d}{dt}\left(\frac{\partial \mathcal{L}}{\partial u_{i,j}} \right) - \frac{\partial \mathcal{L}}{\partial u_i} = 0, \qquad i = 1, 2, 3. \quad (4.96)$$

For $\mathcal{L}$ given by (4.90) we get the scalar wave equations for each component of $\mathbf{u}$

$$c^2 \nabla^2 u_i - u_{i,tt} = 0, \quad (4.97)$$

where the Laplacian $\nabla^2 = \partial/\partial x_i \, \partial/\partial x_i$, and c is the transverse wave speed corresponding to transverse waves if we identify $\mathbf{u}$ with $\mathbf{E}$ and $\mathbf{H}$ for an electromagnetic medium.

Summary and Extensions. We extended Lagrange's equations to a continuous medium by extending Hamilton's principle. To this end, we introduced the Lagrange density function and derived Euler's equations (4.96) in terms of this function. We then showed that the three-dimensional wave equations for the u_i's were obtained from Euler's equations, by constructing the appropriate Lagrange density function. In order to go from a discrete model to a continuum we replaced the generalized coordinates $\mathbf{q}$ and virtual displacements $\delta\mathbf{q}$ by $\mathbf{u}$ and $\delta\mathbf{u}$, respectively. We could have used another approach: We could decompose $\mathbf{u}$ into the gradient of a scalar potential ϕ and the curl of a vector potential $\mathbf{\psi}$. We would then have arrived at the appropriate wave equations for ϕ and $\mathbf{\psi}$. In addition, we could extend the Lagrange approach for a continuum to include nonconservative forces such as those satisfying the Rayleigh dissipation function. This would lead to damped wave equations, which involve the $\mathbf{u}_t$ term expressing the fact that the damping force is proportional to the particle velocity. If we had an external force $f(\mathbf{x}, t)$ applied to the system then we must add the terms $-fu_i$, so that Euler's equations (4.96) would yield the nonhomogeneous wave equation

$$\nabla^2 \mathbf{u} - \mathbf{u}_{tt} = f(\mathbf{x}, t).$$

Whichever way we look at Lagrange's equations we must remember that they are based on energy concepts; since kinetic and potential energy and the velocity potential or Rayleigh's dissipation function are involved in constructing the appropriate Lagrange density function. Energy, being the first integral of the equations of motion, has the advantage of giving a global approach rather than a local one; also local errors arising in numerical calculations tend

---

† These are sometimes called the *Euler–Lagrange equations* for $\mathcal{L}$.

to be smoothed out. Another important advantage of this integral approach vis-à-vis Hamilton's principle is that, since the kinetic and potential energies and velocity potential are *scalar invariants* with respect to a coordinate system, Lagrange's formulation (as well as Hamilton's) is *invariant with respect to a choice of coordinate systems*. This aspect is important to electromagnetic theory as well as to quantum mechanics and relativity theory.

Since the Euler equations are second-order PDEs, the wave equation for **u** derived from them is also second order. We also learned that each scalar wave equation can be reformulated into a pair of first-order PDEs by defining new dependent variables.

In using variational approaches to obtain equations of motion, we saw in the discrete model that Hamilton developed a pair of first-order equations of motion for each (q_i, p_i), i.e., his canonical equations of motion. We saw some advantages of his method over Lagrange's method, especially in the use of cyclic coordinates which we take advantage of when we treat the Hamilton–Jacobi theory. In accordance with our program of relating wave propagation phenomena to the variational treatment in a continuum, we extend Hamilton's formulation to a continuum in the next section.

4.14. Hamilton's Equations of Motion for a Continuum

In this section we reinvestigate wave propagation in a three-dimensional continuous medium from the point of view Hamilton's formulation for a continuum. We use the same notation as in the previous section.

As a motivation for the extension of Hamilton's equations to a continuum we start with a one-dimensional discrete model, namely, an n degree of freedom system. Let L_i be the Lagrangian for the ith particle. Then

$$L = \Delta x \sum_{i=1}^{n} L_i, \tag{4.98}$$

where $\Delta x = x_{i+1} - x_i$. If u_i is the displacement of the ith particle (the generalized coordinate q_i) then the conjugate momentum p_i is given by

$$p_i = \frac{\partial L}{\partial \dot{u}_i} = \Delta x \frac{\partial L_i}{\partial \dot{u}_i}, \tag{4.99}$$

where $\dot{u}_i$ is the velocity of the ith particle. The Hamiltonian can be written as

$$H = \sum_{i=1}^{n} (p_i \dot{u}_i - \Delta x L_i) = \sum_i \left(\dot{u}_i \frac{\partial L_i}{\partial \dot{u}_i} - L_i \right) \Delta x. \tag{4.100}$$

To obtain H for a continuous medium (say a one-dimensional bar or length L) we pass to the limit as $\Delta x \to dx$, $u_i \to u(x, t)$, $\dot{u}_i \to u_t(x, t)$. The sum tends to an integral over x, and H tends to

$$H = \int_0^t \left[\left(\frac{\partial \mathcal{L}}{\partial u_t} \right) u_t - \mathcal{L} \right] dx = \int_0^L \mathcal{H} \, d\tau, \tag{4.101}$$

where $\mathscr{L}$ is the Lagrange density function of the one-dimensional continuum and $\mathscr{H}$ is called the one-dimensional *Hamiltonian density function*. Referring to (4.99) for the definition of p_i for a discrete medium in terms of L_i we note that there is a singularity in p_i when we tend to a continuum, in the sense that $\partial L_i/\partial u_i$ becomes infinite when $\Delta x \to dx$. However, we can get around this difficulty by introducing the *momentum density function* π defined in terms of the Lagrange density function

$$\pi = \frac{\partial \mathscr{L}}{\partial u_t}. \tag{4.102}$$

Then the Hamiltonian density function $\mathscr{H}$ becomes

$$\mathscr{H} = \pi u_t - \mathscr{L}. \tag{4.103}$$

The extension to three dimensions is straightforward. Recall that $\mathbf{u} = (u_1, u_2, u_3)$ and $\mathbf{v} = (u_{1,t}, u_{2,t} u_{3,t})$. The momentum density function is a vector whose ith component is

$$\pi_i = \frac{\partial \mathscr{L}}{\partial u_{i,t}} \qquad i = 1, 2, 3. \tag{4.104}$$

The three-dimensional generalization for the Hamiltonian is

$$H = \int \mathscr{H} \, d\tau = \int (\pi_i u_{i,t} - \mathscr{L}) \, d\tau, \tag{4.105}$$

where $d\tau$ is the volume element in three-space.

We now obtain Hamilton's canonical equations of motion for a continuum. We first expand dH (as usual, we use the tensor summation convention).

$$dH = \int d\mathscr{H} \, d\tau = \int \left(\frac{\partial \mathscr{H}}{\partial \pi_i} \, d\pi_i + \frac{\partial \mathscr{H}}{\partial u_{i,j}} \, du_{i,j} + \frac{\partial \mathscr{H}}{\partial u_{i,t}} \, du_{i,t} \right) d\tau.$$

We now perform an integration by parts on integrals of the form

$$\int \left(\frac{\partial \mathscr{H}}{\partial \pi_i} \, du_{i,j} \right) d\tau.$$

The integrated term can be made to vanish by making the volume of integration so large that the values of $\mathbf{u}$ and $\mathscr{H}$ can be made to vanish at infinity. As a result, we write dH as

$$dH = \int \left(\frac{\partial \mathscr{H}}{\partial \pi_i} \, d\pi_i \right) - \frac{d}{dx_j} \left(\frac{\partial \mathscr{H}}{\partial u_{i,j}} \, du_i \right) d\tau. \tag{4.106}$$

Forming the differential on the right-hand side of (4.105) gives

$$dH = \int \left[\pi_i \, du_{i,t} + u_{i,t} \, d\pi_i - \frac{\partial \mathscr{L}}{\partial u_i} \, du_i - \frac{\partial \mathscr{L}}{\partial u_{i,t}} \, du_{i,t} \right] d\tau. \tag{4.107}$$

Now the generalization of (4.103) for a three-dimensional medium is

$$\mathscr{H} = \pi_i u_{i,t} - \mathscr{L} \quad \text{(summed over } i\text{)}. \tag{4.108}$$

Using (4.108), the Euler–Lagrange equations given by (4.96), and equating like coefficients from (4.106) and (4.107), we get Hamilton's canonical equations of motion for a three-dimensional continuum in terms of the Hamilton density function.

$$\partial \mathcal{H}/\partial u_i = \frac{d}{dx_j}\left(\frac{\partial \mathcal{H}}{\partial u_{i,j}}\right) = -\pi_{i,t}, \qquad \frac{\partial \mathcal{H}}{\partial \pi_i} = u_{i,t}. \tag{4.109}$$

For the general case, where H is an explicit function of t, we have the additional equation

$$\frac{\partial \mathcal{H}}{\partial t} = -\frac{\partial \mathcal{L}}{\partial t}, \tag{4.110}$$

which is analogous to (4.77) for the relation between H and L for the time-varying functions.

We now apply Hamilton's equations given by (4.109) to wave propagation in three dimensions. We first construct $\mathcal{H}$ by inserting (4.90) into (4.108). We obtain

$$\mathcal{H} = \pi_i u_{i,t} - \tfrac{1}{2}\rho u_{i,t} u_{i,t} + \tfrac{1}{2}\mu u_{i,j} u_{i,j}. \tag{4.111}$$

The second equation of (4.109) becomes an identity. Since $\mathcal{H}$ is independent of the u_i's, $\partial \mathcal{H}/\partial u_i = 0$. Also $\partial \mathcal{H}/\partial u_{i,j} = -\mu u_{i,j}$. Using these expressions the first equation of (4.109) becomes

$$\frac{du_{i,j}}{dx_j} = \mu u_{i,jj} = \pi_{i,t} = \rho u_{i,tt},$$

which gives the wave equation for the ith component of $\mathbf{u}$:

$$c^2 u_{i,jj} = c^2 \nabla^2 u_i = u_{i,tt}, \qquad c^2 = \mu/\rho. \tag{4.112}$$

General Remarks about Hamilton's Equations. We noticed above that if we apply Hamilton's equations of motion in a straightforward manner we wind up with practically the same differential equations to be solved as those provided by the Lagrange formulation. Therefore, the advantages of Hamilton's formulation lie not in its use as a method of calculation, but rather in the deeper insight into the fundamental aspects of physics it offers by way of Hamilton's variational principle. But Lagrange's method is also based on this variational principle. What makes Hamilton's formulation fundamentally more useful than Lagrange's? The answer is that the equal status given to generalized coordinates and momenta as independent variables gives us more freedom in selecting those physical quantities which we designate as "coordinates" and "momenta". As a result we are better able to consider the fundamental aspects of physics, especially the modern theories of matter.

CHAPTER 5

Canonical Transformations and Hamilton–Jacobi Theory

Introduction

This chapter is divided into two parts: "Canonical Transformations" and "Hamilton–Jacobi Theory". Concerning the subject of canonical transformations, we have seen in our study of cyclic coordinates that the integration of a dynamical system can generally be effected by transforming it into another dynamical system with fewer degrees of freedom by the use of *cyclic coordinates*. We also saw that, in the Hamiltonian formulation, the Hamiltonian does not contain the cyclic coordinates and the corresponding conjugate momenta are conserved. Such transformations that decrease the number of degrees of freedom of the system and leave Hamilton's canonical equations of motion invariant, are called *canonical transformations*. We shall investigate these ideas in the context of the general theory that underlies the solutions of dynamical systems and is the basis for the Hamilton–Jacobi theory. This theory, to be discussed in Part II, gives the foundation for the modern theory of partial differential equations (PDEs) as applied to wave propagation phenomena.

PART I. CANONICAL TRANSFORMATIONS

5.1. Equations of Canonical Transformations and Generating Functions

Suppose we have an n degree of freedom conservative system so that the Hamiltonian is a constant of the motion (the total energy). Further, suppose we can find a transformation into another set of q_i's and p_i's where *all* the q_i's are cyclic. As we saw, the Hamiltonian is then independent of $\mathbf{q}$, and all the conjugate p_i's are also cyclic so that they are all constant, or

$$p_i = \alpha_i, \qquad i = 1, 2, \ldots, n.$$

Since $H = H(\alpha_1, \alpha_2, \ldots, \alpha_n)$ we can use Hamilton's equations of motion to solve easily for the q_i's from the differential equations

$$\dot{q}_i = \frac{\partial H}{\partial \alpha_i} = \omega_i, \qquad i = 1, 2, \ldots, n, \tag{5.1}$$

where the ω_i's are only functions of the α_i's. These differential equations immediately yield the solutions

$$q_i(t) = \omega_i t + \beta_i. \tag{5.2}$$

The β_i's are easily seen to be the initial values of q_i or $q_i(0) = \beta_i$. Since $\alpha_i = m_i \dot{q}_i(0)$ (where m_i is the mass of the ith particle), it follows that the α_i's are the initial momenta or $p_i(0) = \alpha_i$.

EXAMPLE. Consider an n degree of freedom conservative system. The Hamiltonian becomes

$$H(q_1, \ldots, q_n, p_1, \ldots, p_n) = \frac{1}{2} \sum_{i=1}^{n} \frac{p_i^2}{m_i} + \frac{k}{2} \sum_{i=1}^{n} q_i^2.$$

If we transform to cyclic coordinates then the Hamiltonian is transformed into

$$H(\alpha_1, \alpha_2, \ldots, \alpha_n) = \frac{1}{2} \sum \frac{\alpha_i^2}{m_i}.$$

The system (5.1) becomes

$$\dot{q}_i = \frac{\partial H}{\partial \alpha_i} = \frac{\alpha_i}{m_i} = \omega_i.$$

The solutions are given by (5.2) where $\omega_i = \alpha_i/m_i$.

The above ideas lead to the general concept of *canonical transformations* within the framework of Hamilton's canonical equations of motion. First, we note that $\mathbf{q}$ and $\mathbf{p}$ are called canonical coordinates because they satisfy Hamilton's canonical equations of motion. Canonical transformations are defined as a set of transformations that simultaneously transform the generalized coordinates $\mathbf{q}$ and momenta $\mathbf{p}$ to a new set $\mathbf{Q} = (Q_1, Q_2, \ldots, Q_n)$ and $\mathbf{P} = \mathbf{P}(P_1, P_2, \ldots, P_n)$, such that $\mathbf{Q}$ and $\mathbf{P}$ are also canonical in the sense that they also satisfy Hamilton's equations. The equations of transformation are

$$\begin{aligned}
Q_i &= Q_i(q_1, q_2, \ldots, q_n, p_1, p_2, \ldots, p_n, t) &&\text{or} && \mathbf{Q} = \mathbf{Q}(\mathbf{q}, \mathbf{p}, t), \\
P_i &= P_i(q_1, q_2, \ldots, q_n, p_1, p_2, \ldots, p_n, t) &&\text{or} && \mathbf{P} = \mathbf{P}(\mathbf{q}, \mathbf{p}, t).
\end{aligned} \tag{5.3}$$

Note that the new coordinates and momenta are each defined in terms of the old coordinates and momenta.

Since $\mathbf{Q}$ and $\mathbf{P}$ are canonical, as mentioned above, they must also satisfy Hamilton's equations of motion. In order for this to be, we must define the Hamiltonian in these new variables: $K = K(\mathbf{Q}, \mathbf{P}, t)$. Note that we allow for the more general case where the Hamiltonian may be time-dependent so that the system is not conservative. Recall that Hamilton's equations in the old coordinates and momenta are given by (4.76). With respect to $\mathbf{Q}$ and $\mathbf{P}$ these equations take the form

$$\dot{Q} = \frac{\partial K}{\partial P_i}, \qquad \dot{P}_i = -\frac{\partial K}{\partial Q_i}, \qquad i = 1, 2, \ldots, n. \tag{5.4}$$

As pointed out, transformations given by (5.3), for which (5.4) are valid, are valid, are said to be canonical,† and K is the Hamiltonian in the transformed variables.

Hamilton's principle may be written as

$$\delta \int_{t_1}^{t_2} \left(\sum_i p_i \dot{q}_i - H(\mathbf{q}, \mathbf{p}, t) \right) dt = 0. \tag{5.5}$$

This is easily seen by recalling that Hamilton's principle is $\delta \int_{t_1}^{t_2} L(\mathbf{q}, \mathbf{p}, t) \, dt = 0$ and $H(\mathbf{q}, \mathbf{p}, t) = \sum_i p_i \dot{q}_i - L(\mathbf{q}, \mathbf{p}, t)$. Using the canonical transformations, (5.5) is transformed into Hamilton's principle with respect to the $(\mathbf{Q}, \mathbf{P})$ or transformed system.

$$\delta \int_{t_1}^{t_2} \left(\sum_i P_i \dot{Q}_i - K(\mathbf{Q}, \mathbf{P}, t) \right) dt = 0. \tag{5.6}$$

Equation (5.6) is therefore Hamilton's principle in the transformed canonical coordinates. Equations (5.5) and (5.6) are, respectively, of the form

$$\delta \int_{t_1}^{t_2} g(\mathbf{q}, \mathbf{p}, t) \, dt = 0,$$

$$\delta \int_{t_1}^{t_2} G(\mathbf{Q}, \mathbf{P}, t) \, dt = 0,$$

which yields

$$\delta \int_{t_1}^{t_2} [g(\mathbf{q}, \mathbf{p}, t) - G(\mathbf{Q}, \mathbf{P}, t)] \, dt = 0.$$

This equation does not mean that $g - G = 0$. It does mean that there exists a function $F = F(t)$ such that $g - G = dF/dt$. In other words,

$$\int_{t_1}^{t_2} \left[\frac{dF(t)}{dt} \right] dt = F(t_2) - F(t_1),$$

---

† These canonical transformations are sometimes called *contact transformations*. Although a distinction is sometimes made between these two terms, the literature on this point is not uniform. A. Sommerfeld confines contact transformations to those in which (5.3) do not involve time. Others give a more general definition of contact transformations which allows for time in (5.3), in keeping with the covariant transformations of relativity theory. Contact transformations originated in projective geometry in the study of the transformation of surfaces. We shall pick up on this concept when we investigate the transformation of wave fronts. However, in keeping with the practice of most physicists we shall treat these two terms as synonymous.

and the variation of this integral is automatically zero for any $F(t)$ since the variation vanishes at the end points. We therefore have

$$\sum_i p_i \dot{q}_i - H - \sum_i P_i \dot{Q}_i + K = \frac{dF}{dt}. \tag{5.7}$$

The arbitrary F is an important function;[†] it is called the *generating function* for the canonical transformation since, as we shall see, once F is determined, the transformation equations (5.3) are completely specified. But this means that F must depend on $(\mathbf{q}, \mathbf{p}, \mathbf{Q}, \mathbf{P}, t)$, or F must depend on the $4n$ old and new generalized coordinates and momenta as well as time. Since the system (5.3) consists of $2n$ transformation equations, it follows that only $2n$ of these coordinates and momenta are independent. This means that F can be written as a function of $\mathbf{q}, \ldots, \mathbf{P}, t$ in one of the following four combinations: $(\mathbf{q}, \mathbf{Q}, t)$, $(\mathbf{q}, \mathbf{P}, t)$, $(\mathbf{p}, \mathbf{Q}, t)$, $(\mathbf{p}, \mathbf{P}, t)$. It is clear that each combination such as $(\mathbf{q}, \mathbf{Q})$ represents $2n$ independent variables. The circumstance of the problem will determine which combination to choose. For example, suppose we have $\mathbf{q} = \mathbf{q}(\mathbf{Q}, t)$, then the first combination is ruled out, since $\mathbf{q}$ and $\mathbf{Q}$ are not independent.

Suppose $F = F_1(\mathbf{q}, \mathbf{Q}, t)$. We can expand the total derivative as

$$\frac{dF_1}{dt} = \sum_i \left(\frac{\partial F_1}{\partial q_i} \right) \dot{q}_i + \sum_i \left(\frac{\partial F_1}{\partial Q_i} \right) \dot{Q}_i + \frac{\partial F_1}{\partial t}.$$

Substituting this expansion into (5.7) gives

$$\sum_i p_i \dot{q}_i - H - \sum_i P_i \dot{Q}_i + K = \sum_i \left(\frac{\partial F_1}{\partial q_i} \right) \dot{q}_i + \sum_i \left(\frac{\partial F_1}{\partial Q_i} \right) \dot{Q}_i + \frac{\partial F_1}{\partial t}.$$

Since $\mathbf{q}$ and $\mathbf{Q}$ are independent variables the above equation can be valid only if the coefficients of the $\dot{q}_i$'s and $\dot{Q}_i$'s vanish separately. This yields

$$p_i = \frac{\partial F_1}{\partial q_i}, \qquad P_i = -\frac{\partial F_1}{\partial Q_i}, \qquad i = 1, 2, \ldots, n, \qquad K = H + \frac{\partial F_1}{\partial t}. \tag{5.8}[‡]$$

The first set of equations of (5.8) are n relations involving $(\mathbf{q}, \mathbf{p}, \mathbf{Q}, t)$. They can be solved for $\mathbf{Q} = \mathbf{Q}(\mathbf{q}, \mathbf{p}, t)$ yielding the first half of the transformation equations (5.3). From the second set we get $\mathbf{P} = \mathbf{P}(\mathbf{q}, \mathbf{Q}, t)$, finally obtaining the second half of the transformation equations.

Let $F_1(\mathbf{q}, \mathbf{Q}, t) = F_2(\mathbf{q}, \mathbf{P}, t)$. Then the generating function expressed by F_2 can be related to that represented by F_1 by the following equation which, in a sense, defines F_2.

$$F_2(\mathbf{q}, \mathbf{P}, t) = F_1(\mathbf{q}, \mathbf{Q}, t) + \sum_i P_i Q_i. \tag{5.9}$$

[†] This treatment of the generating function follows [22, Chap. 8].

[‡] These are called the *transformation equations* for F_1.

From (5.9) we easily see that F_2 satisfies the system

$$p_i = \frac{\partial F_2}{\partial q_i},\tag{5.10}$$

which is the same set of equations for F_2, as the first set of (5.8) is for F_1. This is obvious since $\partial F_1/\partial q_i = \partial F_2/\partial q_i$. Setting $F = F_1$ in (5.7), solving (5.9) for F_1, and substituting into (5.7) gives

$$\sum_i p_i \dot{q}_i - H + K = \sum_i P_i \dot{Q}_i + \frac{d}{dt}\left[F_2(\mathbf{q}, \mathbf{P}, t) - \sum_i P_i Q_i \right]$$

$$= \frac{dF_2}{dt} - \sum_i Q_i \dot{P}_i$$

$$= \sum_i \left(\frac{\partial F_2}{\partial q_i}\right)\dot{q}_i + \sum_i \left(\frac{\partial F_2}{\partial P_i}\right)\dot{P}_i + \frac{\partial F_2}{\partial t} - \sum_i Q_i \dot{P}_i.$$

Equating the coefficients of $\dot{q}_i$ and $\dot{P}_i$ we obtain (5.10) and

$$Q_i = \frac{\partial F_2}{\partial P_i}, \qquad K = H + \frac{\partial F_2}{\partial t}.\tag{5.10'}$$

Since F_2 depends on the transformed P_i's, (5.10) can be used to solve for $\mathbf{P} = \mathbf{P}(\mathbf{q}, \mathbf{p}, t)$. We can then solve the first equation of (5.10') for $\mathbf{Q} = \mathbf{Q}(\mathbf{q}, \mathbf{p}, t)$, thus yielding the transformations (5.3).

We consider the third combination $(\mathbf{p}, \mathbf{Q}, t)$ by setting the generating function $F = F_3(\mathbf{p}, \mathbf{Q}, t)$. Using the first equation of (5.8) we can relate F_3 to F_1 by

$$F_3(\mathbf{p}, \mathbf{Q}, t) = F_1(\mathbf{q}, \mathbf{Q}, t) - \sum_i q_i p_i.\tag{5.11}$$

Note that (5.11) satisfies (5.8). Solving (5.11) for F_1 and substituting into (5.7) (where $F = F_1$), expanding dF_3/dt, and equating coefficients of $\dot{p}_i$ and $\dot{Q}_i$ yields

$$q_i = -\frac{\partial F_3}{\partial p_i}, \qquad P_i = -\frac{\partial F_3}{\partial Q_i}, \qquad K = H + \frac{\partial F_3}{\partial t}.\tag{5.12}$$

The first equation of (5.12) gives $\mathbf{Q} = \mathbf{Q}(\mathbf{q}, \mathbf{p}, t)$ and the second gives $\mathbf{P} = \mathbf{P}(\mathbf{q}, \mathbf{p}, t)$.

And, finally, $F_4(\mathbf{p}, \mathbf{P}, t)$ is used when $(\mathbf{p}, \mathbf{P})$ are taken as independent canonical variables. F_4 may be related to F_1 by

$$F_4(\mathbf{p}, \mathbf{P}, t) = F_1(\mathbf{q}, \mathbf{Q}, t) + \sum_i P_i Q_i - \sum_i p_i q_i.\tag{5.13}$$

Using the same procedure as in the above three cases we obtain

$$q_i = -\frac{\partial F_4}{\partial p_i}, \qquad Q_i = \frac{\partial F_4}{\partial P_i}, \qquad K = H + \frac{\partial F_4}{\partial t}.\tag{5.14}$$

5.2. Some Examples of Canonical Transformations

Suppose the generating function $F_1(\mathbf{q}, \mathbf{Q}, t)$ is given by

$$F_1 = \sum_i q_i Q_i. \tag{5.15}$$

Then the transformation equations for F_1 given by (5.8) yield

$$p_i = Q_i, \qquad q_i = -P_i. \tag{5.16}$$

This transformation interchanges the coordinates and momenta in the following sense: The new coordinates are the old momenta and the new momenta are the old coordinates (with a change in sign).

Suppose the generating function $F_2(\mathbf{q}, \mathbf{P}, t)$ is given by

$$F_2 = \sum_i q_i P_i. \tag{5.17}$$

Using (5.10) and (5.10′) we get the identity transformation

$$P_i = p_i, \qquad Q_i = q_i, \qquad K = H. \tag{5.18}$$

More generally, suppose F_2 is given by

$$F_2 = \sum_i f_i(\mathbf{q}, t) P_i. \tag{5.19}$$

Then the first equation of (5.10′) yields

$$Q_i = f_i(\mathbf{q}, t), \tag{5.20}$$

which tells us that, with this expression for F_2, the new generalized coordinates depend only on the old coordinates and time, and not on the old generalized momenta.

If F_2 is given by

$$F_2 = \sum_i G_i(\mathbf{P}, t) q_i, \tag{5.21}$$

then (5.10) gives

$$p_i = G_i(\mathbf{P}, t), \tag{5.22}$$

so that the old momenta depend only on the new momenta and time.

Suppose $\mathbf{Q}$ is connected to $\mathbf{q}$ by an *orthogonal transformation*. If $F_2 = \sum_i f_i(\mathbf{q}, t) P_i$, then

$$f_i = Q_i = \sum_j a_{ij} q_j,$$

where the a_{ij}'s are the direction cosines of the transformation. We then have

$$F_2 = \sum_{i,j} a_{ij} q_j P_i. \tag{5.23}$$

From (5.10) we obtain

$$p_j = \sum_i a_{ij} P_i, \qquad j = 1, 2, \ldots, n. \tag{5.24}$$

Using the orthogonality conditions on the a_{ij}'s we may solve this set of equations for **P**, obtaining the set of equations

$$P_i = \sum_j a_{ij} p_j. \tag{5.25}$$

This tells us that the momenta also transform orthogonally with respect to this form of F_2.

With this mathematical apparatus behind us we are now ready to investigate the Hamilton–Jacobi theory from the point of view of the PDEs of wave propagation.

PART II. HAMILTON–JACOBI THEORY

Introduction

This division is a modification and extension of some of the material contained in [15, Chap. 9], for the convenience of the reader. The treatment presented here will emphasize wave propagation in electromagnetic media. This theory is a very powerful method of solving PDEs.

Hamilton–Jacobi theory was developed by the great Irish mathematical physicist Sir W.R. Hamilton, in the early part of the nineteenth century, primarily to solve problems in analytical dynamics. Hamilton then went further by attempting to tie in the fields of optics and mechanics by relating the variational principle he developed in mechanics with Fermat's principle in optics. He therefore seized upon the analogy between these two variational principles, and used them as guides in the development of the optical theory and the theory of analytical dynamics by the use of his canonical equations of motion. These equations were then extended by the German mathematician Jacobi to first-order PDEs. This approach of Jacobi gives a powerful insight into the nature and geometry of PDEs which are the basis for wave propagation.

The Hamilton–Jacobi theory is based on the Hamilton–Jacobi PDE for *Hamilton's principle function*, which will be described below. Solutions of the PDE yield integral surfaces. It turns out that the Hamilton canonical equations of motion (which are ordinary differential equations (ODEs)) are the characteristics of this PDE. Thus far, in our treatment of first-order PDEs, we have used the characteristic ODEs to solve these equations. Jacobi's ingenious method was to turn this procedure around and use the Hamilton–Jacobi PDEs to solve Hamilton's canonical equations of motion. In this part we shall show how this is done with the aid of Hamilton's principle function, making use of the canonical transformations and generating functions described above.

5.3. Derivation of the Hamilton–Jacobi Equation for Hamilton's Principle Function

We learned in Part I that the canonical transformations given by (5.3) transform the Hamiltonian $H(\mathbf{q} \cdot \mathbf{p}, t)$ into the transformed Hamiltonian $K(\mathbf{Q}, \mathbf{P}, t)$. Suppose the transformed coordinates and moments $\mathbf{Q}$ and $\mathbf{P}$ are constant in time, as they would be if they were the initial conditions. We can automatically ensure that $\mathbf{Q}$ and $\mathbf{P}$ are constant in time by requiring that K be identically zero. For, if $K = 0$, then (5.4) tells us that

$$\dot{Q}_i = \frac{\partial K}{\partial P_i} = 0 \quad \text{and} \quad \dot{P}_i = -\frac{\partial K}{\partial Q_i} = 0.$$

It is convenient to let the generating function depend on $\mathbf{q}$, $\mathbf{P}$, and t, since we want to deal with the transformed momenta $\mathbf{P}$. We shall therefore consider F_2 which we set equal to $S(\mathbf{q}, \mathbf{P}, t)$. Then for $K = 0$, the transformation equations (5.10) and (5.10′) become

$$p_i = \frac{\partial S}{\partial q_i}, \quad Q_i = \frac{\partial S}{\partial P_i}, \quad H + \frac{\partial S}{\partial t} = K = 0. \tag{5.26}$$

From the first and third equations of (5.26) we obtain

$$H\left(q_1, q_2, \ldots, q_n, \frac{\partial S}{\partial q_1}, \frac{\partial S}{\partial q_2}, \ldots, \frac{\partial S}{\partial q_n}\right) + \frac{\partial S}{\partial t} = 0. \tag{5.27}$$

Equation (5.27) is called the *Hamilton–Jacobi equation*. It is to be solved for the generating function S as a function of $(\mathbf{q}, \mathbf{P}, t)$. $S(\mathbf{q}, \mathbf{P}, t)$ is called *Hamilton's principle function*. From the structure of (5.27) we note that the Hamilton–Jacobi equation is a first-order PDE that does not depend directly on the generating function S. Rather, it depends on the partial derivatives $\partial S/\partial q_i$ or **grad** S and $\partial S/\partial t$.

Solving the Hamilton–Jacobi equation for Hamilton's principle function S means that we can then obtain $\mathbf{Q}$ and $\mathbf{P}$ from the first two of the transformation equations (5.26). The first equation can be inverted to get $\mathbf{P} = \mathbf{P}(\mathbf{q}, \mathbf{p}, t)$. Combining this result with the second equation yields $\mathbf{Q} = \mathbf{Q}(\mathbf{q}, \mathbf{p}, t)$. Therefore, as mentioned previously, solving the Hamilton–Jacobi equations allows us to solve the canonical transformation equations (5.3) for $\mathbf{Q}(\mathbf{q}, \mathbf{p}, t)$ and $\mathbf{P}(\mathbf{q}, \mathbf{p}, t)$. It is clear that the Jacobian of the mapping from $(\mathbf{q}, \mathbf{p}, t)$ to $(\mathbf{Q}, \mathbf{P}, t)$ will have a nonzero determinant in order for these transformation equations to have a unique inverse.

There are $(n + 1)$ independent variables $q_1, q_2, \ldots, q_n, t$ upon which S depends. This means that a complete solution of (5.27) must involve $(n + 1)$

constants of integration. Since $\mathbf{P}$ is constant we may choose the first n constants as the n components of the initial momenta given by the α_i's, meaning that $\mathbf{P} = (\alpha_1, \alpha_2, \ldots, \alpha_n)$. As mentioned, the Hamilton–Jacobi equation does not depend on S directly. This means that if S is a solution then $S + \alpha_n$ is also a solution, where α_n is an additive constant. To within this additive constant we may write the solution of the Hamilton–Jacobi equation in the form

$$S = S(q_1, q_2, \ldots, q_n, \alpha_1, \alpha_2, \ldots, \alpha_n). \tag{5.28}$$

Suppose we have solved the Hamilton–Jacobi equations. The functional form for S given by (5.28) can then be used in the transformation equations (5.26). From the first set of n equations of (5.26) we solve for the p_i's as functions of the q_i's and the α_i's. These solutions are inserted into the second set of n equations. Setting $Q_i = \beta_i$, the initial generalized coordinates, we then solve this second set for the q_i's as functions of the initial conditions $(\alpha_1, \ldots, \alpha_n, \beta_1, \ldots, \beta_n)$. This procedure gives $\mathbf{q}$ and $\mathbf{p}$ as functions of the initial coordinates and momenta (initial conditions), or

$$\begin{aligned} \mathbf{q} &= \mathbf{q}(\alpha_1, \ldots, \alpha_n, \beta_1, \ldots, \beta_n), \\ \mathbf{p} &= \mathbf{p}(\alpha_1, \ldots, \alpha_n, \beta_1, \ldots, \beta_n). \end{aligned} \tag{5.29}$$

Indeed, this is within the spirit of the solution of any of the dynamical field equations of mathematical physics, where the position and momenta of a system of particles at any time is given in terms of their initial conditions.

The third equation of (5.26) then allows us to solve for the Hamiltonian, since

$$H(\mathbf{q}, \mathbf{p}, t) = -\frac{\partial S}{\partial t}. \tag{5.30}$$

We repeat the important point that Hamilton's principle function S is the generating function (generates the canonical transformations from $(\mathbf{q}, \mathbf{p}, t)$ to $(\mathbf{Q}, \mathbf{P}, t)$), and S can be obtained by solving the Hamilton–Jacobi equation.

5.4. *S* Related to a Variational Principle

We shall now show that Hamilton's principle function S is related to Hamilton's variational principle, thus showing the physical significance of S. To this end, we expand the total time derivative of S and obtain

$$\frac{dS}{dt} = \sum_i \left(\frac{\partial S}{\partial q_i} \right) \dot{q}_i + \frac{\partial S}{\partial t},$$

since the P_i's are constant. Using the first and third equations of (5.26), the above expansion becomes

$$\frac{dS}{dt} = \sum_i p_i \dot{q}_i - H = L. \tag{5.31}$$

Integrating from t_1 to t_2 gives

$$S = \int_{t_1}^{t_2} L \, dt. \tag{5.32}$$

This gives the important result that Hamilton's principle function is the functional which satisfies Hamilton's variational principle. Hamilton recognized that S is a solution to the Hamilton–Jacobi equation, and Jacobi realized that S could be considered a generating function used to generate the proper canonical transformations, and thereby solve problems in dynamics as well as optics. In actual calculations the result given by (5.32) is of no help, because we cannot integrate the Lagrangian without knowing $\mathbf{q}$ and $\mathbf{p}$ as functions of time (but this means solving the problem for the trajectories).

5.5. Application to Harmonic Oscillator

At this stage we give a simple example, that of a simple harmonic oscillator, to illustrate the theory developed thus far. The Hamiltonian $H(\mathbf{q}, \mathbf{p}, t)$ becomes

$$H = \frac{p^2}{2m} + \left(\frac{k}{2}\right) q^2,$$

where as usual m is the mass and k is the spring constant. By setting $p = \partial S / \partial q$ the Hamilton–Jacobi equation (5.27) becomes

$$\left(\frac{1}{2m}\right)\left(\frac{\partial S}{\partial q}\right)^2 + \left(\frac{k}{2}\right) q^2 + \frac{\partial S}{\partial t} = 0, \tag{5.33}$$

where $S = S(q, \alpha, t)$ and the cyclic coordinate α is the initial value of the momentum. Since a simple harmonic oscillator is an example of a conservative system we must have $H = E$, the total energy of the system. We therefore may separate out the time-dependent part of S by introducing a function $W = W(q, \alpha)$ such that

$$S(q, \alpha, t) = W(q, \alpha) - Et. \tag{5.34}$$

Inserting this expression into (5.33) yields

$$\left(\frac{1}{2m}\right)\left(\frac{\partial W}{\partial q}\right)^2 + \left(\frac{k}{2}\right) q^2 = E, \tag{5.35}$$

which is actually an ODE for W since α is constant. Integrating with respect to q gives

$$W = \sqrt{mk} \int \sqrt{\frac{2E}{k} - q^2} \, dq = S + Et.$$

We are not interested in W as such, but in its partial derivatives with respect to α and E. Differentiating this expression with respect to E gives

$$\frac{\partial W}{\partial E} = \sqrt{\frac{m}{k}} \int \frac{dq}{\sqrt{(2E/k) - q^2}} = -\sqrt{\frac{m}{k}} \cos^{-1} q \sqrt{\frac{k}{2E}} + \tau = t,$$

where τ is a constant having the dimension of time. Solving this equation for q gives

$$q = \sqrt{\frac{2E}{k}} \cos \omega(t - \tau), \qquad (5.36a)$$

where it is clear that $\omega\tau$ is the phase angle. The conjugate momentum becomes

$$p = -\sqrt{2Em} \sin \omega(t - \tau). \qquad (5.36b)$$

Since $\alpha = p(0)$ we obtain $P = \alpha = \sqrt{2Em} \sin \omega\tau$. Note that we used the expression $\omega = \sqrt{k/m}$ for the natural frequency of the oscillator.

This example shows it is possible to integrate the Hamilton–Jacobi equation because S was separated into a time-independent part $W(q, \alpha)$ and a part proportional to time. This decomposition of S is always possible if H is independent of time, since the only time-dependent part of the Hamilton–Jacobi equation is the $\partial S/\partial t$ term. The function W is called *Hamilton's characteristic function*. Its significance will be shown in the next section.

5.6. Hamilton's Characteristic Function

We now consider the case where $\mathbf{q}$ and $\mathbf{p}$ are n-dimensional vectors. We also consider a conservative system so that the additive constant $\alpha_{n+1} = E$. Generalizing (5.34) we decompose S into a time-independent and a time-dependent part.

$$S(\mathbf{q}, \boldsymbol{\alpha}, t) = W(\mathbf{q}, \boldsymbol{\alpha}) - Et, \qquad (5.37)$$

where $\boldsymbol{\alpha} = (\alpha_1, \ldots, \alpha_n)$ is the transformed momentum $\mathbf{P}$ or the initial momentum. $W(\mathbf{q}, \boldsymbol{\alpha})$ is *Hamilton's characteristic function* for this more general case. Inserting (5.37) into the Hamilton–Jacobi equation (5.27) gives

$$H\left(q_1, \ldots, q_n, \frac{\partial W}{\partial q_1}, \ldots, \frac{\partial W}{\partial q_n}\right) = E. \qquad (5.38)$$

Note that we set

$$p_i = \frac{\partial W}{\partial q_i}. \qquad (5.39)$$

We now assume that W can be decomposed as follows:

$$W(q_1, \ldots, q_n, \alpha_1, \ldots, \alpha_n) = \sum_{i=1}^{n} W_i(q_i, \alpha_1, \ldots, \alpha_n), \qquad (5.40)$$

where each W_i is a function of only one q_i and the constant vector $\boldsymbol{\alpha}$. If this decomposition is valid then the q_i's occurring in (5.38) are said to be *separable* in the sense that (5.38) can be split into n equations, the ith equation being

$$H_i\left(q_i, \frac{\partial W_i}{\partial q_i}\right) = \alpha_i.$$

The reason for setting this constant to α_i is as follows: Suppose H depends only on one q_i and t so that $H = H_i$ and $S = S_i$, where

$$H_i\left(q_i, \frac{\partial W_i}{\partial q_i}\right) + \frac{\partial S}{\partial t} = 0.$$

Setting

$$S_i = W_i(q_i) - \alpha_i t, \tag{5.41}$$

we get

$$H_i\left(q_i, \frac{\partial W_i}{\partial q_i}\right) - \alpha_i = 0. \tag{5.42}$$

Now t may be considered to be a *generalized coordinate*. For a conservative system (which we have been considering all along), H is conserved so that t may be treated as a *cyclic coordinate*. From (5.4.1) we get $-\partial S/\partial t = \alpha_i$. Comparing this expression to the first equation of (5.26) we see that $p_i = \alpha_i$, the cyclic momentum conjugate to the cyclic coordinate t. This is the reason for setting the constant in (5.41) equal to $\alpha_i = P_i$.

A separation of variables in Hamilton's characteristic function of the form (5.40) can be accomplished when all but one of the q_i's are cyclic. Suppose q_1 is the only noncyclic coordinate. Since the p_i's conjugate to the q_i's (for $i \neq 1$) are cyclic, we must have

$$\frac{\partial W_i}{\partial q_i} = \alpha_i, \qquad W_i = \alpha_i q_i, \qquad i \neq 1. \tag{5.43}$$

The Hamilton–Jacobi equation for W_1 then reduces to

$$H\left(q_1, \frac{\partial W_1}{\partial q_1}, \alpha_2, \ldots, \alpha_n\right) = \alpha_1, \tag{5.44}$$

which is an ODE for W_1 which is first order and second degree in $\partial W_1/\partial q_1$, and hence can be integrated. We can then separate out W_1 and write W as

$$W = W_1 + \sum_{i=2}^{n} \alpha_i q_i,$$

so that W is decomposed into $W_1(q_1)$ plus separable functions of the q_i's, and the Hamiltonian corresponding to W_1 only involves q_1, as shown by (5.44). We can perform a similar transformation that allows q_2 to be noncyclic and demands that the q_i's ($i = 3, \ldots, n$) be cyclic. We get

$$W = W_1 + W_2 + \sum_{i=3}^{n} \alpha_i q_i,$$

and the Hamilton–Jacobi equation becomes

$$H\left(q_2, \frac{\partial W}{\partial q_2}, \alpha_1, \alpha_3, \ldots, \alpha_n\right) = \alpha_2,$$

so that the Hamiltonian depends only on q_2. We now repeat this process until all the q_i's are separable. This allows the Hamiltonian to be decomposed into the sum of Hamiltonians such that the ith one depends only on q_i: $H = \sum_i H_i$.

5.7. Application to n Harmonic Oscillators

We now consider another case where we can separate the q_i's in W, namely, n uncoupled simple harmonic oscillators all having the same mass and spring constant. The energy equation becomes

$$H = \frac{1}{2m} \sum_i (p_i)^2 + \frac{k}{2} \sum_i (q_i)^2 = E.$$

Using (5.39) this equation becomes the Hamilton–Jacobi equation

$$\frac{1}{2m} \sum_i \left(\frac{\partial W}{\partial q_i}\right)^2 + \frac{k}{2} \sum_i (q_i)^2 = E. \tag{5.45}$$

Since $\partial W / \partial q_i = dW_i / dq_i$ we can rewrite (5.45) in the following form:

$$\frac{1}{2m} \left(\frac{dW_1}{dq_1}\right)^2 + \left(\frac{k}{2}\right)(q_1)^2 = F(q_2, \ldots, q_n).$$

The left-hand side depends on q_1 and the right-hand side is independent of q_1 so that $F = \text{const}$, since the q_i's are independent coordinates. We therefore obtain

$$\frac{1}{2m} \left(\frac{dW_1}{dq_1}\right)^2 + \left(\frac{k}{2}\right)(q_1)^2 = H_1 = \alpha_1.$$

Continuing this procedure, we write F in a form that separates out the q_2 dependence. By the same argument we obtain

$$\frac{1}{2m} \left(\frac{dW_2}{dq_2}\right)^2 + \left(\frac{k}{2}\right)(q_2)^2 = H_2 = \alpha_2.$$

By continuing this process we finally arrive at

$$\frac{1}{2m} \left(\frac{dW_n}{dq_n}\right)^2 + \left(\frac{k}{2}\right)(q_n)^2 = H_n = \alpha_n.$$

This procedure shows that the decomposition given by (5.40) is valid for this application.

5.8. Hamilton–Jacobi Theory Related to Characteristic Theory

We previously mentioned that Hamilton's canonical equations of motion are the characteristic differential equations for the Hamilton–Jacobi equation. We shall now prove this statement by exploring this important correspondence between the Hamilton–Jacobi equation, which is a first-order PDE, and the canonical equations of motion, which are first-order ODEs in the q_i's and p_i's.

Our starting point is the Hamilton–Jacobi equation given by (5.27). For ease of comparison with the theory of PDEs we change the notation to conform with standard PDE theory. We revert to using subscripts for partial differentiation and use the following correspondence between the notation of Hamilton–Jacobi theory and that of PDEs.

$$W \sim u, \qquad S \sim u - Et, \qquad q_i \sim x_i, \qquad H - E = F,$$

$$\frac{\partial S}{\partial q_i} = \frac{\partial W}{\partial q_i} \sim u_x = p_i, \qquad \frac{\partial S}{\partial t} = -E.$$

(5.46)

The Hamilton–Jacobi equation (5.27) can then be written in the general form

$$F(x_1, \ldots, x_n, p_1, \ldots, p_n, t, \alpha_1, \ldots, \alpha_n) = 0. \tag{5.47}$$

Two Degree of Freedom System

The reader is referred to Section 2.12 of Chapter 2 where nonlinear PDEs of the hyperbolic type in two independent variables were discussed in detail. The notation used for this case is: $q_1 = x$, $q_2 = y$, $S = u$, $S_q = u_x = p$, $S_q = u_y = q$ (not to be confused with the generalized coordinate). The Hamilton–Jacobi equation (5.47) becomes

$$F(x, y, u, p, q, t) = 0. \tag{5.48}$$

Note again that u does not appear explicitly in (5.48), but only in terms of the partial derivatives p, q. The characteristic ODEs for the fully nonlinear PDE (5.48) was given by (2.11). For convenience we repeat them.

$$\frac{dx}{dt} = F_p, \qquad \frac{dy}{dt} = F_q,$$

$$\frac{dp}{dt} = -(F_x + pF_u), \qquad \frac{dp}{dt} = -(F_y + qF_u), \tag{5.49}$$

$$\frac{du}{dt} = pF_p + qF_q + F_t,$$

where we added the term F_t to the right-hand side of the last equation, to allow for the fact that H may depend explicitly on time. It is easy to show that the first four equations of the system (5.49) are Hamilton's canonical equations of motion for the two degree of freedom case. Reverting back to the notation of Hamilton–Jacobi theory by inserting (5.46) into (5.49) we get

$$\frac{dq_1}{dt} = H_{p_1}, \qquad \frac{dq_2}{dt} = H_{p_2},$$

$$\frac{dp_1}{dt} = -H_{q_1}, \qquad \frac{dp_2}{dt} = -H_{q_2}, \tag{5.50}$$

$$\frac{dS}{dt} = \left(\frac{\partial S}{\partial q_1}\right)\dot{q}_1 + \left(\frac{\partial S}{\partial q_2}\right)\dot{q}_2 + \frac{\partial S}{\partial t}.$$

The first four equations of (5.50) are Hamilton's equations of motion for the two degree of freedom case, and the last equation represents the expansion of S (recognizing that $S = S(q, \alpha, t)$).

Two Degree of Freedom Simple Harmonic Oscillator. We now apply (5.49) to this example. Since the Hamiltonian or total energy is conserved, the Hamilton–Jacobi equation becomes

$$H - E = F(x, y, p, q) = \left(\frac{1}{2m}\right)(p^2 + q^2) + \left(\frac{k}{2}\right)(x^2 + y^2) - E = 0. \tag{5.51}$$

The characteristic ODEs (5.49) become

$$\frac{dx}{dt} = \frac{p}{m}, \qquad \frac{dy}{dt} = \frac{q}{m},$$

$$\frac{du}{dt} = \left(\frac{1}{m}\right)(p^2 + q^2) \tag{5.52}$$

$$\frac{dp}{dt} = -kx, \qquad \frac{dq}{dt} = -ky.$$

The first two equations give the definition of the x and y components of the momentum. The last two equations give the x and y components of the equations of motion for the simple harmonic oscillator, relating the time rate of change of momentum to the spring force. The third equation tells us that $du/dt = 2T$. This is related to the principle of least action which tells us that

$$\Delta S = \Delta \int_{t_1}^{t_2} 2T\, dt = 0. \tag{5.53}$$

Manipulating the system (5.52) we get

$$\left(\frac{1}{m}\right)\left(p\frac{dp}{dt} + q\frac{dq}{dt}\right) + k\left(x\frac{dx}{dt} + y\frac{dy}{dt}\right) = 0.$$

Integrating this equation yields the Hamilton–Jacobi equation (5.51) where the constant of integration E is interpreted as the total energy. The principle of the decomposition of the Hamiltonian ($H = \sum_i H_i$) applies here. Setting $F = F_1 + F_2$ we have

$$F_1 = \left(\frac{1}{2m}\right)p^2 + \left(\frac{k}{2}\right)x^2, \qquad F_2 = \left(\frac{1}{2m}\right)q^2 + \left(\frac{k}{2}\right)y^2.$$

We then set $u = u_1 + u_2 = W_1(x) + W_2(y)$, so that $p = u_{1,x}$ and $q = u_{2,x}$. This reduces our two degree of freedom system to two uncoupled simple harmonic oscillators.

n Degree of Freedom System

We now generalize characteristic theory in relation to the Hamilton–Jacobi theory, to an n degree of freedom system. Still using the notation of PDEs we let $\mathbf{x} = (x_1, \ldots, x_n)$ and $\mathbf{p} = (p_1, \ldots, p_n)$ be the generalized coordinates and momenta, respectively. (Note that $p_i = \partial u / \partial x_i \equiv u_{x_i}$.) For the two-dimensional case $p_1 = p$, $p_2 = q$, $p_i = 0$ for $i = 3, \ldots, n$. The Hamilton–Jacobi equation (5.47) can be written in vector form as

$$F(\mathbf{x}, \mathbf{p}, t, \boldsymbol{\alpha}) = 0.$$

The characteristic equations for the two-dimensional case (5.49) are generalized to

$$\frac{dx_i}{dt} = F_{p_i}, \qquad i = 1, \ldots, n,$$

$$\frac{dp_i}{dt} = -F_{x_i} + p_i F_u, \tag{5.54}$$

$$\frac{du}{dt} = \sum_i p_i F_{p_i} + F_t.$$

The system (5.54) is a set of $(2n + 1)$ equations for the $(2n + 1)$ variables $(x_1, \ldots, x_n, u, p_1, \ldots, p_n)$.

Since $H = T + V$, H is a quadratic function of the q_i's and p_i's. In addition, if H is conserved $(H = E)$ then it is independent of time, so that the Hamilton–Jacobi equation is simplified to that form

$$F(\mathbf{p}, \boldsymbol{\alpha}) = 0. \tag{5.55}$$

Reverting to the physical notation of Hamilton–Jacobi theory, the first two sets of equations of (5.54) then become

$$\frac{dp_i}{dt} = H_{p_i}, \qquad i = 1, \ldots, n,$$

$$\frac{dp_i}{dt} = -H_{q_i}, \tag{5.56}$$

which are Hamilton's canonical equations of motion for a conservative system of n degrees of freedom. We have thus shown that the characteristic equations associated with the Hamilton–Jacobi PDE are indeed the Hamilton equations of motion.

5.9. Hamilton–Jacobi Theory and Wave Propagation

The analogy between the variational principle of geometrical optics (Fermat's principle) and that of dynamics (principle of least action) was seized upon by Hamilton and used as a guide in the development of both optical and dynamical theory. The origin of Hamilton's method is to be found in a celebrated memoir on optics which was presented to the Royal Irish Academy in 1824. This seminal paper was the basis of the Hamilton–Jacobi theory of PDEs. As pointed out previously it was left to Jacobi to show that the Hamilton–Jacobi PDE could be used to solve the characteristic equations which are Hamilton's canonical equations of motion, vis-à-vis Hamilton's principal function. Up until de Broglie and Schrödinger came on the scene, the correspondence between the above-mentioned variational principles was considered as an analogy between geometric optics and dynamics. But these founders of the science of quantum mechanics showed that these two variational principles may be used directly as a stepping stone for the development of a true quantum or wave mechanics, similar in form to *physical* rather than *geometric* optics. This gave a correspondence between electromagnetic theory and dynamics compatible with the concept of quantized energies developed by Planck and Einstein. We shall show below how this correspondence leads, in a formal way, to the Schrödinger equation of quantum mechanics.

We now show how the Hamilton–Jacobi theory can be applied to wave propagation phenomena. We shall continue to investigate only those systems where the Hamiltonian is conserved so that $H = E$. Recall that Hamilton's principal function S and his characteristic function W are connected by (5.37) (which we repeat, for convenience)

$$S(\mathbf{q}, \mathbf{P}, t) = W(\mathbf{q}, \mathbf{P}) - Et. \tag{5.37}$$

We now consider the n-dimensional *configuration space* given by $\mathbf{q} = (q_1, \ldots, q_n)$. Both S and W are embedded in this space. Since W is independent of time each surface, characterized by a constant value of W, is fixed in configuration space. However, for a fixed value of W, the corresponding surface of constant S moves with time according to (5.37).

Suppose at some time t the surface of constant S corresponds to the surface of constant W in configuration space. At time $t + dt$ that surface coincides with the surface for which $W = S + E\,dt$. This means that during the time interval dt, the surface $S = $ const. travels from the surface on which $W = $ const. to a new surface given by $W + dW$ where $dW = E\,dt$. The surface $S = $ const. is interpreted as the *wave front*, so that surfaces of constant S are considered as wave fronts propagating in configuration space with a characteristic wave velocity. The outward drawn normal at each point on the surface $S = $ const. gives the direction of this wave or phase velocity. If S is a planar surface when the wave velocity is constant and a plane wave is thereby propagated (the electromagnetic medium is homogeneous so that the refractive index is constant). However, in general, the medium is not homogeneous (the index of refraction is variable), so that the wave velocity is not constant in space. In any case, at each point on the surface (wave front), the wave velocity c is given by the normal distance ds which the wave front travels in the time interval dt, so that

$$c = \frac{ds}{dt}.$$

Since W also moves with a normal velocity equal to c, and using (5.37), we obtain

$$\frac{dW}{dt} = \textbf{grad } W \cdot \textbf{c} = E,$$

so that

$$c = \frac{ds}{dt} = \frac{E}{|\textbf{grad } W|}. \tag{5.57}$$

EXAMPLE (A Single Particle Without Constraints). The configurations space reduces to three dimensions so that $\textbf{q} = (x, y, z)$, where we may use rectangular coordinates. $\textbf{grad } W$ is furnished by applying the Hamilton–Jacobi equation to a single particle. We get

$$\left(\frac{1}{2m}\right)[(W_x)^2 + (W_y)^2 + (W_z)^2] = \left(\frac{1}{2m}\right) \textbf{grad } W \cdot \textbf{grad } W$$

$$= \left(\frac{1}{2m}\right)(\textbf{grad } W)^2 = E - V. \tag{5.58}$$

From (5.57) and (5.58) we get the following expression for c:

$$c = \frac{E}{\sqrt{2m(E - V)}} = \frac{E}{\sqrt{2mT}} = \frac{E}{p}, \tag{5.59}$$

which also holds for the general case of an n degree of freedom system.

n Degree of Freedom System

We again consider the surfaces W and S (related to W) embedded in n-dimensional configuration space $\textbf{q}$. Recall that $p_i = \partial W/\partial q_i$ $(i = 1, \ldots, n)$. This gives the following relation between $\textbf{p}$ and W:

$$\textbf{p} = \textbf{grad } W. \tag{5.60}$$

Inserting (5.60) into (5.57) tells us that the reciprocal relation between c and p, as given by (5.59), holds in the general case. Equation (5.60) clearly means that both $\textbf{grad } W$ and $\textbf{p}$ are in the direction normal to the surface of constant W, thus giving the direction of wave propagation. The trajectory in configuration space of a system of particles is always normal to these surfaces. Recall that a point in configuration space represents a particular value of the generalized coordinates of the n-particle system at a given time, so that a single trajectory in this n-dimensional space represents the motion of the system. As the system of particles moves along the trajectory, the surface S generating the motion will also travel in this space. But the motions of the system of particles and the surface do not keep in step. In fact, when the surface moves faster the particles slow down, and conversely. This is seen from the reciprocal relation between c and p given by (5.59).

In the section on the principle of least action in Chapter 4 we introduced the metric tensor m_{ij} in terms of the arc length $d\rho$ given by (4.87). In configuration space the wave speed of the moving surfaces (wave fronts) becomes

$$c = \frac{d\rho}{dt}. \tag{5.61}$$

Therefore, instead of obtaining (5.59) for the relation between c and p for a single particle, we get the following relation which holds in configuration space:

$$c = \frac{E}{\sqrt{2(E - V)}} = \frac{E}{2T}. \tag{5.62}$$

The velocity of the system point in configuration space is proportional to $T^{1/2}$ where T is the kinetic energy of the system. Thus the reciprocal relation between the wave velocity c and the velocity of the system point is preserved in configuration space, as shown by (5.62). Also, as seen above, the trajectories of the system point were found to be normal to the surfaces of constant W or S. Thus we see that the transition from a single particle to an n-particle system introduces no new physical results.

Wave Propagation in Electromagnetic Media

The Hamilton–Jacobi equation tells us that the geometrical optics limit of wave motion corresponds to classical mechanics, in the sense that light rays are orthogonal to the wave fronts of the electromagnetic waves corresponding to the surfaces of constant S. By the "geometrical optics limit" we mean the limit of an electromagnetic or light wave as the frequency becomes infinite. This procedure of passing asymptotically to the infinite frequency case allows us to go from physical optics, where wave properties such as diffraction phenomena are considered, to geometrical optics (infinite frequency case), where *light rays* are considered so that wave properties are ignored. This is best seen by an asymptotic expansion process which will be discussed below.

Recall that the solution of the Hamilton–Jacobi equation (5.38) for W allows us to obtain Hamilton's characteristic function S from (5.37). The surfaces of constant S are characterized as wave fronts because they propagate in space in the same manner as wave surfaces with the wave speed c.

We now discuss the propagation of electromagnetic waves in the context of the Hamilton–Jacobi theory. From Chapter 1 let us recall some of the properties of electromagnetic waves: An electromagnetic wave is characterized by the time and space oscillations of both the electric field vector **E** and the magnetic flux density or magnetic induction vector **B**. **E** and **B** are normal to each other and lie in the plane normal to the direction of wave propagation given by the wave vector or wave number **k**. This means that they are transverse waves. **E** and **B** satisfy Maxwell's equations which is a system of coupled first-order PDEs. By some mathematical manipulations it was shown, in Chapter 1, that each component of **E** and **B** satisfies the wave equation. It was also shown that a scalar potential Φ and a vector potential **A** can be introduced, both of which satisfy the wave equation. We also know that light waves are electromagnetic waves.

We now turn to an investigation of the three-dimensional scalar wave equation

$$\nabla^2\phi - \left(\frac{n^2}{c_0}\right)\phi_{tt} = 0, \qquad \nabla^2 = \frac{\partial^2}{\partial x^2} + \frac{\partial^2}{\partial y^2} + \frac{\partial^2}{\partial z^2}, \qquad (5.63)$$

where ϕ is a scalar quantity. Equation (5.63) is a generic equation so that ϕ may stand for a component of **E**, **B**, **A** or the scalar potential Φ. c_0 is the velocity of an electromagnetic wave in a vacuum and n is the index of refraction defined by

$$n = \frac{c_0}{c}, \qquad (5.64)$$

where c is the velocity of an electromagnetic wave in the medium. In general, n is a function of space since it depends on the optical density of the medium which may be nonhomogeneous. The electromagnetic or optical medium is assumed to be isotropic, the scalar n is to be replaced by a vector for a nonisotropic medium. If n is constant (so that the medium is homogeneous) then (5.64) is satisfied by a plane wave solution of the form

$$\phi = \phi_0 \exp[i(\mathbf{k}\cdot\mathbf{r} - \omega t)], \qquad (5.65)$$

where ϕ_0 is the complex amplitude and, in the usual manner, we take the real part of the right-hand side. **r** is the radius vector to a point on the wave surface in (x, y, z) or Cartesian three-space. Equation (5.65) represents a progressing wave. The general solution of (5.63) in an extended medium is a linear combination of such a progressing wave and a regressing wave, obtained by replacing the minus sign in the exponent with a plus sign. The wave number k, which is the magnitude of the wave vector **k**, is given by

$$k = \frac{2\pi}{\lambda} = \frac{n\omega}{c_0} = \frac{\omega}{c}, \qquad k_0 = \frac{\omega}{c_0}, \qquad (5.66)$$

where ω is the frequency and λ is the wave length. The term $\mathbf{k}\cdot\mathbf{r}$ in the exponent of (5.65) is the projection of **r** in the direction of the propagating wave. As pointed out above, if n is constant, we have a plane wave. This means that we have a plane wave if **k** is independent of **r**.

Almost Plane Waves. By an "almost" plane wave we mean one whose wave front varies slightly from a planar surface. We know that n and ϕ_n are independent of **r** on a planar wave front so that (5.65) is a plane wave solution of the wave equation. If $n = n(\mathbf{r})$ and $\phi_0 = \phi_0(\mathbf{r})$ then (5.65) is no longer a solution of the wave equation. We define an almost plane wave as one where n and ϕ_0 in (5.65) vary very slowly with **r**. We now replace the plane wave solution by the more general solution (resembling it)

$$\phi = \exp[A(\mathbf{r}) + ik_0(\psi(\mathbf{r}) - c_0 t)], \qquad (5.67)$$

where the amplitude ϕ_0 is replaced by $e^{A(\mathbf{r})}$ for convenience, and k_0 is the wave number *in vacuo* (where $n = 1$) so that $k_0 = \omega/c_0$. (We assume $\omega_0 = \omega$.) For a plane wave solution ψ is proportional to **r** and $k_0(\psi - c_0 t) = \mathbf{k} \cdot \mathbf{r} - \omega t$, where the magnitude of **k** is given by (5.66). For example, if **k** is in the x direction then the exponent of the plane wave solution becomes $ik_0(nx - \omega t)$. At this stage (5.67) is to be interpreted as an exact solution to the wave equation (5.63), since no approximations of slowly varying $A(\mathbf{r})$ and $\psi(\mathbf{r})$ have been made. The function $\psi(\mathbf{r})$ is called the *phase* of the wave or the *eiconal*. For a light wave, $\psi(\mathbf{r})$ is called the *optical path length*.

Since (5.67) is assumed to be a solution of the wave equation we insert (5.67) into (5.63). To this end, we successively apply the **grad** operator to (5.67) and obtain

$$\mathbf{grad}\ \phi = \mathbf{grad}(A + ik_0\psi),$$

$$\nabla^2\phi = \phi[\nabla^2(A + ik_0\psi) + (\mathbf{grad}(A + ik_0\psi)^2]$$

$$= \phi[\nabla^2 A + ik_0\nabla^2\psi + (\mathbf{grad}\ A)^2 - k_0^2(\mathbf{grad}\ \psi)^2$$

$$+ 2ik_0\ \mathbf{grad}\ A \cdot \mathbf{grad}\ \psi].$$

The wave equation (5.63) becomes

$$ik_0[2\ \mathbf{grad}\ A \cdot \mathbf{grad}\ \psi + \nabla^2\psi]\phi$$

$$+ [\nabla^2 A + (\mathbf{grad}\ A)^2 - k_0^2(\mathbf{grad}\ \psi)^2 + n^2 k_0^2]\phi = 0. \qquad (5.68)$$

Since A and ψ are real, the real and imaginary parts of (5.68) must separately be set equal to zero. This gives

$$\nabla^2 A + (\mathbf{grad}\ A)^2 + k_0^2[n^2 - (\mathbf{grad}\ \psi)^2] = 0,$$

$$2\ \mathbf{grad}\ A \cdot \mathbf{grad}\ \psi + \nabla^2\psi = 0, \qquad \phi \neq 0. \qquad (5.69)$$

Both equations of (5.69) are exact since no approximations have been made.

We now invoke the hypothesis that n varies very slowly with distance, which we take as λ. Let a characteristic distance be designated by L, the distance traveled by the electromagnetic wave in one second in a medium where $n = 1$. Then L is defined by

$$L = \frac{2\pi c_0}{\omega},$$

since c_0 is the wave velocity in a medium of unit index of refraction and $\omega/2\pi$ is the frequency in cycles per second. From this definition of L and (5.66) we get

$$n = \frac{L}{\lambda}.$$

From this expression we get (upon fixing L)

$$\frac{dn}{d\lambda} = -\frac{L}{\lambda^2}.$$

Now $dn/d\lambda \ll 1$ since n is slowly varying with respect to distance. Fixing λ, this inequality must hold only if $\omega \gg 1$, using the definition of L. We have therefore shown that ω must be very large in order for n to be slowly varying with respect to distance (either L or λ, which are of the same order of magnitude). Equation (5.66) then tells us that k_0 is also very large. For the limiting case ($k_0 \to \infty$) the first equation of (5.69) yields

$$(\mathbf{grad}\ \psi)^2 = n^2. \tag{5.70}$$

Equation (5.70) is an important equation in optics and electromagnetic theory in general. It is called the *eiconal equation*. This limiting case of infinite frequency is the assumption of geometrical optics.† However, in the high frequency spectrum (where k_0 is not infinite), the eiconal equation is also valid if we assume that the amplitude A is constant in space, as shown by the first equation of (5.69). Then the second equation tells us that ψ is a harmonic function. Therefore the eiconal equation holds for physical optics and electromagnetic wave propagation in general (diffraction phenomena considered) as long as A is constant.

Recall that in Chapter 2, Example 2, "Geometrical Optics", (5.70) was discussed from the point of view of characteristic theory. It was pointed out that surfaces of constant optical path length or eiconal ψ represent wave fronts and the characteristics represent the light rays. (If n is constant then the wave fronts are planar surfaces.) The important statement: "Surfaces of constant eiconal determined from (5.70) represent wave fronts", allows us to relate geometrical optics to classical mechanics, by comparing the eiconal equation with the Hamilton–Jacobi equation for Hamilton's characteristic function W given by (5.58). Then ψ is analogous to W, and n to $\sqrt{2m(E - V)}$. The Hamilton–Jacobi equation therefore tells us that classical mechanics corresponds to the geometrical optics limit of wave motion, in which the particle trajectories, orthogonal to the surfaces of constant S, correspond to light rays (characteristics) which are orthogonal to surfaces of constant ψ. We can also see the correspondence between analytical dynamics and geometrical optics

---

† Strictly speaking, geometrical optics deals with light phenomena as light rays, neglecting its wave properties such as diffraction effects.

by relating the principle of least action (in dynamics), given in the form

$$\Delta \int \sqrt{2mT} \, ds = 0,$$

to Fermat's principle of least time (in optics). This is easily seen since $\sqrt{2mT} = \sqrt{2m(E - V)}$ is analogous to n, as previously mentioned, and the principle of least action can then be written as

$$\Delta \int n \, ds = \Delta \int \frac{ds}{c} = \Delta \int \frac{ds}{\lambda(\omega, \mathbf{r})} = 0,$$

which are well-known variations of Fermat's principle. The local velocity of light c is a function of ω and $\mathbf{r}$. As the frequency is treated as constant in varying the functional and as the local wave length $\lambda = 2\pi c/\omega$, we may substitute λ for c in the statement of Fermat's principle and hence obtain the last equation.

5.10. Hamilton–Jacobi Theory and Quantum Mechanics

As mentioned above, the Hamilton–Jacobi equation tells us that analytical dynamics or classical mechanics corresponds to the geometrical optics limit of wave propagation, in which the particle trajectories orthogonal to the surfaces of constant S correspond to light rays orthogonal to the wave fronts. This explains why Huyghen's theory of wave propagation[†] and Newton's corpuscular theory were equally able to account for the phenomena of reflection and refraction (but not diffraction!) which occur in geometrical optics. We also showed the formal relationship between the principle of least action in mechanics and Fermat's principle in optics, where $\sqrt{2m(E - V)}$ was seen to be analogous to n. The physics underlying this correspondence, which indicate the *duality of particles and waves*, cannot be adequately explained within the framework of classical physics. All that classical physics does is connect mechanics with optics in the limiting case of very high frequency, which means we cannot explain any of the phenomena of *physical optics*, since this branch of optics requires an understanding of the wave properties of light necessary to explain the phenomena of diffraction.

We must therefore abandon classical physics and appeal to modern physics, in the form of quantum mechanics, in order to get an adequate understanding of the phenomenon of light waves. We are reminded of the great contribution of Maxwell who showed that light waves are electromagnetic waves; this theory being within the framework of classical physics. However, this theory does not relate wave motion with the corpuscular theory of light, which quantum mechanics does. However (still within the framework of classical

---

† See [15, Chap. 6], Sound Waves (Huyghen's Principle).

physics), if we speculate about what form the optical wave equation for the potential would take, for which the Hamilton–Jacobi equation represents the short wave length limit, we arrive at the astonishing conclusion that this wave equation becomes the *Schrödinger equation* of quantum mechanics. In order to reach this result we use the following argument: The similarity of the eiconal equation (5.70) with the Hamilton–Jacobi equation (5.58) tells us that the optical path length or eiconal ψ is proportional to Hamilton's characteristic function W, as mentioned above. If W corresponds to ψ then $S = W - Et$ must be equal to a constant times the "total phase" of the light wave which is given by the exponent in the plane wave solution (5.65), so that we have

$$W - Et = (2\pi h)k_0(\psi(\mathbf{r}) - c_0 t), \tag{5.71}$$

where h is a constant. (The factor 2π is given for convenience.) We have anticipated a result in quantum mechanics: h is called *Planck's constant* and is associated with *quantized energy levels* which is fundamental to quantum theory. Equation (5.71) tells us that the total energy $E = (2\pi h)k_0 c_0$. Since $k_0 c_0 = \omega = v/2\pi$, where v is the frequency cycles/sec, the energy becomes

$$E = hv. \tag{5.72}$$

Equation (5.72) is of fundamental importance in quantum mechanics. It applies on a molecular level and is the basis for the modern theory of atomic structure. It tells us that the total energy of an atomic or molecular system is "quantized" in the sense that the energy depends *only* on the frequency in a *discrete manner*. For example, if the electron of a hydrogen atom is excited by an external energy source then it emits a quantum (discrete bit) of energy equal to hv (proportional to the frequency of excitation). Planck's constant h is a universal constant (independent of the coordinate system) and equals 6.625×10^{-22} joul/sec, thus having units of (energy) $\times$ (time). We now relate wave length to momentum in the following way: We have the relation

$$c = \lambda v.$$

Hamilton–Jacobi theory yields the result (5.59). Combining these two expressions we get

$$\lambda = \frac{h}{p}. \tag{5.73}$$

Equation (5.73) is another fundamental equation of quantum mechanics. To get a feel for the fact that (5.73) only applies to small-scale phenomena, let us apply it to a body of macroscopic dimensions such as a golf ball which weighs 47 grams and travels with a speed of 1 meter/sec. Then $\lambda = 1.41 \times 10^{-20}$ meter. This indicates that diffraction effects are hopelessly beyond the reach of measurement in case of large-scale bodies. On the other hand, the wave length becomes appreciable if we apply (5.73) to atomic phenomena. For example, an oxygen molecule with a speed corresponding to a mean thermal energy of 300 K (temperature of 300 degrees Kelvin) has a wave length of approximately 1.5×10^{-8} cm, which is a dimension of the order of magnitude of an atomic diameter and is in the X-ray region and is thus subject to measurement.

We now show how the Hamilton–Jacobi theory leads, in a formal sense, to the Schrödinger equation of quantum mechanics. The optical wave equation for the potential $\phi(\mathbf{r}, t)$ is

$$c^2 \nabla^2 \phi - \phi_{tt} = 0, \tag{5.74}$$

where $c^2 = c_0^2/n^2$ (using the definition of n). We now separate variables in (5.74) by letting

$$\phi(\mathbf{r}, t) = u(\mathbf{r}) \exp(i\omega t) = u \exp\left(\frac{2\pi i c t}{\lambda}\right). \tag{5.75}$$

Inserting this expression for the potential into (5.74) yields

$$\nabla^2 u + \left(\frac{4\pi^2}{\lambda^2}\right) u = 0. \tag{5.76}$$

Equation (5.76) is called the *reduced wave equation* corresponding to (5.74). According to Hamilton–Jacobi theory, $p = \sqrt{2m(E - V)}$. From this expression we get $\lambda = h/\sqrt{2m(E - V)}$. Inserting this expression for the wave length into the reduced wave equation yields

$$\nabla^2 u + \left(\frac{8\pi^2 m}{h^2}\right)(E - V(\mathbf{r}))u = 0. \tag{5.77}$$

Equation (5.77) is the Schrödinger equation or reduced wave equation (time-independent or steady state equation) of quantum mechanics. It is sometimes called the *first form of Schrödinger's wave equation*. The function $u(\mathbf{r})$ is called the amplitude or "space factor" of the *wave function* which is a fundamental variable in quantum mechanics.

CHAPTER 6

Quantum Mechanics—A Survey

Introduction

The phenomenon of wave propagation in electromagnetic media has been presented from a macroscopic viewpoint; this neglects the atomic nature or the structure of matter. The relationship between classical mechanics and quantum mechanics, vis-à-vis Hamilton–Jacobi theory, was also presented from this viewpoint. The electromagnetic nature of wave propagation in continuous media has, as its counterpart, the electromagnetic force which is one of the four forces that are involved in understanding the nature of the atom. Therefore a proper understanding of the nature of electromagnetic involves getting some insight into the physics of matter from an atomic viewpoint using the modern methods of quantum mechanics. We shall also show that the microcosmos is intimately related to the macrocosmos, meaning that quantum mechanics is related to astronomical or cosmological events. This ties up events involving picosecond (one trillionth of a second) time intervals and distances of 10^{-12} cm (diameter of an atom) occurring in particle physics with billions of years (age of the universe) and distances measured in light years.

Although a detailed study of this subject would distort the balance of this book, we feel that the reader should be given an overview of quantum mechanics and cosmology. The vast scope of these fields precludes our giving a detailed discussion; we, no doubt, shall omit many important developments. The reader interested in more technical details may consult the many books on these subjects, some of which are listed in the references.

Planck's Quantized Black-Body Radiation

In 1900 the great German physicist, Max Planck formulated the quantum theory of radiation in the course of his investigations of the laws of thermal or black-body radiation. He set himself the problem of attempting to explain the distribution of energy in the continuous spectrum of a heated black body as a function of temperature. The experimental fact is that the intensity of black-body radiation per unit frequency interval rises from zero at very low frequencies to a maximum whose position and magnitude depend on the temperature—then falls, approaching zero magnitude at very high frequencies. This drop in intensity at the high-frequency range is in sharp conflict with a result previously obtained by Lord Rayleigh and Jeans on the basis of the *equipartition theorem* of classical statistical mechanics, which says that the mean kinetic energy of a system of particles in thermal equilibrium depends only on the temperature. The classical law of black-body radiation is called the *Rayleigh–Jeans law* and agrees well with experiment for low frequencies or high temperatures. But it cannot be right for the whole frequency spectrum, for it implies an ever-increasing energy density as the frequency is increased, and leads to the catastrophe of infinite total energy density (when integrated over the complete frequency spectrum). This is sometimes called the *ultraviolet catastrophe.*

Planck realized that this infinite total-energy density paradox was based on the classical equipartition of the energy theorem. He suggested that it broke down when applied to the high-frequency spectrum. He therefore abandoned this approach and made the brilliant hypothesis that, if the vibrating particles which emit radiation have motions restricted to certain *discrete* (*quantized*) *energy levels*, there would be a departure from the laws of classical statistical mechanics which would be in accordance with the experimental facts. Using this hypothesis, Planck was able to derive a formula for the intensity of radiation in terms of temperature *and frequency* which fits the empirical data within the bounds of experimental error—the classical equipartition principle says the thermal energy density depends only on the temperature.

This hypothesis of Planck was indeed revolutionary since it was incompatible with both Newton's corpuscular theory and the electromagnetic theory of light. It was deduced from Planck's work that the mechanics of the interaction between atoms and radiant energy is completely nonclassical, in the sense that light is emitted or absorbed in discrete packets called "quanta". An inference is that radiant energy is corpuscular in nature, but in a different sense than visualized by Newton. Planck was very unhappy about this result for he saw that it led to a complete revolution in the foundations of physics. He spent much time in an unsuccessful attempt to rescue classical physics, especially the classical wave theory of light, but to no avail.

Photoelectric Effect

Albert Einstein showed, in his investigations of the photoelectric effect, that Planck's fruitful hypothesis can be applied to the structure of matter. It is interesting that he received the Nobel prize in 1921 for his investigations of the photoelectric effect rather than for his revolutionary contribution to physics, namely, the general theory of relativity. The photoelectric effect was discovered in 1887 by Heinrich Hertz, who noticed that a spark would jump more readily between two charged spheres when their surfaces were illuminated by light. This showed that electrons can acquire sufficient energy to escape from the surface of cold metal if it is illuminated by light. But it was left to Einstein to correctly interpret the experimental fact that the maximum speed of the emitted electrons depends only on the frequency of the incident light. He applied Planck's quantum theory of radiation and thereby showed that light from the metal surface is absorbed or emitted in quanta. This treatment of the photoelectric effect in terms of quantized energy levels generalized Planck's theory of quantized black-body radiation to the quantization of matter, and thereby laid the foundations of quantum mechanics.

Rutherford's Atom

We know that an atom of any element is composed of a positively charged nucleus around which electrons rotate (the orbital or planetary electrons). The nucleus has a mass almost one thousand times that of an electron. This concept of the structure of an atom was due to Ernest Rutherford, the great English physicist. It was developed before quantum mechanics, and it was left to Niels Bohr to apply quantum mechanics to Rutherford's model. At the Cavendish Laboratory of Cambridge University in 1899 Rutherford discovered that there are three kinds of rays associated with an atom:

(1) α rays which turned out to be helium ions (having a double positive charge).
(2) β rays which could pass through aluminum foils. They turned out to be streams of high-velocity electrons.
(3) γ rays which could penetrate lead shields many centimeters thick. They are similar to X-rays but have shorter wave lengths.

When beams of these rays pass through a magnetic field the α rays bend in a direction showing them to consist of positively charged particles. The β rays bend in the opposite direction thus showing them to have a negative charge. And, finally, the γ rays do not deflect, showing them to be neutral.

In 1911 Rutherford proved that α particles exist when he carried out scattering experiments, wherein he bombarded thin foils of various metals with a high-velocity beam of α particles. From an analysis of the scattering pattern of these particles he deduced the *nuclear model of the atom*, wherein an atom of an element is composed of a positive nucleus, composed of protons, and surrounded by revolving electrons. Since an uncharged atom is neutral there are as many electrons orbiting the nucleus as there are protons making up the nucleus.

Rutherford and his collaborator, Frederick Soddy, soon came to the conclusion that the phenomenon of radioactivity is the result of the spontaneous transformation of one element into another by emission or absorption of these rays by the nucleus. Thus, unlike chemical reactions which do not involve the nucleus but only involve interactions with the electrons surrounding the nucleus, radioactivity involves transformations of the nucleus. For example, suppose a radioactive element emits an alpha particle which has a $+2$ charge and mass of four units (two protons and two neutrons). (A neutron is an uncharged particle in the nucleus having essentially the same mass as a proton.) Then the transformed element will have an atomic number two less, and an atomic weight four less, than the untransformed element. The atomic number of an atom is equal to the number of orbital electrons, and the atomic weight is equal to the sum of the protons and neutrons making up the nucleus. (Since the mass of an electron is almost two thousand times less than that of a proton, we can neglect it in calculating the atomic weight.)

At the onset of World War I Rutherford became director of the Cavendish Laboratory. The British Admiralty asked him to transform the Cavendish Laboratory into a war research institute for developing antisubmarine warfare methods against the Germans. Rutherford refused saying that he had the more important job of breaking the atomic nucleus. Even though his pioneering work on the structure of the atom paved the way for the development of the atomic and hydrogen bombs, it is ironic that he did not foresee these events. In fact in 1937, just before his death, Rutherford had a strong debate with the Hungarian physicist, Leo Szilard—Rutherford taking the stand that large-scale liberation of nuclear enery was not feasible. Szilard, to back up his argument, went to the patent office and took out a patent on a large-scale nuclear reaction method. Three years later the fission of the uranium nucleus was discovered and six years after that the first atomic bomb was exploded over Hiroshima. As Gamow [21] points out: Rutherford, in sitting on a celestial cloud, undoubtedly thought "So what? Now these damn chaps are using my discoveries for killing one another."

Bohr's Atom

In 1913 Niels Bohr united the Rutherford concept of the atom with the quantized energy level hypothesis to formulate his famous theory of the structure of the hydrogen atom. This gave the correct theoretical explanation for the hydrogen spectrum. In 1922 he was awarded the Nobel Prize for "studies of the structure of atoms and their radiation".

It is interesting to get a glimpse of Bohr's personality by way of an amusing anecdote given by George Gamow [21]. Gamow mentions that perhaps Bohr's most outstanding characteristic was his slowness of thinking. He goes on to say that this same slowness of comprehension was brought out at scientific meetings. Once, when a young physicist was giving a visiting lecture at Copenhagen, everybody in the audience would understand the argument quite clearly (or thought they did), except Bohr. Then his coworkers would explain to Bohr the simple point he had missed, "and in the resulting turmoil everybody would stop understanding anything". After a considerable time Bohr would finally understand the argument. But, what he understood about the problem was quite different from what the lecturer meant. What Bohr understood was correct, while the visitor's interpretation was wrong! I imagine that this incident illustrates that, unlike a 100-meter dash, the race for the Nobel prize does not always go to the swift. Bohr, I am sure, would not have thought of the Nobel prize as a race, but rather as the international recognition of his deep and passionate quest for an understanding of the nature of matter.

Bohr's theory of the atom, published in 1913, was based on Rutherford's model of the atom. In studying the Rutherford model, Bohr realized that atoms would be unstable in the sense that they could not exist any longer than a time interval of the order of 10^{-8} sec. Bohr showed that the system involving an electron revolving around a proton (a hydrogen atom) represents a miniature antenna which radiates energy in the form of electromagnetic waves. The electron would thereby lose energy and rapidly spiral into the nucleus leading to the paradox of an unstable atom. To get around this paradox, Bohr used the analogy with Planck's treatment of black-body radiation. If black-body radiation is quantized, why not quantize the energy of an electron orbiting the proton? Bohr then made a series of hypotheses. The first was that only a discrete set of orbits were possible for the electron, namely, those orbits that have the property that no electromagnetic radiation takes place when the electron is in one of them. The criterion for such an orbit is that the phase integral J is quantized so that

$$J = \oint \left(\frac{dS}{dt}\right) dq = nh,$$

and the integration is taken over a period of q taken as an angle variable) for a circular orbit. Planck's constant h is sometimes called the "quantum of action". Note the connection with Hamilton–Jacobi theory. The second hypothesis of Bohr is that the emission or absorption of electromagnetic energy (light) takes place as the atom passes from a higher- to a lower-energy state or vice versa. The energy difference between these two states must equal the quantum $h\nu$ of emitted or absorbed energy. These two hypotheses allowed Bohr to determine the rules of quantization of a hydrogen atom. From them he was able to give a theoretical interpretation of the spectral lines of hydrogen that were known from spectroscopic experiments.

Bohr's concept of the atom was limited, in that it did not lead to any satisfactory way of accounting for the wave phenomena of interference and diffraction on the basis of a purely particle theory. The duality of light led to conflicts between the proponents of corpuscular and wave theory. Experiments showed this duality in a confusing way.

De Broglie Waves

It was left to Louis de Broglie, the great French physicist, to come up with the extraordinary concept that the motion of material particles is guided by *pilot waves* which propagate through space along with the particles. Then Bohr's model of the hydrogen atom could be interpreted as one in which the quantized orbits of the electron satisfy the condition that their lengths (the phase integral) contain an *integral number of these pilot waves*, which are standing wave solutions of the wave equation for quantized energy levels. The wave length is given by $\lambda = h/p$. These de Broglie waves were demonstrated experimentally by the American physicists C.J. Davisson and L.H. Germer who directed a beam of electrons (accelerated in an electric field) at a crystal. The result was a diffraction pattern. The wave length estimated from the diameters of the diffraction rings coincided with that given by de Broglie's theory. It was thus demonstrated experimentally by Davisson and Germer that electrons can show the same sort of diffraction pattern that light shows, being diffracted by crystals and even by ruled gratings.

As an example of the use of de Brogile's formula $\lambda = h/p$, let us calculate the wave length associated with an electron moving with a velocity of 10^8 cm/sec. The mass of the electron is 9×10^{-28} gm. Using de Broglie's formula we get $\lambda = 7.3 \times 10^{-8}$ cm, a quantity of atomic dimensions. Thus if the electron passed through an aperture of atomic size it could easily be diffracted through a large angle. It is thus demonstrated that diffraction of electrons on at atomic scale is measurable.

De Broglie's pilot waves, which are standing waves associated with electron orbits, were developed further by Schrödinger by the use of his wave equation. We showed above that the time-independent Schrödinger equation resulted from the analogy between classical mechanics and geometrical optics vis-à-vis Hamilton–Jacobi theory, thus showing a duality between the corpuscular and wave theory of light. In the same sense we showed above that there is a duality between the particle nature of an electron and its wave properties. Thus we have the same sort of duality between the wave and corpuscular nature of particles of atomic dimensions. This duality, as we already mentioned, has a firm basis in many experiments. From a quantum mechanical point of view, it can best be understood by recognizing the statistical relation between the intensity of the wave (say the de Broglie pilot wave) and the *probability of finding the electron at a given point with a given momentum.*

Heisenberg's Uncertainty Principle

This probabilistic concept regarding the relation between **q** and **p** associated with an electron (or any particle of atomic dimensions) and its wave nature was correctly answered by Werner Heisenberg. He set himself the problem of attempting to answer the question: What is the physical meaning (if any) of de Broglie's waves? Are they real waves like electromagnetic waves or are they only mathematical fictions that were introduced just for convenience in describing the physics of atomic phenomena? In general, Heisenberg asked how the quantum laws, developed by Planck, Einstein, Bohr, and others, affect the basic concepts of classical physics. He went to the heart of the problem by constructing a "thought experiment" which involved attempting to determine the position and momentum of a stream of light photons as they impinged on an electron. Specifically, he designed an ideal laboratory in which an electron gun could shoot a single electron in a completely evacuated chamber. He also had an ideal light source which emits photons of light of any desired number and wave length. He was also equipped with an ideal microscope tuned over the complete frequency spectrum with which he could watch the motion of the electron. Clearly, the photons must impinge on the electron in order for it to be visible by the microscope. According to classical mechanics, the electron, when fired from the gun, would follow a deterministic parabolic trajectory until it was impacted by a photon, at which time it will recoil thus changing its velocity. Observing the motion of the electron we would have found it to take a zigzag path due to repeated impacts by the stream of photons. Now suppose we decrease the photon's energy in order to lessen the affect of their impact on the electron. This is done by lowering the frequency or increasing the wave length. However, we pay a price for minimizing the erratic trajectory of the electron due to repeated impacts. For as the wave length gets longer, the less we are able to define the position of the electron because of the diffraction effect. If we attempt to remedy this uncertainty in measuring its position by shortening the wave length, the electron's trajectory becomes more erratic and therefore its momentum becomes more erratic so that there is more error in its measurement. Summarizing: For long wave lengths there is uncertainty in measuring the electron's position, for short wave lengths there is uncertainty in measuring its momentum. Heisenberg analyzed these effects and came up with his *uncertainty principle* which stated that $\Delta p \Delta q \geq h$, meaning that the product of the uncertainties in the measurement of the momentum and position can never be less than Planck's constant.

Heisenberg's uncertainty principle is a fundamental principle of quantum mechanics. From a philosophical point of view it states that nature on a molecular scale is not deterministic. Einstein, who was one of the founders of quantum mechanics, was convinced that the fundamental probabilistic nature of quantum mechanics was wrong. Both his special and general relativity theories were firmly based on determinism. He said, "God does not play dice with man", and spent the last years of his life in a vain attempt to rescue quantum mechanics from the "tyrany" of probability theory by searching for a unified field theory that would wed quantum mechanics and general relativity theory in one grand design of the universe. However, his failure was actually not in vain for it spurred modern research into string theory, which is an esoteric attempt at a grand unifying theory wherein astronomical and atomic phenomena would both obey the same set of universal laws. This has yet to be accomplished. How does one reconcile the physics of nuclear particles with their spectacularly confusing array of colorful particles and short-time reactions with the gigantic picture that problems in cosmology present with its stupendously large space and time? Nobody knows, yet.

Coming back to Heisenberg's uncertainty principle: as mentioned, this principle is the backbone of quantum mechanics. That is why solutions of Schrödinger's equation are given in terms of the wave function (to be precise its square). The square of the wave function is probabilistic in nature. Solving Schrödinger's equation for the hydrogen atom gives the trajectories of the electron orbiting the nucleus as a set of quantized orbits that are not deterministic, but exist as a smeared-out set of orbits satisfying the wave function where the eigenvalues are quantized observables which yield the appropriate energy levels. The electron trajectories are not mathematically determined lines infinitely thin and deterministic as they would be in classical mechanics. Moreover, Heisenberg's uncertainty principle implies that there is an interaction between the phenomenon measured in the laboratory and the measuring equipment which yields an uncertainty in measurement of position and momentum. In this sense, quantum mechanics is a subjective science. It tells us that there is no absolutely "true" set of measurements because of this interaction between the object measured and the measuring device. Clearly, this principle of uncertainty holds only on a micro scale.

Particle Accelerators

In 1939 Gamow, working with Rutherford at the Cavendish Laboratory, calculated that protons would make much better projectiles for penetrating atomic nuclei than α particles. Noting this, Rutherford asked his students, J. Cockcroft and E.T.S. Walton to construct an apparatus for producing beams of protons. Thus the first "atom smasher" went into operation in 1931.

The pioneering work of Cockcroft and Walton led to the development of larger and larger particle accelerators. This started with the Van de Graaf machine based on the principle of electrostatics. Then Ernest Lawrence, using the principle of the cyclotron, developed a more powerful nuclear accelerator. The cyclotron utilizes the energy obtained from the multiple acceleration of charged particles moving in a circular trajectory in a powerful magnetic field. Specifically, the ions of the element to be used as atomic projectiles have their trajectories bent into small circles by an alternating magnetic field, in such a manner that the ions will accelerate along an unwinding spiral path and hit the target with very high velocities. More and more powerful accelerators are being constructed. Their purpose is to impinge very high-speed particles onto nuclear targets in order to obtain more and more information about the structure of the nucleus.

Pauli's Exclusion Principle

The electron orbits obey the quantum statistics dictated by Schrödinger's equation within the framework of certain rules governing the quantum numbers. These rules were formulated by the great Austrian physicist Wolfgang Pauli in the form of his exclusion principle. This states that no two electrons can be in the same state. The state of an electron is characterized not only by the nodal surface of the wave function but also by the spin which can assume two different values. It turns out that four quantum numbers are used to determine the state of an electron. Thus Pauli's exclusion principle tells us that no two electrons can have all four quantum numbers the same. This guides one in selecting the appropriate wave functions. From this principle a satisfactory interpretation of the periodic table for the elements was developed, yielding the correct electronic orbits.

As an aside: There are many stories of the clumsiness of the theoretician in the laboratory. Pauli fits this description very nicely. In his book, Gamow [21] gives a charming insight into the personality of Pauli whose "corpulent and jovial figure was a familiar and welcome sight in Bohr's Institute of Theoretical Physics". Pauli was a tremendously gifted theoretical physicist. But surrounding him, as Gamow puts it, was a mysterious phenomenon known as the "Pauli effect". This effect illustrates the well-known axiom that all good theoretical physicists are to the laboratory what a bull is to a china shop! Pauli was no exception. He was such a good theoretical physicist that "things broke down when he merely walked into a laboratory". Gamow goes on to illustrate the most persuasive case of this effect by an incident that happened in Professor James Frank's laboratory at the Physics Institute at the University of Göttingen. One day the equipment in the laboratory unexpectedly blew up and went to pieces without any apparent reason. An investigation showed that this catastrophe took place at the exact time when a train carrying Pauli from Zurich to Copenhagen stopped for five minutes at the Göttingen railroad station!

One fact that Gamow does not point out is the exception to the rule that every good theoretical physicist is, *a fortiori*, clumsy in the laboratory. And that glaring exception is Enrico Fermi. He was a great theoretician who was equally at home in the laboratory. His pioneer work in the development of the atomic bomb necessitated that he become quite familiar with laboratory methods. In fact, Fermi supervised the construction of the first atomic pile (under the grandstand of Stagg Field at the University of Chicago) which started operating on December 2, 1941.

Radioactive Decay

The story of nuclear fission goes back to the discovery of radioactive decay discovered in 1896 by the French physicist Henri Becquerel. He placed a crystal of the mineral uranyle (the sulfate salts of uranium and potassium) on a photographic plate wrapped in black paper. After a few hours of exposure to sunlight it showed a dark spot where the crystal had been placed. After repeating the experiment several times the dark spot remained even though the plate was wrapped in black paper. The story goes that he put his freshly wrapped photographic plate with the crystal in a desk drawer waiting for better weather to develop the plate. When the sun finally appeared after almost a week Becquerel again exposed his contraption to the sun's rays and then examined the plate in a dark room. To his amazement he noted that the darkening of the plate due to the crystal was going on uninterruptedly all the while it was sitting in the desk drawer in the absence of sunlight. He thus discovered that the uranyle crystal was emitting penetrating radiation similar to the X-rays discovered previously by Roentgen. He tried to do everything he could think of (heating, chilling, etc.) to the crystal but the intensity of the radiation remained constant. It became clear that this was a new property of matter which received the name of *radioactivity*. Then Marie Sklodowska Curie, the great Polish chemist and wife of the physicist Pierre Curie, carried out extensive tests on many elements and compounds for their radioactivity. She observed that uranium ores are about five times as much radioactive than would be expected from their uranium content, thus indicating that those ores must contain small amounts of some other radioactive substances much more active than uranium. With great labor and using tons of uranium ore, Madame Curie succeeded in separating a new radioactive element with properties similar to bismuth, which she called *polonium* in honor of her native country. She continued her pioneering work and managed to separate another radioactive element with properties similar to barium. She called this element *radium*. It is two million times more radioactive than uranium. The discovery of polonium and radium was followed by the discoveries of more radioactive elements, *actinium* which is a close relative of the fissionable uranium, and the family of thoriums which was separated by Otto Hahn, who forty years later discovered that the unranium nucleus can be split (the phenomenon of nuclear fission).

Mass Defect

Radioactive disintegration can be understood by invoking Einstein's famous law $E = mc^2$ which shows the *equivalence of mass and energy*. This principle is used for calculating the *mass defect* of an atomic nucleus. This may be illustrated by considering an oxygen atom whose nucleus is composed of eight protons and eight neutrons (giving an atomic weight of 16 and an atomic number of 8). Suppose these nuclear particles are brought together from some distance. Then nuclear attractive forces are brought into play making a stable nucleus. As the particles accelerate toward each other the potential energy due to the nuclear forces is changed into kinetic energy. In order that the particles remain stable (not fly apart) this energy must be given out as electromagnetic or γ radiation. This amount of energy is now lacking in the atomic nucleus (the mass defect). This is precisely the amount of energy needed to separate the nucleus again into its parts. Thus the atomic nucleus is lighter than the sum of its parts by the amount of energy calculated by Einstein's equation (the mass defect). This mass defect is usually expressed in electron volts. (One electron volt (eV) is defined as the energy imparted to an electron when accelerated from rest by a potential of 1 volt.) The energies involved in chemical reactions are of the order of 1 eV. However, the energy with which a single proton or neutron is bound to the nucleus is six to eight million eV. The energy of radioactive α particles is about five million eV. Their disadvantage as projectiles is that only one in about a million particles hit the target (say aluminum foil). The others are lost in passing through the atomic shells, for they impart energy to the planetary electrons, thus losing energy. Protons and deuterons are generally better projectiles than α particles, since they can be produced with energies far surpassing those particles.

Development of Atomic Energy

At this point it is of interest to give a brief description of the development of atomic energy starting with a story, told again by Gamow, concerning how the field of nuclear fission, which led to the development of the atomic bomb, got started. On January 27, 1939, Niels Bohr was a visiting dignitary at a small conference on theoretical physics at Washington, DC, jointly sponsored by the George Washington University and the Carnegie–Mellon Institute. On that day Bohr mentioned that he received a letter from a German physicist, Lisa Meitner, who at that time (because of Hitler) was working in Stockholm. She mentioned that she got word from her former colleague, Otto Hahn in Berlin, telling her that he and his assistant, Fritz Strassman, bombarded uranium with neutrons and found the presence of barium. Her nephew Otto Frisch, a physicist who was with her, thought that it might be the result of fission wherein the uranium nucleus splits upon being bombarded by neutrons. As soon as Bohr imparted this information to the participants at the conference there was a sudden shift in the discussion, from the rather unexciting conference to a heated argument, as to whether the fission of the uranium nucleus could possibly lead to large-scale liberation of nuclear energy. Gamow was at that conference and he tells how Enrico Fermi went to the blackboard and wrote formulas pertaining to the fission process. He goes on to say that a correspondent from a Washington newspaper suddenly woke up and furiously began taking notes, but Merle Tuve, a nuclear physicist from the Carnegie–Mellon Institute, quickly ushered him out with the excuse that the discussion was too technical for him. This was probably the first application of security regulations which thereafter monitored atomic energy developments. However, an article about this development got into the newspaper and Robert Oppenheimer called Gamow wanting to know what it was all about. In any case, Bohr published an article on the theory of nuclear fission in the September, 1939, issue of *The Physical Review*. This was the first and last article published on that subject before the AEC† brought down the security curtain.

Fermi–Dirac and Bose–Einstein Statistics

We now explain briefly these types of statistics which are involved in the statistical mechanical interpretation of matter from a quantum mechanics point of view. In general, statistical mechanics is concerned with the statistical

---

† Atomic Energy Commission.

behavior of an ensemble of particles (molecules, atoms, protons, neutrons, etc.) in phase space, $(\mathbf{q}, \mathbf{p})$ space. This space is divided up into n cells of equal volume. In classical statistical mechanics this ensemble is described by *Maxwell–Boltzmann statistics*. Simply stated, this statistics allows us to calculate the probability of putting r indistinguishable particles into the n above-mentioned cells in phase space. There are n^r possible placements of equal probability. The probability that r_1 particles can be placed in cell number $1, \ldots, r_n$ particles in cell number n, is given by

$$\left(\frac{r!}{r_1! \, r_2! \ldots r_n!} \right) n^{-r}.$$

However, quantum mechanics tells us that the Maxwell–Boltzmann statistics is wrong. It was this formulation of statistical mechanics that Planck rejected in investigating black-body radiation, since his assumption of quantized energy levels negated the Maxwell–Boltzmann statistics. The *Fermi–Dirac* statistics, which is designed to take into account quantized energy states, is based on the following hypotheses:

(1) It is impossible for two or more particles to be in the same cell in phase space. This requires that $r \leq n$.
(2) All distinguishable arrangements satisfying the first hypothesis have equal probabilities.

An arrangement of particles in phase space is then completely described by stating which of the n cells contain a particle. Since there are r particles the corresponding cell can be chosen in $\binom{n}{r} = \dfrac{n!}{(n-r)! \, r!}$ ways which is the number of possible arrangements, each having probability $\binom{n}{r}^{-1}$.

Bose–Einstein statistics also discusses the probability of putting r particles into n cells in phase space. But in this statistics only distinguishable arrangements in the cells are considered, rather than at most one particle in a cell. If r_i is the number of particles in the ith cell, so that

$$\sum_{i=1}^{n} r_i = r,$$

where r is the total number of particles, then the number of distinguishable distributions (the number of different solutions of the above equation) is given by $A_{r,n}$ where

$$A_{r,n} = \binom{n+r-1}{r}.$$

Nuclear Particles

We now give a brief survey of the current state of nuclear physics by describing the zoo of nuclear particles. Experiments with particle accelerators have led to the classification of nuclear particles into the following two categories:

(1) *fermions*, which make up matter; and
(2) *bosons*, which are the carriers of force between nuclear particles.

Fermions obey Fermi–Dirac statistics, while bosons obey Bose–Einstein statistics.

Fermions, which are the nuclear particles, are further divided into two subclasses: *quarks* and *leptons*. The word "quark"† comes from a passage in *Finnegan's Wake* by James Joyce, "Three quarks for Muster Mark!". Why this curious term is used in nuclear physics is not known to the author. Quarks come in at least six "flavors": up, down, strange, charmed, bottom, and top. Each flavor comes in three "colors": red, blue, green. Actually, quarks are not colored since they are much smaller than the wave length of visible light. A proton contains two up and one down quark. A neutron contains two down and one up quark. Particles made up of other quarks can be created; but they all have a much greater mass and rapidly decay into protons and neutrons.

The word "lepton" comes from the Greek "leptos" meaning small particle. Leptons may carry an electric charge. For example, an electron is a lepton. Some leptons are uncharged. A neutrino is a lepton. A neutrino has no discoverable mass and can interact with practically nothing. Each nuclear particle has an *antiparticle* with the same mass and lifetime and opposite electrical properties.

The interactions among these various particles are governed by four fundamental forces, each of which is carried by a separate set of bosons. These forces are: *electromagnetic, gravitational, strong nuclear, weak nuclear.*

The electromagnetic force interacts with electrically charged particles such as electrons and charged quarks, but clearly does not interact with uncharged particles such as neutrons and gravitons (which are associated with the gravitational force). The electromagnetic force is much stronger than the gravitational force between atomic particles. For example, the electromagnetic repulsive force between two electrons is 10^{42} times as strong as the gravitational force of attraction. The electromagnetic attraction of the electrons with the positively charged nucleus causes them to orbit the nucleus, just as the gravitational attraction causes the earth to orbit the sun. But the mechanism is different. The electromagnetic attraction of the electrons to the nucleus is pictured as being caused by the exchange of photons. When an electron changes from one quantized or allowed orbit to another one nearer to the nucleus, energy is released and a photon is emitted (which can be detected). Conversely, when an electron absorbs a photon it goes to an orbit further away from the nucleus (this can also be detected experimentally). A photon, which is a quantum of light, obeys Bose–Einstein statistics and is therefore a boson and has a spin 1. It travels with the speed of light and is

---

† Named by the California Institute of Technology physicist Murray Gell-Mann, who won the Nobel price in 1969 for his work in nuclear physics.

therefore massless (if it had mass it would be infinite according to Einstein's special relativity theory).

The gravitational force is universal in the sense that every particle is affected by it since it has mass. Gravity is the weakest force when we deal with the atom. The gravitational force is always attractive and is carried by the *graviton* which is a boson having spin 2.† It has not been discovered as yet. The gravitational attraction between any two particles is due to the virtual exchange of gravitons between the particles. Associated with the gravitons are gravity waves, which have not yet been discovered.

The strong force is the nuclear force which keeps the nucleus stable. It is carried by eight *gluons* which are bosons, also having a spin 1. These gluons interact with themselves and with the quarks. The strong nuclear force has the property of *confinement* meaning that it always binds the quarks together into combinations that have no color (white). This means that quarks of three different colors must combine (red + blue + green = white) to give a triplet, which is a proton or a neutron. Another possible set of combinations consists of three pairs each of which is a quark and its corresponding *antiquark*. An antiquark has an "anticolor". This means we can have red + antired, blue + antiblue, or green + antigreen, all of which equals white. Such combinations make up *mesons*. A quark and its antiquark combine to annihilate each other, producing electrons and other particles. Therefore a meson is unstable since it decays into other particles. Gluons also have color. Confinement prevents one having a single gluon. Instead, we must have a triplet of gluons of different colors adding up to white. This is called a *glueball*. Experiments with high-energy accelerators show that, at high energies the strong nuclear force becomes weaker so that the quarks and gluons behave almost like free particles. This is the phenomenon of "asymptotic freedom".

The weak nuclear force is responsible for radioactivity. It acts on all particles having a spin $\frac{1}{2}$. It does not act on particles of other spin quantum numbers such as photons and gravitons. This force was not clearly understood until 1967, when Abdus Salam at the Imperial College in London and Steven Weinberg at Harvard proposed theories that unified the weak nuclear force with the electromagnetic force, analogous to the unification of electricity and magnetism done by Maxwell about one hundred years earlier. The Weinberg–Salam theory postulates that, in addition to the photon, there are three other spin 1 particles known as *massive vector bosons* that carried the weak nuclear force. These are called W^+, W^-, and Z^0, and each has a mass of about 100 GeV (100 billion eV). This theory has the property of *spontaneous symmetry breaking*. This means that the nuclear particles that appear in different states at low-energy levels coalesce into a single state if the energy level is high enough (much greater than 100 GeV. In 1979 Salam and Weinberg were awarded the Nobel prize in physics for this work, together with Sheldon Glashow, of Harvard, who posed a similar theory which unified the electro-

---

† See below, under "Spin".

magnetic and weak nuclear forces. In 1983 these three massive vector bosons were discovered by investigators at CERN (European Center for Nuclear Research). The leader of the group of physicists, Carlo Rubbia, together with Simon van der Meer, the CERN engineer, received the Nobel prize for this experimental work in 1984.

The current state of knowledge tells us that all interactions in the universe can be reduced to combinations of these four forces.

Spin

It was mentioned above that some particles have spin $\frac{1}{2}$, gluons have spin 1, gravitons have spin 2, etc. What does this mean? We will give a semi-quantitative discussion of spin, eliminating many technical details. We first consider an electron which can be thought of as a tiny spinning permanent magnet. It thus has a magnetic moment which can be oriented in certain allowable directions according to the rules of quantum mechanics. Each spin axis can be oriented either parallel to or opposite to an arbitrary direction in space. The spin has a quantized angular momentum vector which depends on the quantum number $l = \frac{1}{2}$ (l is called the "azimuthal" quantum number), and has two possible orientations given by the 'magnetic" quantum number $m = \pm\frac{1}{2}$,† parallel or opposite to the spin axis. In a sense, the spin may be considered to be a coordinate. Four quantities are needed to describe the position of an electron: the three coordinates of its center of gravity and its spin. It turns out that the wave function (solution to the Schrödinger equation) for a single electron depends on these four coordinates. Since there are two possible values for the spin, this means that there are two possible values for the wave function which characterizes a single electron. Of course, the appropriate solutions of the Schrödinger equation must, as mentioned above, satisfy Pauli's exclusion principle.

The spin of a particle is related to the idea of symmetry in the sense that the spin tells us what the particle looks like from different directions. A particle of spin 0 is like a tiny sphere looking the same from every direction; it has spherical symmetry. A particle of spin 1 is like an arrow; it must be turned through a complete revolution for it to look the same. A particle of spin 2 is like a double-headed arrow; it looks the same if it is turned through half a revolution. And finally, a particle of spin 1/2 must be rotated through two complete revolutions in order for it to look the same.

All particles of the universe belong to two classes:

(1) Those particles making up the matter of the universe have spin 1/2.
(2) Those particles of spin 0, 1, or 2 which are carriers of the forces between matter.

---

† This was the Uhlenbeck–Goudsmit hypothesis. They introduced the concept of electron spin in 1926.

One of the benefits of Pauli's exclusion principle is that it explains why matter particles do not collapse to a state of very high density under the influence of the force carriers. Without the exclusion principle, instead of having well-structured atoms, the quarks and electrons making up these atoms would all collapse to a highly dense "soup".

Antimatter

In 1928 Paul Dirac supplied the proper understanding of spin $\frac{1}{2}$ particles by a theory that combined quantum mechanics with the special theory of relativity. His theory also predicted that the electron should have a partner which has the same mass and an opposite charge, a *positron*. The positron was discovered in 1931 by an American physicist, Carl Anderson, while he was studying the tracks of high-energy electrons produced in a cloud chamber due to cosmic ray showers. He placed the cloud chamber in a strong magnetic field and photographed the tracks of the electrons. To his surprise, he observed that one-half of the electrons were deflected one way while the other half were deflected in the opposite direction. Thus there was a mixture of half positively charged and half negatively charged electrons, both having the same mass. This discovery of positrons supplied the experimental verification of Dirac's theory. A positron–electron pair can be produced by the impact of high-energy light quanta (cosmic rays or gamma rays) against the nuclei, and the probability of these events occurring coincides exactly with the values calculated on the basis of Dirac's theory. When a positron and an electron collide they annihilate each other. The energy equivalent to their mass (Einstein's equivalence of mass and energy) is liberated in the form of high-energy photons. Dirac was awarded the Nobel prize for physics in 1933. Dirac's theory, which is essentially a relativistic quantum field theory, leads to the mathematical possibility of another "negative" world in which all matter in our world would have *negative mass*. This has not been observed. Maybe it is because we can only perform experiments in this world, and there can be no communication between our world and the antiworld.

Particle Physics and Cosmology

We now briefly discuss the relationship between particle physics and cosmology. Cosmology attacks such problems as the age of the universe and hence deals with billions of years and with astronomical distances, while particle physics deals with reaction times of microseconds and less and with atomic distances. These fields of physics therefore appear to be at opposite ends of the scientific spectrum. However, as pointed out above, the four forces that govern interactions in the nucleus also govern cosmological events. The real marriage between these two seemingly disparate fields lies with the time period supposedly associated with the creation of the universe according to the *big bang theory*. Cosmologists appear to have traced the physics of matter as early as 10^{-43} to 10^{-35} sec after the primordial explosion known as the big bang. Within this time interval a super colossal accelerator appears to have existed far beyond the capacity of the experimental physicist to design. This supplies a conceptual laboratory for the nuclear physicist. The tremendous temperatures and pressures that must have existed produced highly energetic radiation which, upon cooling, became the prototype for the formation of the elements and hence all the matter of the universe. The analysis of the physics that occurred during the big bang, within the framework of the ideal laboratory of this super accelerator, is within the scope of the particle physicist.

David N. Schramm and Gary Steigman[†] have been pioneers in developing the interface between cosmology and particle physics. In their article they claim that important clues to the nature of physics during the time interval of the big bang may come from a current theory known as "the ultimate theory of everything" (the TOE theory). This theory, which is yet to be developed, would describe all the interactions among the fundamental particles in a single stroke, and a single *super force* would describe the interactions of all the particles of the universe. It would appear that the TOE theory would fulfill the dream of Einstein of a unified field theory. However, Einstein's theory does not involve quantum mechanics.

Modern studies in particle physics show that at very high temperatures the four forces controlling all reactions begin to unify. This was shown in various laboratories. For example, at CERN it was shown that the weak and electromagnetic forces merge into a single *electroweak* force at energies greater than 100 GeV. This enormous energy corresponds to the temperature of the universe about 10^{-10} sec after the big bang (more than 10^{15} times room temperature). The experiments at CERN give promise of a "grand unified field theory" (GUT) wherein the strong force will merge with the electroweak force at roughly 10^{15} GeV, and by 10^{19} GeV the force of gravity will join this super force to yield a TOE.

The authors of the above-mentioned article state that their analysis of the nuclear reactions that occurred when the universe was about one second old leads to the prediction that the number of fundamental types of elementary particles must be small. Their studies involved *primordial nucleosynthesis* which is the process of attempting to determine how the elements were formed from the big band. Temperatures nearly 100 million times as great as room temperature are needed to forge the elements from elementary particles such as neutrons and protons. Such temperatures supposedly occurred about one second after the big bang (when the universe is said to have been created). They claim that, by measuring the relative abundance of the elements, they can probe conditions as far back as one second after the big bang. The current microwave background radiation also serves as a probe of the universe, but only back to about 10^5 years after the big bang. At that time temperatures

---

[†] Article entitled, "Particle Accelerators Test Cosmological Theory", by David N. Schramm and Gary Steigman, in *Scientific American*, June, 1988.

were only about 3,000 K. The theory of nucleosynthesis has predicted the abundance of several light elements and their isotopes. They used as a model a gas of neutrons and protons (called a *nucleon gas*, since neutrons and protons are called *nucleons*) in an environment simulating an expanding and cooling universe. At less than a microsecond after the big band when temperatures were greater than 10 billion degrees Kelvin, protons and neutrons were in equilibrium in roughly equal numbers. As the temperature fell to 10 billion degrees the mass difference between neutrons and protons became more significant. Since a neutron is slightly heavier than a proton, as the temperature decreased more neutrons changed into protons (the ratio of neutrons to protons decreased with decreasing temperature). When the average temperature of the universe dropped to a billion degrees this ratio was low enough to allow for protons and neutrons to begin to fuse into the nucleus of deuterium consisting of a neutron and a proton. Deuterium is an isotope of hydrogen. But at these temperatures electrons were stripped away leaving the ionic form (just the nucleus). Then, as the temperature continued to drop, deuterium reacted with a neutron to produce tritium (one proton and two neutrons). Then helium-3 (two protons and one neutron) was produced. Then helium-4 (two protons and two neutrons) was produced with a small amount of beryllium-7 (four protons and two neutrons). The reactions essentially stopped producing more complex nuclei. The majority of the other elements were produced inside stars which have conditions allowing three helium-4 nuclei to combine to produce carbon-12. The authors go on to argue that the theory of big bang nucleosynthesis sets limits on the allowed number of families of elementary particles. Their argument is that, if the number of families of nuclear particles exceeded three of four, the predicted abundance of helium-4 would exceed the observed abundance. The details of their argument are given in the above-mentioned article, which should be of great interest to those readers interested in this aspect of the creation of matter.

Mention was made of the big bang theory of the formation of the universe. The other rival theory of cosmology was the *steady-state* theory developed by in 1948 by Hermann Bondi, Thomas Gold and Fred Hoyle. It says that matter is continuously being created (creation of protons), so that as the universe expands its density remains constant. As previously mentioned, the big bang theory asserts that the universe was once very hot and enormously dense. It then expanded, cooled and became less dense.

These theories attempted to account for the fundamental discovery of Edwin P. Hubble in 1929 that the universe is expanding according to the law that galaxies are receding from each other at a speed proportional to their distance from each other, *Hubble's law*. This discovery, that the universe is expanding, was one of the great revolutions of science. Looking back (with 20–20 vision!) it is difficult to know why nobody thought of this before. Clearly Newton, Lagrange, and others should have realized that a nonexpanding universe would eventually start to collapse under the gravitational attraction of the galaxies. If the universe were expanding at a slow rate (the galaxies receding from each other) then the force of gravity would tend to make the universe stop expanding and eventually contract. If it were expanding at a great rate then the force of gravity would never allow it to contract and it would expand forever. Therefore, there must be a *critical rate of expansion* such that above this rate it would expand forever and below this rate it would eventually contract. This critical rate is similar to what occurs when a rocket is fired away from the earth's surface. A low initial velocity will cause the rocket to come back to earth. There is a critical velocity (about 7 miles/sec), above which it will escape from the earth's gravitational attraction. This concept of a static universe was ingrained in the minds of physicists working in this field. Even Einstein, when he developed his general theory of relativity in 1915, used a static model of the universe. He used a rather artificial idea that space–time had an inherent tendency to expand, and this could be made to balance exactly the attraction of the galaxies, thus resulting in a static universe. However, the Russian mathematical physicist, Alexander Friedmann (who was a teacher of George Gamow) took relativity theory at its face value by avoiding the static model of the universe. Friedmann assumed the principle of an isotropic universe both locally and globally. This means that the universe looks the same in any direction at a given observation point, and this statement would be true at any observation point in the universe. Friedmann then predicted an expanding universe from these two principles. This was in 1922, several years before Hubble discovered that the universe was expanding from his analysis of the red shift of stars.

The following argument supports the big bang over the steady-state theory: According to the steady-state theory the density of the universe is constant in time, so that it has always been what it is today. As the galaxies move away from each other (the expanding universe principle) new galaxies would continuously be formed from the continual creation of protons. However, the rate of formation of protons was very low, about one proton per cubic kilometer. This did not conflict with experiment and only involved a slight modification of general relativity theory. It followed that the universe never existed in a dense hot state, and hence there would be no residual thermal radiation. On the other hand, the big band theory holds that the universe was once sufficiently so dense and hot that the matter at the time during the big bang process could only exist as radiant energy. Therefore the universe would generate a characteristic spectrum of thermal radiation whose residual effect should be seen today. Indeed, this was verified experimentally by Arno Penzias and Robert W. Wilson, of Bell Telephone Laboratories, who were awarded the Nobel prize in 1978 for their work. They discovered a microwave background radiation that is consistent with the prediction of the big bang theory.

The big bang theory, as with all cosmological concepts, can only be understood in the context of Einstein's space–time continuum. Since we cannot visualize this four-dimensional space, we shall give a demonstration in a three-dimensional space–time setting of the big bang theory. As our model we take a partially blown-up balloon whose surface is a good approximation to a spherical surface. We paint dots uniformly distributed on this surface. Each dot represents a galaxy, so that the surface represents the universe in two spatial dimensions at a particular time. If we further blow up the balloon at a constant rate the galaxies will move away from each other at a constant rate in accordance with Hubble's principle. This is a two-dimensional model of the expanding universe. At each instant the distribution of dots represents the state of the universe at that time. To represent the big bang phenomenon we let air out of the balloon in such a manner that it collapses at a constant rate. This reverses time so that the universe is getting younger and the galaxies are moving toward each other at a constant rate. As the balloon continues to collapse its surface gets smaller and smaller and the density distribution of the galaxies gets larger and larger; also the curvature of the surface gets larger and larger. Finally, the balloon collapses into a "point" where the sphere has zero radius, infinite curvature, and zero surface area so that the density of the galaxies is infinite. Clearly this is a singularity, since physically a point has zero dimensions while the balloon's surface always has two dimensions however small the surface. This singularity represents the beginning of the universe and the time at which it occurs, the beginning of time. The big bang starts at this singularity when all matter of the universe is concentrated at the "singular point" so that it has infinite density (occupying zero surface area). All of Friedmann's solutions to his variation of general relativity theory have the property that between ten and twenty billion years ago the distance between neighboring galaxies must have been zero thus giving the above-mentioned singularity which is the start of the big bang process. This is a singularity in another sense: That point in time corresponding to the singularity, which we call the creation of the universe, separates events before from events after the big bang. We cannot "cross the divide" in an attempt to determine what happened before the big bang from our understanding of events after the big bang, and conversely.

In the neighborhood of this singularity (time interval up to about 10^{-40} sec after the singularity) the physics must be different so that general relativity theory, or its variations, cannot be used. Stephen W. Hawking, in his book, [24], describes his pioneering work in attempting to understand the physics of this time interval.

Hawking is a remarkable person. He suffers from Lou Gehrig's disease, is confined to a wheelchair, and uses a speech synthesizer as a result of a tracheostomy operation which removed his ability to speak. He once jokingly remarked that it has one fault: "It gave me an American accent!" Anyhow, his affliction has not prevented him from becoming a world class physicist (some authorities say, the most brilliant theoretical physicist since Einstein), having made outstanding contributions to the field of cosmology. His courage, in the face of this debilitating disease, which allows him to focus his brilliant and pentrating mind on problems in cosmology of first-rate importance, has earned him, among other honors, the prestigious chair of Lucasian Professor of Mathematics at Cambridge University, a post once held by Newton and Dirac. As a serendipitous footnote: he was born on the anniversary of the death of Galileo.

Returning to the singularity mentioned above. Einstein's general theory of relativity predicted that Minkowski space† began at the big bang singularity and would come to an end at the *big crunch* where the entire universe would collapse to a singularity, or locally where a star would collapse to a black hole causing a singularity to occur inside it. These results were finally proved by Hawking and Penrose. Any matter that fell into the black hole would be destroyed at the singularity. Hawking and his coworkers investigated the nature of this singularity and attempted to devise a theory that would obviate it. In 1965 Roger Penrose, an English mathematical physicist, investigated the nature of a collapsing star. Using the nature of light cones‡ in general relativity he showed that a star which collapses under its own gravitational field is trapped in a region whose surface shrinks to zero size, thus yielding a singularity. This means that all the matter in the star would be contained in a region of zero volume thus producing a condition of infinite density which is a singularity. This is a black hole. (A general description of a black hole will be given below.) In general, Penrose's theorem states that any body undergoing gravitational collapse must form a singularity. Hawking [24] points out that, at the time Penrose produced his theorem, he was a graduate student looking for a problem with which he could complete his Ph.D. thesis. He had been diagnosed as suffering from Lou Gehrig's disease and was given two years to live so he didn't have the heart to work on his doctorate. However, two years passed and he didn't become much worse. Besides he was engaged to "a very nice girl", so he needed the Ph.D. to get a job! He then read Penrose's theorem, expanded on it and obtained his doctorate. He and Penrose then wrote a joint paper in 1970, wherein they expanded their ideas to the nature of the big bang. They finally proved that there must have been a big bang singularity within the framework of relativity theory. Hawking says that later he changed his mind, and he is now attempting to convince other physicists that there is no singularity, provided that quantum effects are taken into account.

In the broader sense, this is an attempt to design a quantum theory of gravitation, which has not yet been achieved. Quantum theory is concerned with a study of particles of atomic dimensions and is a probabilistic or stochastic science which has as its backbone the uncertainty principle. On the other hand, the gravitational fields produced by stars are best attacked by general relativity theory which is deterministic, meaning that it does not consider the stochastic methods that must occur in quantum mechanics. In

---

† See Chapter 8 for an explanation of Minkowski's four-dimensional space–time used in both the special and general theory of relativity.

‡ See Chapter 8 for a description of light cones in the setting of the special theory of relativity.

fact, Einstein, who was essentially the founder of quantum mechanics, in his latter days, rejected the principles of quantum mechanics. As he said: "I refuse to believe that God plays dice with the universe". Therefore, we have two theories which are inconsistent with each other. For the micro domain of particle physics, gravitational effect may be neglected so that quantum mechanics holds away. For problems of cosmological scope where quantum effects may be neglected the general theory of relativity is very successful. However, the time interval during the big bang is a shadowy area where, as Hawking points out, quantum mechanics must be used in conjunction with relativity theory to get a more realistic picture of the physics of a very highly dense high-energy plasma gas. This essentially means constructing a quantum theory of gravity which would wed quantum mechanics and general relativity theory into one grand theory encompassing the "micro" domain of particles and the "mega" domain of cosmology.

At this point we digress to give a somewhat historical description of black holes. The term *black hole* was invented in 1969 by the American physicist John Wheeler to describe a phenomenon which John Michell, a Cambridge don, wrote about in the *Philosophical Transactions of the Royal Society of London* in 1783. Michell pointed out that a sufficiently massive and dense star could have a strong enough gravitational field to prevent light from escaping and thus could not be seen, although its gravitational attraction could be felt. These are black holes. Laplace made a similar suggestion (apparently independent of Michell) which he incorporated in the first two editions of his book *The System of the World*. However, he left it out of the later editions, apparently not believing that light could be affected by a gravitational field, according to the wave theory of light. The first successful determination of the speed of light was reported by the Danish astronomer Olaus Roemer in 1675, when he announced the calculation of the speed of light from observations of the irregularities in the time between successive eclipses of the innermost moon of Jupiter by that planet. Before this, it was thought that light traveled with infinite speed. A consistent theory of how gravity affects light did not come along until Einstein proposed his general relativity theory in 1915.

In order to understand how a black hole is formed it is first necessary to see how a star is born. A star is formed when a large amount of hydrogen gas collapses on itself due to its gravitational attraction. As time goes on the gravitational attraction makes the collection of gas molecules more and more dense, and hotter. Eventually, the temperature is great enough to start a controlled thermonuclear reaction and helium is formed. The heat released in this reaction (equivalence of mass and energy) gives off light thus making the star shine. The pressure of the star is also increased until it is sufficient to balance the gravitational attraction, and the gas therefore stops contracting. Stars will remain stable like this for a long time. For example, our sun has got probably enough nuclear fuel to last for another five billion years or so.

When a star runs out of fuel it starts to contract. Then several things may happens. The various possibilities were worked out by the Indian mathematical physicist Subrahmanyan Chandrasekhar, who studied with Sir Arthur Eddington, an authority on general relativity theory. As an aside: Once in the 1920s a journalist told Eddington that he heard there were only three people who understood general relativity theory. Eddington paused and then said: "I am trying to think who the third person is!" Chandrasekhar realized that when a start becomes smaller the molecules get very near to each other, and by the Pauli exclusion principle they must have very different velocities giving tremendous temperature gradients. The result is that the star expands until an equilibrium condition occurs which gives a balance between the gravitational attraction and the force of repulsion that arises from Pauli's principle. The constraint on the maximum velocity of the star particles is, of course, the velocity of light (according to relativity theory). This gives a limit to the expansion of the star due to the exclusion principle. On the other hand, when a star gets sufficiently dense this expansion process is overbalanced by the gravitational attraction, and the star collapses into a black hole. In 1931 Chandrasekhar made the momentous discovery that white-dwarfs had a maximum mass equal to about 1.4 times the mass of the sun, the exact value depended on the composition of the star. In 1934 he calculated the maximum possible mass a stable cold star must have above which it would collapse into a black hole. This mass is called the *Chandrasekhar limit*, and is the same as the above-mentioned maximum mass of a white-dwarf. He wrote: "The life history of a star of small mass must be essentially different from the life history of a star of large mass. For a star of small mass the natural *white-dwarf* stage is an initial step toward initial extinction (conversion to a black hole). A star of large mass cannot pass into the white-dwarf stage and one is left speculating on the other possibilities." A similar discovery was independently made by L.D. Landau in 1932. Chandrasekhar received much hostility from other astronomers, especially Eddington who thought it was not possible for a star to collapse to a singularity. Because of this attitude Chandrasekhar abandoned this speciality and turned to the study of star clusters. However, he was awarded the Nobel prize in 1983 partly for his early work on the limiting mass of cold stars.

Robert Oppenheimer in 1939 was the first one to solve the problem of understanding what would happen to a white-dwarf according to general relativity theory. His work lay dormant in the 1940s and 1950s due to the pressure of his work on the atomic bomb. But in the 1960s his work was rediscovered and extended by others. We get the following picture from Oppenheimer's work: According to general relativity theory the gravitational field of the white-dwarf changes the paths of the light rays in Minkowski space. The light cones are bent inward near the surface of the star. As the star contracts, the gravitational field at its surface gets stronger and the light cones are bent inward more which makes it more difficult for light to escape. Thus the light appears dimmer to the observer and there is a corresponding red shift. Eventually, when the star has shrunk to a very small critical volume, the gravitational field at the surface becomes so strong that no light can escape to be observed. This gives a set of events in Minkowski space from which it is not possible to reach a distant observer. The boundary of this region is called the *event horizon*—it coincides with the paths of the light rays that just fail to escape from the black hole, which is thus formed.

Hawking† proved an important theorem about black holes which has far-reaching consequences: In any interaction the surface area of a black hole can never decrease. This theorem arose as a consequence of defining the event horizon as the boundary of the black hole, which has the property that the light cones just fail to get away from the black hole. He proved that the paths of the light rays could never approach one another, so that these paths had always to be moving parallel to or away from each other. Thus the event horizon or boundary of the black hole can never have its surface area decrease. Following up this discovery a research student at Princeton, named Jacob Beckenstein,‡ suggested that the area of the event horizon was a measure of the entropy of the black hole. As matter carrying entropy fell into the black hole, this area would increase so that the sum of the entropies of the matter outside the black hole and the area of the event horizon would never decrease.

Hawking [24] goes on to say that in 1973, while visiting Russia, he discussed black holes with two leading Soviet experts, Yakov Zeldovich and Alexander Starobinsky, both of whom convinced him that rotating black holes should create and emit particles, according to the uncertainty principle. Hawking devised a better mathematical treatment, and, to his surprise he found that even a nonrotating black hole would create and emit particles just as if it were a black body. It was previously thought that nothing can escape from within the event horizon of a black hole. But, Hawking points out, quantum mechanics tells us that the particles do not come from within the black hole, but from the "empty" space just outside the event horizon. The way this happens is that the electromagnetic and gravitational fields in this space, just outside the black hole, are not quite identically zero, since Heisenberg's uncertainty principle implies that the more accurately we know the position of a particle that may exist (which must cause the fields) the less accurately we would know its momentum. Therefore, the supposition of exactly zero particles in this region (same as zero fields) is a deterministic statement and thus violates the uncertainty principle. This means that there must be a minimum amount of quantum fluctuations in the values of these fields. We may consider these fluctuations as pairs of particles or photons that appear,

---

† See Hawking, S.W. and G.F.R. Ellis, *The Large Scale Structure of Space–Time*, Cambridge University Press, 1973. Other articles by Hawking that are of interest to the reader who desires a more detailed discussion of this topic are: Hawking, S.W., "Gravitational Collapsed Objects of Very Low Mass", *Mon. Not. Roy. Astron. Soc.*, **152** (1971), 75. Ibid., "Black Hole Explosions?", *Nature*, **248** (1974), 30. Ibid., "Particle Creation by Black Holes", *Commun. Math. Phys.*, **43** (1975), 199.

‡ See Beckenstein, J.D., "Black Holes and Entropy", *Phys. Rev.* D **7**, 2333 (1973). Also, ibid "Statistical Black Hole Thermodynamics", *Phys. Rev.*, D **12** (1975), 3077.

come together, and then annihilate each other. These are *virtual particles* and cannot be directly observed with a particle detector.

Continuing with Hawking's discussion of the role of quantum mechanics, in allowing for radiation from a black hole: We know that one of the partners in a particle–antiparticle pair will have positive energy, and the other, negative energy. The negative energy particle is a short-lived and clearly a virtual particle. It wants to seek out its positive energy partner and annihilate it. It is possible for a virtual particle (either particle or antiparticle) to fall into the black hole, whereupon it becomes a real particle (or antiparticle). Then its forsaken partner outside the black hole may also fall into the black hole. Or, if it has enough positive energy, it will escape from the vicinity of the black hole. To a distant observer it will appear to have been emitted from the black hole. The smaller the black hole, the shorter the distance the virtual particle (negative energy) will have to go before it becomes a real particle, and thus the greater the emission, raising the apparent temperature of the black hole. The positive energy of the emitted radiation would be balanced by the flow of negative energy particles into the black hole, according to Einstein's principle of this equivalence of mass and energy given by $E = mc^2$. Therefore, the flow of negative energy into the black hole reduces its mass, thus decreasing the area of the event horizon and decreasing the entropy. But this decrease in entropy of the black hole is more than compensated for by the entropy of the emitted radiation, so that the second law of thermodynamics is never violated.

We have given the reader just a taste of the fascinating field of cosmology in its relation to quantum mechanics vis-à-vis the study of black holes. We invite the reader to discover for himself or herself this fascinating field of cosmology as described so eloquently in Hawking's book. The references in the footnotes will give the reader a more detailed technical discussion of this subject.

CHAPTER 7

Plasma Physics and Magnetohydrodynamics

Introduction

In this chapter we take up wave propagation in an electromagnetically conducting fluid which we call a *plasma*. Conduction occurs when there are free
electrons or other charged particles (ions) that can move under the action of
applied forces. We need a fluid (liquid or gas) for the mass motion of these
charged particles, since in a solid the electrons are bound and thus suffer a
smaller motion on a macro scale so that there is practically no mass motion.
However, in a solid the electrons can move on an atomic scale within the
lattice structure; thus there are dynamic effects, such as the Hall effect, which
are observed when external fields are applied. These fields cause stresses to
occur in the lattice structure. We do not consider the effect of external fields
on the electromagnetic properties of solids, but we concentrate on fluids or
plasmas. For a plasma, the effect of an external field on electrons and other
charged particles is to produce bulk motion of the fluid and other dynamic
effects associated with the reaction of plasmas to oscillating electromagnetic
fields. We are therefore concerned with the fluid dynamics of a fluid medium
containing charged particles. The equations of fluid dynamics must contain the electromagnetic properties of the fluid as expressed by Maxwell's
equations.

The propagation of electromagnetic waves in plasma is divided into two
domains: *plasma physics* and *magnetohydrodynamics*. To understand this division it is first necessary to see how electrons and ions collide in a plasma under
the action of an external electromagnetic field. In a simple model the electrons
and ions are accelerated by the applied oscillating fields, but change direction
as a result of the collisions. Thus their motion in the direction of the external
field is retarded by an effective frictional force equal to $vm\mathbf{v}$ where v is the
collision frequency. Ohm's law represents the balance between the applied
force and the frictional drag. Let ω be the frequency of the applied field. When
ω is comparable to v the charged particles have time to accelerate and
decelerate between collisions, thus causing inertial effects to take over so that

the conductivity becomes complicated. Also, if these frequencies are comparable, the description of the collision process in terms of an effective friction force tends to become invalid. If $\omega \gg \nu$ then there is the phenomenon of *charge separation* which means that the positive ions and electrons are accelerated in opposite directions and thus tend to separate. As a result of this charge separation, strong electrostatic restoring forces are set up and characteristic high-frequency oscillations are produced. These are called *plasma oscillations*. This is the domain of plasma physics. The domain in which we have magnetohydrodynamic electromagnetic waves occurs at the other end of the spectrum, where $\omega \ll \nu$. These low-frequency oscillations do not involve charge separation, just the motion of the fluid.

Propagation of electromagnetic waves in plasmas is encountered in a number of cases. We mention a few:

(1) Propagation of radio waves in the higher layers of the earth's atmosphere (ionosphere).
(2) Propagation of low-frequency magnetohydrodynamic waves in space (the ionosphere and regions of interplanetary space).
(3) Propagation of radio waves of cosmic origin in the solar atmosphere, in nebulae, and interstellar space, all of which are investigated by the methods of radio astronomy. This includes radio communication with artificial satellites, etc.
(4) Propagation of plasma waves in space, the solar corona, and the ionosphere.
(5) Propagation of various types of electromagnetic waves in plasmas created in the laboratory, such as controlled nuclear reactions. Plasma physics plays a key role in nuclear fission and fussion processes.

7.1. Fluid Dynamics Equations—General Treatment

In [15, Chap. 7], we treated, in some detail, the fluid dynamics equations both for an inviscid as well as a viscous fluid, from the point of view of wave propagation in a nonconducting medium. Here we briefly review this treatment in preparation for the study of wave propagation in a plasma.

The equations of fluid dynamics are called the *field equations*. In formulating these field equations we use the *three conservation laws of physics*. They are: *conservation of mass* which yields the *continuity equation*, *conservation of momentum* which gives *Euler's equations of motion*, and *conservation of energy* from which we obtain the *equation of state* which defines the medium. The equations that arise from the conservation laws are the field equations for the fluid. The solution of these equations yields the state of the medium with respect to the applied forces and the boundary and initial conditions. This state is completely defined when we know the dynamic variables (particle velocity, stress, etc.) and the thermodynamic variables (temperature, entropy, etc.) associated with each particle of the medium as a function of space and time.

There are two ways of representing the state of each particle: The Lagrange and Euler representations. (Historically speaking, both of them are essentially due to Euler.) In the *Lagrange representation*, coordinates are assigned to each particle, and are called the Lagrange coordinates which are the independent variables. The dependent variables (dynamic and thermodynamic) are functions of the Lagrange coordinates and time. This representation is useful in describing the dynamics of solids since it is more appropriate for small particle motion. The *Euler representation* is more useful for the dynamics of fluids since this involves bulk motion or large particle velocities. The independent variables are called the *Euler coordinates*. Consider an inertial reference system (one not accelerating or rotating with respect to an arbitrary reference system), and consider a moving fluid embedded in this coordinate system. If an observer stations himself at a given point (field point) in this reference frame he can observe the time-varying state of the fluid particles as they flow past him. Each particle is identified by its Lagrangian coordinates and its dynamic and thermodynamic variables. The set of all such field points is called the *Eulerian coordinates*. As mentioned above, in studying the flow of fluids, we use the Eulerian representation. Another point discussed in [15] is that the difference between the Lagrangian and Eulerian representations occurs only in the nonlinear terms of the field equations. If we consider small amplitude particle motion then we can linearize the field equations so that there is no difference between the Lagrangian and Eulerian representations.

We now discuss the general field equations for a fluid under the action of an arbitrary system of external forces as a basis for applications to magneto-hydrodynamics and plasma physics. Let ρ be the density; v, the particle velocity; p, the pressure; and F the resultant of the external forces acting on the fluid other than the pressure force which is given by $\mathbf{grad}\ p$.

The continuity equation is

$$\rho_t + \operatorname{div} \rho\mathbf{v} = 0. \tag{7.1}$$

Let (u, v, w) be the (x, y, z) components of v. Expanding div ρv:

$$\operatorname{div} \rho\mathbf{v} = (\rho u)_x + (\rho v)_y + (\rho w)_z$$

$$= \operatorname{div} \mathbf{v} + \mathbf{v} \cdot \mathbf{grad}\ \rho,$$

where $\mathbf{grad}\ \rho = \mathbf{i}\rho_x + \mathbf{j}\rho_y + \mathbf{k}\rho_z$. $(\mathbf{i}, \mathbf{j}, \mathbf{k})$ are unit vectors in the (x, y, z) directions. Then (7.1) can be written as

$$\rho_t + \operatorname{div} \rho\mathbf{v} + \mathbf{v} \cdot \mathbf{grad}\ \rho = 0 \tag{7.1'}$$

The conservation of momentum leads to Euler's equations of motion, which we write in the following vector form:

$$\rho \frac{d\mathbf{v}}{dt} = -\mathbf{grad}\ p + \mathbf{F}, \tag{7.2}$$

where $\mathbf{F} = \mathbf{i}X + \mathbf{j}Y + \mathbf{k}Z$. (X, Y, Z) are the (x, y, z) components of the external force not derivable from the potential p.

Kinematics. We can discuss the physical meaning of the particle acceleration $d\mathbf{v}/dt$ that occurs in the left-hand side of (7.2), without recourse to the system of forces that produces it. This is the branch of mechanics called kinematics. Surround any field point P: $(\mathbf{x})$ by a spherical surface of very small radius. As P moves with the fluid the volume enveloped by this surface contains the same fluid particles. Associated with this tiny spherical volume are all the dynamic and thermodynamic variables which are assumed to be constant at a particular time within that volume. Let f be a generic dynamic or thermodynamic variable. Then $f = f(\mathbf{x}, t)$ where $\mathbf{x}$ is the set of Eulerian coordinates in three dimensions (rectangular coordinates) so that f_P depends only on t. Consider a neighboring point $P'(\mathbf{x} + \mathbf{v}\Delta t)$. Clearly, the distance $\overline{PP'}$ equals $\mathbf{v}\Delta t$. We now calculate the rate at which f varies for the moving particle. As $P' \rightarrow P$, $\Delta t \rightarrow dt$ so that $(f_{P'} - f_P)/\Delta t \rightarrow df/dt$. We therefore obtain the following expansion for df/dt at P:

$$\frac{df}{dt} = f_t + uf_x + vf_y + wf_z = f_t + \mathbf{v} \cdot \mathbf{grad}\, f.$$

The differential operator $d/dt = \partial/\partial t + \mathbf{v} \cdot \mathbf{grad}$ is sometimes called the *Stokesian derivative* after Sir G.G. Stokes who was one of the founders of fluid dynamics. It is clear that this operator denotes a differentiation following the motion of the fluid. Setting f equal to $\mathbf{v}$ (f may also stand for a vector quantity) we get the following expansion for the particle acceleration:

$$\frac{d\mathbf{v}}{dt} = \mathbf{v}_t + u\mathbf{v}_x + v\mathbf{v}_y + w\mathbf{v}_z = \mathbf{v}_t + \mathbf{v} \cdot \mathbf{grad}\, \mathbf{v}. \tag{7.3}$$

The term $\mathbf{v}_t$ represents the part of the acceleration of the particle due to the change in time at a fixed point. The term $\mathbf{v} \cdot \mathbf{grad}\, \mathbf{v}$ is called the *convective term*. It represents the portion of the acceleration due to the infinitesimal motion of the particle at a fixed time. We note that $\mathbf{grad}\, \mathbf{v}$ is the gradient of a vector which is a tensor of rank two (we need two indices to describe each component of $\mathbf{grad}\, \mathbf{v}$). The scalar product of this tensor with $\mathbf{v}$ (from the left) yields a vector. Writing out the vector equations (7.3) in component form, we get

$$\frac{du}{dt} = u_t + uu_x + vu_y + wu_z,$$

$$\frac{dv}{dt} = v_t + uv_x + vv_y + wv_z, \tag{7.3'}$$

$$\frac{dw}{dt} = w_t + uw_x + vw_y + ww_z.$$

Inserting (7.3) into (7.2) gives

$$\rho(\mathbf{v}_t + \mathbf{v} \cdot \mathbf{grad}\, \mathbf{v}) = -\mathbf{grad}\, p + \mathbf{F}. \tag{7.4}$$

Written out in extended or component form, we get

$$\rho(u_t + uu_x + vu_y + wu_z) = -p_x + X,$$
$$\rho(v_t + uv_x + vv_y + wv_z) = -p_y + Y, \qquad (7.4')$$
$$\rho(w_t + uw_x + vw_y + ww_z) = -p_z + Z.$$

The energy equation defines the type of fluid under consideration. For an adiabatic gas, where we neglect viscosity, there is no exchange of heat between the fluid and its environment. For this type of gas the energy equation is the *adiabatic equation of state* which gives the following unique relation between the pressure and density:

$$p = A\rho^\gamma, \qquad A = \frac{p_0}{\rho_0^\gamma}, \qquad \gamma = \frac{C_V}{C_p}, \qquad (7.5)$$

where C_V and C_p are the specific heats at constant volume and pressure, and p_0 and ρ_0 are defined at the undisturbed state. For air at room temperature we have $\gamma = 1.4$. Let c_0 be the speed of a wave propagating in the fluid. The wave speed is given by

$$c_0^2 = \frac{dp}{d\rho}.$$

We note that the nonlinear term is the convective term for the particle acceleration in Euler's equations. The field equations consist of a set of three first-order quasilinear hyperbolic partial differential equations (PDEs). Reference [15, Chap. 7 part I] describes the application of characteristic theory to one-dimensional compressible flow, two-dimensional steady state flow, and shock wave propagation.

7.2. Application of Fluid Dynamics Equations to Magnetohydrodynamics†

Magnetohydrodynamic Field Equations

We are now in a position to apply the field equations of fluid dynamics to plasmas—in particular, to the magnetohydrodynamic domain. We first consider the behavior of an electrically neutral plasma, which means that we have a nonionized conducting fluid in an electromagnetic field. This type of fluid has applications in astrophysics, since the plasmas in interstellar space seldom have a net charge (except in the neighborhood of the sun). We assume that the field is incompressible and is acted on by a viscous drag force, the force of gravity, and the electrodynamic forces, in addition to the pressure force

---

† See, for example, [6], [18], and [40].

grad p. Therefore, the external force term **F** in Euler's equation of motion can be written as

$$\mathbf{F} = \left(\frac{1}{c}\right)(\mathbf{J} \times \mathbf{B}) + \eta \nabla^2 \mathbf{v} + \rho g, \tag{7.6}$$

where **J** is the current density, **B** is the magnetic flux density, η is the viscosity coefficient, and g is the acceleration of gravity. The first term on the right-hand side of (7.6) represents the force which the magnetic field exerts on the fluid. We neglect the electric field due to charged particles $q\mathbf{E}$ (q is the charge) since we have a neutral fluid. The second term is the viscous force on the incompressible fluid (the equation of state is not adiabatic), and the last term is the force due to gravity.

$\mathbf{B} = \mu \mathbf{H}$, where μ is the magnetic permeability and **H** is the magnetic field. If the fluid is assumed to be nonmagnetic we have the approximation $\mu = 1$, so that we can replace **H** by **B** in Maxwell's equation (1.4) (in Chapter 1). This becomes

$$\nabla \times \mathbf{B} = \left(\frac{4\pi}{c}\right)\mathbf{J} + \left(\frac{1}{c}\right)\frac{\partial \mathbf{D}}{\partial t}, \tag{7.7}$$

where **D** is the displacement vector. If $|v/c| \ll 1$ we can neglect relativistic effects and the displacement current $\partial \mathbf{D}/\partial t$ can be neglected compared to the other terms in (7.7) so that we get the approximation

$$\nabla \times \mathbf{B} = \left(\frac{4\pi}{c}\right)\mathbf{J}. \tag{7.8}$$

Using (7.8) and the vector identity $(\nabla \times \mathbf{B}) \times \mathbf{B} = (\mathbf{B} \cdot \mathbf{grad})\mathbf{B} - \frac{1}{2}\,\mathbf{grad}\,B^2$, the magnetic force term becomes

$$\frac{1}{2}(\mathbf{J} \times \mathbf{B}) = \left(\frac{1}{4\pi}\right)(\nabla \times \mathbf{B}) \times \mathbf{B}$$

$$= \left(\frac{1}{4\pi}\right)(\mathbf{B} \cdot \mathbf{grad})\mathbf{B} - \left(\frac{1}{8\pi}\right)\mathbf{grad}\,B^2. \tag{7.9}$$

For a fluid moving in a reference frame with velocity **v**, Ohm's law takes the form

$$\mathbf{J} = \sigma\left[\mathbf{E} + \left(\frac{1}{c}\right)(\mathbf{v} \times \mathbf{B})\right]. \tag{7.10}$$

The last term $\sigma\mathbf{E}$ represents Ohm's law for a fluid in a coordinate system at rest. The term $\sigma(1/c)(\mathbf{v} \times \mathbf{B})$ is the contribution to Ohm's law caused by a coordinate system translating with a velocity **v** with respect to an arbitrary fixed coordinate system. If the fluid is infinitely conducting then (7.10) is approximated by

$$\mathbf{E} + \left(\frac{1}{c}\right)(\mathbf{v} \times \mathbf{B}) = 0. \tag{7.11}$$

We mentioned above that the fluid is assumed to be incompressible. This means that the density is constant so that the continuity equation (7.1) reduces to

$$\text{div } \mathbf{v} = 0.$$

We now summarize the above results by giving the field equations for a plasma in the magnetohydrodynamic domain

$$\text{div } \mathbf{v} = 0, \tag{7.12a}$$

$$\rho \frac{d\mathbf{v}}{dt} = -\text{grad } p + \rho g + \eta \nabla^2 \mathbf{v} + \left(\frac{1}{4\pi}\right)(\mathbf{B} \cdot \text{grad})\mathbf{B} - \left(\frac{1}{8\pi}\right)\text{grad } B^2, \tag{7.12b}$$

$$\left(\frac{1}{c}\right)\frac{\partial \mathbf{B}}{\partial t} = -\nabla \times \mathbf{E}, \tag{7.12c}$$

$$\nabla \times \mathbf{B} = \left(\frac{4\pi}{c}\right)\mathbf{J}, \tag{7.12d}$$

$$\mathbf{J} = \sigma\left[\mathbf{E} + \left(\frac{1}{c}\right)(\mathbf{v} \times \mathbf{B})\right]. \tag{7.12e}$$

The system (7.12) must be supplemented by an equation of state for the fluid. If the fluid is inviscid then we may use the adiabatic equation of state given by (7.5) and the term $\eta \nabla^2 \mathbf{v}$ vanishes. Essentially, this is a system of six equations in the six unknowns (ρ, p, $\mathbf{v}$, $\mathbf{B}$, $\mathbf{E}$, $\mathbf{J}$). For an infinitely conducting fluid, (7.12e) is approximated by (7.11). Equation (7.12a) is the continuity equation for an incompressible fluid, (7.12b) is Euler's equation of motion for a viscous incompressible fluid in the presence of electromagnetic fields, (7.12c) is Faraday's law, and (7.12d) is Ohm's law.

Magnetic Diffusion

The right-hand side of (7.12b) represents:

(1) The three mechanical forces which are due to the pressure gradient, the effect of gravity, and viscous damping.
(2) The forces due to the electromagnetic fields for a nonionized fluid.

We shall first consider the effect of the electromagnetic fields. To this end, we combine Faraday's law with Ohm's law. Inserting (7.12e) into (7.12c)† yields

$$\frac{\partial \mathbf{B}}{\partial t} = \left(\frac{c^2}{4\pi\sigma}\right)\nabla^2 \mathbf{B} + \nabla \times (\mathbf{v} \times \mathbf{B}). \tag{7.13}$$

We first consider a fluid at rest. This is the same as introducing a coordinate system moving with the fluid at constant velocity. This means that $\mathbf{v} = 0$ so

---

† We also use the vector identity $\nabla \times (\nabla \times \mathbf{B}) = -\nabla^2 \mathbf{B}$, since div $\mathbf{B} = 0$.

that (7.13) reduces to

$$\frac{\partial \mathbf{B}}{\partial t} = \left(\frac{c^2}{4\pi\sigma}\right) \nabla^2 \mathbf{B}. \tag{7.14}$$

Equation (7.14) is a diffusion equation for $\mathbf{B}$. It is analogous to the three-dimensional unsteady Fourier heat transfer equation. Suppose we have an unsteady magnetic field in a one-dimensional medium of length L with homogeneous boundary conditions at $x = 0$ and $x = L$, and a prescribed initial value for B (the magnitude of the magnetic flux density). This gives the following boundary value problem for B:

$$B_t = \left(\frac{c^2}{4\pi\sigma}\right) B_{xx}, \quad 0 \le x \le L, \quad t > 0,$$

$$B(x, 0) = f(x), \quad 0 \le x \le L; \qquad B(0, t) = B(L, t) = 0, \quad t \ge 0, \tag{7.15}$$

where $f(x)$ is the prescribed initial value of B. It is easily seen (by the method of separation of variables, for example) that the nth mode of the solution is

$$B_n = a_n \exp\left[-\left(\frac{n^2 \pi c^2}{4\sigma L^2}\right) t\right] \sin\left(\frac{n\pi x}{L}\right), \tag{7.16}$$

where a_n is the nth Fourier coefficient. The solution is given by the sum over all modes (it is uniformly continuous in x and t), so that $B = \sum_{n=1}^{\infty} B_n$. The Fourier coefficients are obtained from the initial condition. The decay time for the nth mode is

$$\tau_n = \frac{4\sigma L^2}{n^2 \pi c^2}. \tag{7.17}$$

This is the time for B to decay to e^{-1} of its maximum value (which occurs at $t = 0$). Clearly, the largest decay time is obtained by setting $n = 1$ in (7.17). We shall call this time the *magnetic diffusion time* for our one-dimensional model.

This one-dimensional model indicates that the diffusion time, for the general three-dimensional model is given by

$$\tau = \frac{4\pi\sigma L^2}{c^2}, \tag{7.18}$$

where, for this case, L is a characteristic length associated with the spatial variation of $\mathbf{B}$. Note that (7.18) tells us that the diffusion time is proportional to the conductivity σ. This means that for the asymptotic case of infinite conductivity the diffusion time is also infinite, so that $\mathbf{B}$ does not decay—or there is no dissipation in the magnetic flux density. We also note that, for infinite conductivity, (7.13) reduces to

$$\frac{\partial \mathbf{B}}{\partial t} = \nabla \times (\mathbf{v} \times \mathbf{B}). \tag{7.19}$$

Equation (7.19) means that the magnetic flux through any closed curve moving with the fluid velocity **v** is constant in time. This is the same as saying that $d\mathbf{B}/dt = 0$ at every point on the curve, where $d/dt = \partial/\partial t + \mathbf{v} \cdot \mathbf{grad}$ is the Stokesian derivative operator. This is easily seen by directly expanding the Stokesian derivative of a particular component dB_x/dt (for example) and using (7.19). We could also prove this statement by taking the surface integral of $d\mathbf{B}/dt$ along any closed surface containing the closed curve, using the definition of the Stokesian derivative, (7.19), and invoking Stokes's theorem.

7.3. Application of Characteristic Theory to Magnetohydrodynamics

We now apply the method of characteristics to the magnetohydrodynamic field equations given by the system (7.12). This method gives us a powerful insight into the nature of wave propagation in a plasma. At this stage the reader is advised to review the pertinent material in Chapter 3, especially Sections 3.4, 3.7, and 3.8. For example, we shall make generous use of the characteristic determinant Q given by (5.53), and the concepts of ray cone, normal cone, and bicharacteristics.

The first stage is to recast the field equations in matrix form given by a generalization of (3.50a) (in Chapter 3) to the four-dimensional space–time continuum (x, y, z, t). The matrices A_i $(i = 1, \dots, 4)$, which represent the principal parts of the system of quasilinear first-order PDEs, will then be used to calculate the characteristic determinant and hence obtain the properties of the propagating waves.

We take as our model an incompressible, inviscid fluid acted upon by gravity. Equations (7.12c, d, e) are replaced by (7.19) so that our mathematical model is given by the system (7.12a), (7.12b), and (7.19), where the term $\eta \nabla^2 \mathbf{v}$ is neglected in (7.12b). Writing out this system in component form we obtain

$$u_x + v_y + w_z = 0,$$

$$\rho(u_t + uu_x + vu_y + wu_z)$$
$$= -c_0^2 \rho_x + \left(\frac{1}{4\pi}\right)[B_y(B_{x,y} - B_{y,x}) + B_z(B_{x,z} - B_{z,x})],$$

$$\rho(v_t + uv_x + vv_y + wv_z)$$
$$= -c_0^2 \rho_y + \left(\frac{1}{4\pi}\right)[B_z(B_{y,z} - B_{z,y}) + B_x(B_{y,x} - B_{x,y})], \quad (7.20)$$

$$\rho(w_t + uw_x + vw_y + ww_z)$$
$$= -c_0^2 \rho_z + \left(\frac{1}{4\pi}\right)[B_x(B_{z,x} - B_{x,z}) + B_y(B_{z,y} - B_{y,z})] + \rho g,$$

$$B_{x,t} = B_x u_x + B_y u_y + B_z u_z - (uB_{x,x} + vB_{x,y} + wB_{x,z}),†$$

$$B_{y,t} = B_x v_x + B_y v_y + B_z v_z - (uB_{y,x} + vB_{y,y} + wB_{y,z}),$$

$$B_{z,t} = B_x w_x + B_y w_y + B_z w_z - (uB_{z,x} + vB_{z,y} + wB_{z,z}).$$

The system (7.20) is a set of seven equations in the seven unknowns $\mathbf{v} = (u, v, w)$, $\rho, \mathbf{B} = (B_x, B_y, B_z)$. The first equation is the continuity equation. The next three are Euler's equations of motion for an adiabatic gas where we used the condition $\mathbf{grad}\ p = c_0^2\ \mathbf{grad}\ \rho$, since $c_0^2 = dp/d\rho$. Also gravity acts downward in the z direction. The last three equations are the (x, y, z) components of (7.19). They were easily calculated from the vector identity $\nabla \times (\mathbf{v} \times \mathbf{B}) = \mathbf{B} \cdot \mathbf{grad}\ \mathbf{v} - \mathbf{v} \cdot \mathbf{grad}\ \mathbf{B}$.

We now put the system (7.20) into a generalization of the matrix form given by (3.50a) by introducing the "supervector" $\mathbf{u}$ defined by $\mathbf{u} = (u, v, w, \rho, B_x, B_y, B_z)$. Then (7.20) can be written as

$$\mathbf{A}_1 \mathbf{u}_x + \mathbf{A}_2 \mathbf{u}_y + \mathbf{A}_3 \mathbf{u}_z + \mathbf{A}_4 \mathbf{u}_t + \mathbf{b} = 0, \tag{7.21}$$

where the matrices $\mathbf{A}_1, \dots, \mathbf{A}_4$ and the column vector $\mathbf{b}$ can be obtained by inspecting the system (7.20) in the context of (7.21). Incidentally, $\mathbf{b}$ is not a principle part of (7.21) so that the gravitational term ρg does not affect the results obtained from the method of characteristics. Recall from Chapter 3 that the characteristic surfaces in (x, y, z, t) space are given by $\phi = 0$, or the corresponding families of curves $t = \psi(x, y, z)$ in (x, y, z) space, which we obtain by setting $\psi = t - \phi(x, y, z)$. These characteristic surfaces are the wave fronts associated with wave propagation in the plasma. The characteristic matrix $\mathbf{A}$, as given by (3.52) for our purpose, is written as

$$\mathbf{A} = \sum_{i=1}^{4} \mathbf{A}_i \phi_i, \qquad \phi_1 = \phi_x, \qquad \phi_2 = \phi_y, \qquad \phi_3 = \phi_z, \qquad \phi_4 = \phi_t. \tag{7.22}$$

The determinant of $\mathbf{A}$ is the characteristic determinant and is given by

$$Q(\phi_1, \phi_2, \phi_3, \phi_4) = \det(\mathbf{A}) = 0. \tag{7.23}$$

Setting the characteristic determinant equal to zero gives the characteristic condition (7.23). Having computed the $\mathbf{A}_i$'s by inspection from (7.21), we then multiply $\mathbf{A}_1$ by $\phi_x, \dots, \mathbf{A}_4$ by ϕ_t, add these four terms and set the result equal to zero. We thus obtain Q, which is a (7×7) matrix involving the $\mathbf{A}_i$'s and the ϕ_i's, which we set equal to zero. The calculations are a little complicated but straightforward, and will be reserved for the two-dimensional case discussed below, in order to save space. The result of setting $Q = 0$ is

$$(\mathbf{v} \cdot \mathbf{grad}\ \phi + \phi_t)[4\pi\rho(\mathbf{v} \cdot \mathbf{grad}\ \phi + \phi_t)^2 - (\mathbf{B} \cdot \mathbf{grad}\ \phi)^2]^2 = 0. \tag{7.24}$$

---

† Note that we used the fact that div $\mathbf{v} = 0$ and div $\mathbf{B} = 0$.

We note that

$$\mathbf{v} \cdot \mathbf{grad} \ \phi + \phi_t = \frac{d\phi}{dt}. \tag{7.25}$$

Setting the left-hand factor of (7.24) equal to zero yields $d\phi/dt = 0$ which means that we have a family of characteristic surfaces $\phi = 0$ in (x, y, z, t) space or the corresponding families of characteristic curves $t = \psi(x, y, z)$ in (x, y, z) space, which we obtain by setting $\phi = t - \psi(x, y, z)$. Since $d\phi/dt = 0$, we obtain for the characteristics the surface

$$\phi_t + u\phi_x + v\phi_y + w\phi_z = 0. \tag{7.26}$$

The ray directions are the streamlines of the flow and are given by the velocity components $u = dx/dt$, $v = dy/dt$, $w = dz/dt$.

Setting the right-hand factor of (7.24) equal to zero yields

$$\frac{d\phi}{dt} = \pm(4\pi\rho)^{-1/2}\mathbf{B} \cdot \mathbf{grad} \ \phi. \tag{7.27}$$

Equation (7.25) represents forward and backward *Alfvèn waves* whose speeds are given by the $\pm$ values of $d\phi/dt$, respectively. Their normal surfaces are given by the planes (in $\phi_t, \phi_x, \phi_y, \phi_z$ space)

$$\sqrt{4\pi\rho}(\phi_t + \mathbf{v} \cdot \mathbf{grad} \ \phi) \pm \mathbf{B} \cdot \mathbf{grad} \ \phi = 0.$$

If we introduce a coordinate system moving with the velocity of the fluid then $\mathbf{v} = 0$, so that the Alfvèn wave relative to the fluid has a velocity $\mathbf{v}_A$, where

$$\mathbf{v}_A = \pm(4\pi\rho)^{-1/2}\mathbf{B}. \tag{7.28}$$

Thus a forward Alfvèn wave propagates with a speed relative to the fluid in the $\mathbf{B}$ direction, while a backward Alfvèn wave propagates in the $-\mathbf{B}$ direction.

Two-Dimensional Case. To illustrate the mathematical technique we consider the two-dimensional case in some detail. The space is (x, y, t), the particle velocity is $\mathbf{v} = (u, v, 0)$, and the magnetic flux density is $\mathbf{B} = (B_x, B_y, 0)$. We omit the gravitational field. The field equations (7.20) reduce to

$$u_x + v_y = 0,$$

$$\rho(u_t + uu_x + vu_y) = -c_0^2\rho_x + \left(\frac{1}{4\pi}\right)B_y(B_{x,y} - B_{y,x}),$$

$$\rho(v_t + uv_x + vv_y) = -c_0^2\rho_y + \left(\frac{1}{4\pi}\right)B_x(B_{y,x} - B_{x,y}), \tag{7.29}$$

$$B_{x,t} = B_x u_x + B_y u_y - (uB_{x,x} + vB_{x,y}),$$

$$B_{y,t} = B_x v_x + B_y v_y - (uB_{y,x} + vB_{y,y}).$$

The system (7.29) consists of five field equations in the five unknowns given by the supervector in (x, y, t) space which becomes

$$\mathbf{u} = (u, v, \rho, B_x, B_y). \tag{7.30}$$

This a set of first-order PDEs whose matrix form is given by

$$A_1 \mathbf{u}_x + A_2 \mathbf{u}_y + A_3 \mathbf{u}_t = 0. \tag{7.31}$$

Note that (7.31) is the reduction of (7.21) to our three-dimensional time–space, where the (5×5) matrices A_1, A_2, A_3 become

$$A_1 = \begin{pmatrix} 1 & 0 & 0 & 0 & 0 \\ \rho u & 0 & c_0^2 & 0 & kB_y \\ 0 & \rho u & 0 & 0 & -kB_x \\ -B_x & 0 & 0 & u & 0 \\ 0 & -B_x & 0 & 0 & u \end{pmatrix}, \qquad k = \frac{1}{4\pi}, \tag{7.32a}$$

$$A_2 = \begin{pmatrix} 0 & 1 & 0 & 0 & 0 \\ \rho v & 0 & 0 & -kB_y & 0 \\ 0 & \rho v & c_0^2 & kB_x & 0 \\ -B_y & 0 & 0 & v & 0 \\ 0 & -B_y & 0 & 0 & v \end{pmatrix}, \tag{7.32b}$$

$$A_3 = \begin{pmatrix} 0 & 0 & 0 & 0 & 0 \\ \rho & 0 & 0 & 0 & 0 \\ 0 & \rho & 0 & 0 & 0 \\ 0 & 0 & 0 & 1 & 0 \\ 0 & 0 & 0 & 0 & 1 \end{pmatrix}. \tag{7.32c}$$

The characteristic matrix A is given by

$$A = A_1 \phi_x + A_2 \phi_y + A_3 \phi_t, \tag{7.33}$$

where the scalar $\phi = \phi(x, y, t)$. This is calculated by multiplying A_1 by ϕ_x, A_2 by ϕ_y, A_3 by ϕ_t, and adding the resulting matrices. The characteristic condition is obtained by setting the characteristic determinant $Q = \det(A)$ equal to zero, yielding

$$Q = \begin{vmatrix} \phi_x & \phi_y & 0 & 0 & 0 \\ f & 0 & c_0^2 \phi_x & -kB_y \phi_y & kB_y \phi_x \\ 0 & f & c_0^2 \phi_y & kB_x \phi_y & -kB_x \phi_x \\ -g & 0 & 0 & f & 0 \\ 0 & -g & 0 & 0 & f \end{vmatrix} = 0, \tag{7.34}$$

where

$$f = \rho(u\phi_x + v\phi_y + \phi_t) = \rho(\mathbf{v} \cdot \mathbf{grad} + \phi_t) = \rho \frac{d\phi}{dt},$$

$$g = B_x \phi_x + B_y \phi_y = \mathbf{B} \cdot \mathbf{grad}\ \phi. \tag{7.35}$$

Working out the determinant in (7.34) we obtain the characteristic condition

$$\left(\frac{d\phi}{dt}\right)\left[4\pi\rho\left(\frac{d\phi}{dt}\right)^2 - (\mathbf{B}\cdot\mathbf{grad}\ \phi)^2\right]^2 = 0, \tag{7.36}$$

which is the reduction of (7.24) for (x, y, t) space.

7.4. Linearization of the Field Equations

As our physical model we consider the wave motion of an infinitely conducting compressible, nonviscous fluid in a magnetic field (plasma) in the absence of a gravitational field. We assume an adiabatic fluid so that the pressure–density relationship is given by the adiabatic equation of state. As before, we consider an infinite three-dimensional medium so that the only boundary conditions involve boundedness at infinity. The mathematical model consists of the appropriate field equations, which we repeat (for the readers' benefit).

$$\rho_t + \operatorname{div} \rho\mathbf{v} = 0,$$

$$\rho\mathbf{v}_t + \rho\mathbf{v}\cdot\mathbf{grad}\ \mathbf{v} = -c_0^2\ \mathbf{grad}\ \rho - \left(\frac{1}{4\pi}\right)\mathbf{B}\times(\nabla\times\mathbf{B}), \tag{7.37}$$

$$\mathbf{B}_t = \nabla\times(\mathbf{v}\times\mathbf{B}).$$

The force due to the pressure gradient in Euler's equation of motion (the second equation of (7.37)) is replaced by the square of the sound speed times the density gradient. The reason why we can replace the pressure by the density is easily seen. Pressure is a single-valued function of density so that $p_x = (dp/d\rho)\rho_x = c_0^2\rho_x$, and similarly for the p_y and p_z terms. This means that $\mathbf{grad}\ p = c_0^2\ \mathbf{grad}\ \rho$.

We now linearize the field equations (7.37). This means that we replace $\mathbf{v}$, ρ, and $\mathbf{B}$ by small perturbations about the equilibrium values $\mathbf{v}_0 = 0$ (coordinate system fixed in the uniformly moving fluid), ρ_0, and $\mathbf{B}_0$. We therefore have

$$\mathbf{v} = \mathbf{v}'(\mathbf{x}, t),$$

$$\rho = \rho_0 + \rho'(\mathbf{x}, t), \tag{7.38}$$

$$\mathbf{B} = \mathbf{B}_0 + \mathbf{B}'(\mathbf{x}, t),$$

where $\mathbf{v}'$ is the particle velocity with respect to the uniformly moving fluid, $\mathbf{B}'$ is the perturbed magnetic flux, and ρ' is the perturbed density. We have the following inequalities:

$$|\mathbf{v}'| \ll 1, \qquad \left|\frac{\rho'}{\rho_0}\right| \ll 1, \qquad \left|\frac{\mathbf{B}'}{\mathbf{B}_0}\right| \ll 1,$$

in the sense that we keep only the linear terms (we neglect terms of second and higher order). We neglect the $\text{div}(\rho' \mathbf{v}')$ term in the continuity equation, the convective acceleration term $\mathbf{v}' \cdot \mathbf{grad}\ \mathbf{v}'$ in the equation of motion, and the quadratic terms involving $\mathbf{B}$ in the equation of motion and in Faraday's equation. Using (7.38) the linearized form of (7.37) becomes

$$\rho'_t + \rho_0\ \text{div}\ \mathbf{v}' = 0,$$

$$\rho_0 \mathbf{v}'_t = -c_0^2\ \mathbf{grad}\ \rho' - \left(\frac{1}{4\pi}\right) \mathbf{B}_0 \times (\nabla \times \mathbf{B}'),\qquad(7.39)$$

$$\mathbf{B}'_t = \nabla = (\mathbf{v}' \times \mathbf{B}_0).$$

The sound speed c_0 is constant. We may eliminate the density from the linearized field equations (7.39) as follows: take the gradient of the first equation and obtain

$$\mathbf{grad}\ \rho'_t = \rho_0 \nabla^2 \mathbf{v}'.$$

We then differentiate the second equation partially with respect to t and substitute the above expression into the resulting equation, use the third equation of (7.37) and the definition of the Alfvèn velocity $\mathbf{v}_A$ given by (7.28). We obtain

$$\mathbf{v}'_{tt} - c_0^2 \nabla^2 \mathbf{v}' = -\mathbf{v}_A \times \nabla \times [\nabla \times (\mathbf{v}' \times \mathbf{v}_A)].\qquad(7.40)$$

Equation (7.40) is the vector wave equation for $\mathbf{v}'$. The right-hand side depends, in a complicated way on $\mathbf{v}'$ and $\mathbf{v}_A$. We assume $\mathbf{v}_A$ to be constant. Applying some well-known vector identities, and using the fact that $\mathbf{v}_A$ is constant, we have

$$\nabla \times (\mathbf{v}' \times \mathbf{v}_A) = \mathbf{v}_A \cdot \mathbf{grad}\ \mathbf{v}' - \mathbf{v}_A(\text{div}\ \mathbf{v}'),$$

so that

$$\mathbf{v}_A \times \nabla \times [\nabla \times (\mathbf{v}' \times \mathbf{v}_A)] = \mathbf{grad}(\mathbf{v}_A \cdot (\mathbf{v}_A \cdot \mathbf{grad}\ \mathbf{v}'))$$

$$- \mathbf{v}_A \cdot \mathbf{grad}(\mathbf{v}_A \cdot \mathbf{grad}\ \mathbf{v}')$$

$$- \mathbf{grad}(\mathbf{v}_A \cdot \mathbf{v}_A(\text{div}\ \mathbf{v}'))$$

$$+ \mathbf{v}_A \cdot \mathbf{grad}(\mathbf{v}_A(\text{div}\ \mathbf{v}')).\qquad(7.41)$$

Propagation of Plane Waves

We now consider plane wave solutions of the wave equation (7.40), where the right-hand side is given by (7.41). To this end, we write time-harmonic solutions for $\mathbf{v}'$ in the form

$$\mathbf{v}' = \bar{\mathbf{v}}'\ \exp[i(\mathbf{k} \cdot \mathbf{x} - \omega t)],\qquad(7.42)$$

where $\bar{\mathbf{v}}'$ is a constant vector giving the magnitude and direction of $\mathbf{v}'$. We recall from Chapter 1 that plane waves are defined such that the wave fronts consist of the family of planes $\mathbf{k} \cdot \mathbf{x} = \text{const.}$ which travel in space with a characteristic wave speed. $\mathbf{k}$ is the wave vector whose magnitude is the wave number. Inserting (7.42) into (7.40) and (7.41), after a little manipulation, yields the following equation for $\mathbf{k}$:

$$-\omega^2 \bar{\mathbf{v}}' + (c_0^2 + v_A^2)(\mathbf{k} \cdot \bar{\mathbf{v}}')\mathbf{k} + (\mathbf{k} \cdot \mathbf{v}_A)[(\mathbf{k} \cdot \mathbf{v}_A \bar{\mathbf{v}}') - (\mathbf{v}_A \cdot \bar{\mathbf{v}}')\mathbf{k} - (\mathbf{k} \cdot \bar{\mathbf{v}}')]\mathbf{v}_A = 0. \tag{7.43}$$

If the magnetic field is identically zero then $\mathbf{v}_A = 0$ and (7.43) becomes

$$-\omega^2 \bar{\mathbf{v}}' + c_0^2(\mathbf{k} \cdot \bar{\mathbf{v}}')\mathbf{k} = 0,$$

which yields

$$k = \frac{\omega}{c_0}.$$

Returning to (7.43) we note the following scalar products: $\mathbf{k} \cdot \bar{\mathbf{v}}'$, $\mathbf{k} \cdot \mathbf{v}_A$, and $\bar{\mathbf{v}}' \cdot \mathbf{v}_A$. Recall that $\mathbf{k}$ identifies the direction of the propagating wave. We consider the following cases:

If $\mathbf{k}$ is normal to $\mathbf{v}_A$ then (7.43) reduces to

$$-\omega^2 \bar{\mathbf{v}}' + (c_0^2 + v_A^2)(\mathbf{k} \cdot \bar{\mathbf{v}}')\mathbf{k} = 0.$$

This means that the solution for $\mathbf{v}'$ has a phase velocity given by

$$\mathbf{v}_L = \sqrt{c_0^2 + v_A^2}. \tag{7.44}$$

$\mathbf{v}_L$ is the phase velocity of the *longitudinal magnetosonic* wave. Observe that this phase speed depends on the sum of the squares of the sound speed and the Alfvèn speed. These, in a sense, represent the hydrostatic and magnetic pressures.

If $\mathbf{k}$ is parallel to $\mathbf{v}_A$ and $\mathbf{v}'$ then (7.43) reduces to the case where the phase speed equals the sound speed c. This is the case where the directions of wave propagation, $\mathbf{v}_A$ and $\mathbf{v}'$, are the same.

And finally, if the direction of wave propagation is in the $\mathbf{v}_A$ direction ($\mathbf{k}$ is parallel to $\mathbf{v}_A$) and $\mathbf{v}_A$ is normal to $\mathbf{v}'$, then (7.43) tells us that we have a pure Alfvèn wave whose phase speed is given by

$$v_A = \frac{\omega}{k}.$$

Clearly this is a transverse wave, which is solely a magnetohydrodynamic phenomenon depending only on the magnetic field and the inertia.

We comment briefly on the magnitudes of the Alfvèn velocity. At all laboratory magnetic field strengths the Alfvèn speed is much less than the sound speed. For mercury at room temperature, for example, the Alfvèn speed is $B_0/13.1$ cm/sec compared with the sound speed of 1.45×10^5 cm/sec. In the astrophysical domain, on the other hand, the Alfvèn speed can become very large because of the very small density of the interstellar gas. For example, in the sun's photosphere the density is of the order of 10^{-7} gm/cm^3, which means that there are about 5×10^{16} hydrogen atoms/cm^3. It turns out that $v_A \simeq 10^3 B_0$ cm/sec. Solar magnetic fields are of the order of 1 gauss at the surface, with much larger values around sunspots. As a comparison, the sound speed is the order of 10^6 cm/sec in both the photosphere and chromosphere.

We can determine the magnetic fields for the above cases by appealing directly to the third equation of (7.39) which, upon using a vector identity, may be written as

$$\mathbf{B}'_t = \mathbf{B}_0 \cdot \mathbf{grad}\, v' - \mathbf{B}_0(\mathrm{div}\, v'). \tag{7.45}$$

We again look for time-harmonic plane waves, so that we set

$$\mathbf{B}' = \bar{\mathbf{B}}' \exp[i(\mathbf{k} \cdot \mathbf{x} - \omega t)], \tag{7.46}$$

where $\bar{\mathbf{B}}'$ is a constant vector giving the magnitude and direction of $\mathbf{B}'$. Inserting (7.42) and (7.46) into (7.45) gives

$$\omega\bar{\mathbf{B}}' = -(\mathbf{k} \cdot \mathbf{B}_0)\bar{\mathbf{v}}' + (\mathbf{k} \cdot \bar{\mathbf{v}}')\mathbf{B}_0. \tag{7.47}$$

From (7.47) we obtain the following three possibilities for the perturbed magnetic flux $\mathbf{B}'$:

$$\mathbf{B}' = \begin{cases} \left(\dfrac{k}{\omega}\right)\bar{v}'\mathbf{B}_0 & \text{for } \mathbf{k} \cdot \mathbf{B}_0 = 0 \ (\mathbf{k} \text{ normal to } \mathbf{B}_0), \\[2mm] 0 & \text{for } \mathbf{k} \text{ in the direction of } \mathbf{B}_0,\ \text{longitudinal } \mathbf{k}, \\[2mm] -\left(\dfrac{k}{\omega}\right)\mathbf{B}_0 v' & \text{for } \mathbf{k} \cdot \bar{\mathbf{v}}' = 0 \ (\mathbf{k} \text{ normal to } v'),\ \text{transverse } \mathbf{k}. \end{cases} \tag{7.48}$$

For *the first case* of (7.48) the magnetosonic waves propagating normal to the constant magnetic field $\mathbf{B}_0$ consist of a set of parallel lines in the direction of $\mathbf{B}_0$ which are alternately compressed and rarefied according to the factor (k/ω) (for a given value of v'). *The second case* tells us that there is no resulting magnetic field if the direction of wave propagation in the same as the direction of $\mathbf{B}_0$. This is in keeping with what we learned in Chapter 1, that the magnetic field is normal to the direction of wave propagation. *The third case* tells us

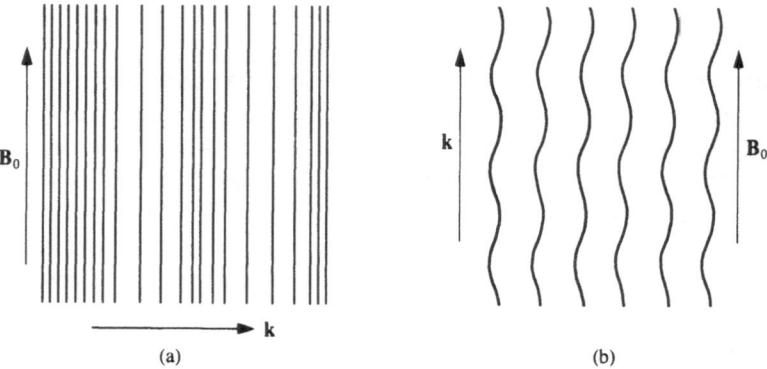

Fig. 7.1. Magnetohydrodynamic waves. (a) Wave propagation normal to $\mathbf{B}_0$; (b) wave propagation parallel to $\mathbf{B}_0$.

that the waves are transverse, or the direction of wave propagation is normal to $\mathbf{v}'$. In this case the perturbed magnetic flux involves Alfvèn waves parallel to $\mathbf{B}_0$ which oscillate back and forth laterally in the $\mathbf{v}'$ direction. Cases one and three are shown in Fig. 7.1(a), (b). Figure 7.1(a) is the case where $\mathbf{k}$ is normal to $\mathbf{B}_0$. The magnetic field lines are parallel straight lines whose distance from each other varies as the reciprocal of the frequency. In Fig. 7.1(b) $\mathbf{k}$ is parallel to $\mathbf{B}_0$. The field lines are oscillatory, are parallel to each other, and are uniformly distributed.

Plane Wave Propagation in a Compressible Viscous Fluid of Finite Conductivity

We now extend the treatment of plane wave propagation in a magnetohydro-dynamic medium to a plasma composed of compressible viscous fluid of finite conductivity σ. We wish to find the effect of viscosity and conductivity on wave propagation in the plasma. We still use the linear approximation to the field equations, which become

$$\rho'_t + \rho_0 \operatorname{div} \mathbf{v}' = 0,$$

$$\rho_0 \mathbf{v}'_t = -c_0^2 \operatorname{\mathbf{grad}} \rho' + \eta \nabla^2 \mathbf{v}' + (\zeta + \tfrac{1}{3}\eta) \operatorname{\mathbf{grad}}(\operatorname{div} \mathbf{v}')$$

$$- \left(\frac{1}{4\pi}\right) \mathbf{B}_0 \times (\nabla \times \mathbf{B}'), \tag{7.49}$$

$$\mathbf{B}'_t = \nabla \times (\mathbf{v}' \times \mathbf{B}_0) + \left(\frac{c^2}{4\pi\sigma}\right) \nabla^2 \mathbf{B}'.$$

The equation of motion (second equation of (7.49)) has two additional viscosity terms compared to the corresponding equation of (7.39). The term $(\zeta + \tfrac{1}{3}\eta) \operatorname{\mathbf{grad}}(\operatorname{div} \mathbf{v}')$ involves the two viscosity coefficients which arise from thermodynamic considerations. It reflects the fact that $\operatorname{div} \mathbf{v}'$ is not zero for a compressible fluid. The reader is referred to [15, chap. 7, Part II] "Viscous Fluids" for a treatment of this and the other viscosity term using a thermo-dynamic treatment of the energy equation. However, from the point of view of plane wave propagation the effect of this additional viscosity term is to merely change the value of the effective viscosity coefficient in determining the relationship amongst the frequency, wave number, and phase velocities. In addition, the term involving $\mathbf{B}'$ in the magnetic force part of the equation of motion must be modified due to the additional term $(c^2/4\pi\sigma)\nabla^2\mathbf{B}'$ in the third equation. This represents the effect of conductivity on $\mathbf{B}'_t$.

We look for time-harmonic solutions for $\mathbf{v}'$ and $\mathbf{B}'$. To do this we form the wave equation for $\mathbf{v}'$ in the usual way by differentiating the second equation of (7.49) and inserting the first equation into the result. We get

$$\rho_0 \mathbf{v}'_{tt} - \rho_0 c_0^2 \nabla^2 \mathbf{v}' - \eta \nabla^2 \mathbf{v}'_t - (\zeta + \tfrac{1}{3}\eta) \operatorname{\mathbf{grad}}(\operatorname{div} \mathbf{v}'_t) = -\left(\frac{1}{4\pi}\right) \mathbf{B}_0 \times (\nabla \times \mathbf{B}'_t).$$

$$\tag{7.50}$$

Inserting (7.42) into (7.50) yields

$$[-\rho_0\omega^2 + \rho_0 c_0^2 k^2 + i\alpha\omega k^2]\bar{\mathbf{v}}' = -\left(\frac{1}{4\pi}\right)\mathbf{B}_0 \times (\nabla \times \mathbf{B}_t'),$$

where $\alpha = \frac{4}{3}\eta + \zeta$ is the "effective" viscosity coefficient. The above equation may be written as

$$\left[1 + i\left(\frac{\alpha k^2}{\rho_0\omega}\right)\right](-\omega^2\rho_0)\bar{\mathbf{v}}' = -\rho_0 c_0^2 k^2\bar{\mathbf{v}}' - \left(\frac{1}{4\pi}\right)\mathbf{B}_0 \times (\nabla \times \mathbf{B}_t'). \quad (7.51)$$

Since $-\omega^2\rho_0\bar{\mathbf{v}}' = \rho_0\mathbf{v}_{tt}'$ (apart from the exponential factor) (7.51) tells us that ω^2 in (7.43) must be multiplied by the factor $[1 + i(\alpha k^2/\rho_0\omega)]$, in order to take into account the viscosity term.

We allow the wave number k to be complex. To get a feel for the physical meaning of a complex wave number we consider the following two special cases:

(1) Effect of the viscous force.
(2) Effect of the magnetic force.

(1) Effect of the Viscous Force

We again consider the linearized field equations given by (7.49). We specialize the equation of motion to the case where the only external force is due to viscous damping. This means that we neglect the **grad** ρ' term and the last term on the right-hand side of (7.49) (meaning that we neglect the magnetic force). There is no loss in generality in assuming a one-dimensional model. To this end, we set $\mathbf{v}' = (u(x, t), 0, 0)$. Then (7.50) becomes

$$\rho_0 u_{tt} = \rho_0 c_0^2 u_{xx} + \alpha u_{xxt}. \quad (7.52)$$

Inserting $u = \bar{u} \exp[i(kx - \omega t)]$ (where $\bar{u}$ is the amplitude of u) into the above equation yields

$$-\rho_0\omega^2 = -\rho_0 c_0^2 k^2 + i\alpha\omega k^2, \quad (7.53)$$

which is the equation relating the wave number to the frequency. It is easily seen that (7.43) reduces to the above equation for this example. The solution for k^2 is

$$k^2 = \frac{\rho_0\omega^2}{c_0^2 - i\alpha\omega} = U + iV,$$

$$U = \frac{\rho_0^2 c_0^2 \omega^2}{(\rho_0 c_0^2)^2 + (\alpha\omega)^2}, \quad (7.54)$$

$$V = \frac{\alpha\rho_0\omega^3}{(\rho_0 c_0^2)^2 + (\alpha\omega)^2}.$$

Set

$$k = k_0 + ik_1. \quad (7.55)$$

We see that the wave number is complex. k_0 is the real part and k_1 is the imaginary part of k. We get

$$k_0^2 - k_1^2 + 2ik_0k_1 = U + iV.$$

Equating real and imaginary parts yields quadratic expressions for k_0 and k_1 whose solutions are

$$k_0^2 = \tfrac{1}{2}(U + \sqrt{U^2 + V^2}) \qquad k_1^2 = \tfrac{1}{2}(-U + \sqrt{U^2 + V^2}).\dagger \qquad (7.56)$$

The solution for $\mathbf{v}'$ becomes

$$\mathbf{v}' = \bar{\mathbf{v}}' \exp(-k_1 x) \exp[i(k_0 x - \omega t)]. \qquad (7.57)$$

The first factor on the right-hand side of (7.57) represents damping in x, since $k_1 > 0$ (we assume $x > 0$).

For very small viscosity we retain only terms linear in α. Then, using (7.54) and (7.56), we get the following approximation for (7.55):

$$k = \frac{\omega}{c_0} + \frac{i\alpha\omega^2}{2c_0^3\rho_0}. \qquad (7.58)$$

Equation (7.58) tells us that the real part of k represents the wave number if the fluid were not viscously damped. The imaginary part of k is a purely viscous damping term proportional to the square of the frequency. To within the approximation given by (7.58), the solution for $\mathbf{v}'$ becomes

$$\mathbf{v}' \simeq \bar{\mathbf{v}}' \exp\left[\left(\frac{-\alpha\omega^2}{2c_0^3\rho_0}\right)x\right] \exp\left[i\left(\frac{\omega}{c_0} - t\right)\right]. \qquad (7.59)$$

Equation (7.59) tells us that, within the approximation of linear α, the damping factor depends quadratically on the frequency. The purely oscillatory term behaves as if there were no viscous damping, meaning that the frequency does not involve a damping term.

Note that if we introduced quadratic and higher-order terms in α, then k_0 would also involve viscous damping. This means that, in this case, the frequency of oscillation depends on the damping.

(2) Effect of the Magnetic Force

Here we are interested solely in the effect of the magnetic force on the wave number. We therefore neglect all the terms on the right-hand side of the second equation of (7.49) except the magnetic force term.

We first consider the special case where we neglect the fluid. This means that we set $\mathbf{v}' = 0$ on the right-hand side of the third equation of (7.49). The resulting equation is (7.14), which was treated in the section entitled "Magnetic Diffusion". In that section we assume the wave number to be real so that there

---

† We take the positive square root since both k_0 and k_1 must be real.

was no spatial damping. There we obtained (7.18) as the expression for the diffusion time. In the treatment given here we relax the restriction of real wave number, and thereby attempt to find the effect of the imaginary part of the wave number on the solution. Again, there is no loss in generality in restricting ourselves to a one-dimensional model, for simplicity. Equation (7.14) becomes

$$B'_t = \left(\frac{c^2}{4\pi\sigma}\right) B'_{xx}. \tag{7.60}$$

where B' is the x component of the magnetic flux density. To obtain plane wave solutions we insert (7.46) into (7.60) (where $\mathbf{k} \cdot \mathbf{x}$ becomes kx) and obtain the following equation relating the wave number to the frequency:

$$k^2 = \left(\frac{4\pi\sigma}{c^2}\right) i\omega. \tag{7.61}$$

from which we obtain

$$k = \frac{(1+i)\sqrt{2\pi\sigma\omega}}{c} = k_0 + ik_1. \tag{7.62}$$

Inserting (7.62) into (7.46) we get a plane wave solution of (7.60)

$$\mathbf{B}' = \bar{\mathbf{B}}' \exp\left[-\frac{\sqrt{2\pi\sigma\omega}}{c} x\right] \exp\left[i\left(\frac{\sqrt{2\pi\sigma\omega}}{c} x - \omega t\right)\right]. \tag{7.63}$$

In this case the damping coefficient depends on the square root of the frequency.

Next we take up the more general case where $\mathbf{v}'$ is not zero. Neglecting the pressure and viscous forces, the second equation of (7.49) becomes

$$\rho_0 \mathbf{v}'_t = -\left(\frac{1}{4\pi}\right)[\text{grad}(\mathbf{B}_0 \cdot \mathbf{B}') - \mathbf{B}_0 \cdot \text{grad } \mathbf{B}']. \tag{7.64}$$

The third equation of (7.49) can be written in the following form:

$$\mathbf{B}'_t = \mathbf{B}_0 \cdot \text{grad } \mathbf{v}' - \mathbf{B}_0(\text{div } \mathbf{v}') + \left(\frac{c^2}{4\pi\sigma}\right) \nabla^2 \mathbf{B}'. \tag{7.65}$$

Inserting (7.42) and (7.46) into (7.65) yields the following relationship between the frequency and wave number:

$$\omega\mathbf{B}' = -(\mathbf{k} \cdot \mathbf{B}_0)\bar{\mathbf{v}}' + (\mathbf{k} \cdot \bar{\mathbf{v}}')\mathbf{B}_0 - i\left(\frac{c^2}{4\pi\sigma}\right) \mathbf{k} \cdot \mathbf{k}\bar{\mathbf{B}}'. \tag{7.66}$$

Note that (7.66) reduces to (7.47) for the special case of infinite conductivity and $\mathbf{k}$ real, so that the last term on the right-hand side of (7.66) represents the effect of conductivity on the magnetic field. The generalization of the three cases given by (7.48) for finite conductivity is given by

$$
\mathbf{B}' = \begin{cases}
\dfrac{(\mathbf{k} \cdot \mathbf{v}')\mathbf{B}_0}{\omega + i(c^2/4\pi\sigma)k^2} & \text{for } \mathbf{k} = \mathbf{B}_0 = 0, \\[3mm]
0 & \text{for } \mathbf{k} \text{ longitudinal}, \\[3mm]
\dfrac{-(\mathbf{k} \cdot \mathbf{B}_0)\mathbf{v}'}{\omega + i(c^2/4\pi\sigma)k^2} & \text{for } \mathbf{k} \cdot \mathbf{v}' = 0,
\end{cases}
\tag{7.67}
$$

where $\mathbf{k}$ is complex so that $\mathbf{k} = \mathbf{k}_0 + i\mathbf{k}_1$. We must set separately the real and imaginary parts of the equations for these cases equal to zero. Note that $\mathbf{k}$ is a complex vector. Written out in full form it becomes

$$
\mathbf{k} = \mathbf{i}(k_{0x} + ik_{1x}) + \mathbf{j}(k_{0y} + ik_{1y}) + \mathbf{k}(k_{0z} + ik_{1z}),
$$

where $(\mathbf{i}, \mathbf{j}, \mathbf{k})$ are unit vectors in the (x, y, z) directions and $k_{0x}, \ldots, k_{1z}$ are the corresponding components of the real/imaginary parts of $\mathbf{k}$.

To complete the picture for the frequency–wave number relationship we need to insert (7.42) and (7.46) into the equation of motion given by (7.64). We thereby obtain

$$
\rho_0 \omega \mathbf{v}' = \left(\frac{1}{4\pi}\right)[(\mathbf{B}_0 \cdot \mathbf{B}')\mathbf{k} - (\mathbf{B}_0 \cdot \mathbf{k})\mathbf{B}'],
\tag{7.68}
$$

where, again, the real and imaginary parts of this equation must be separately set equal to zero.

As an example we discuss the two-dimensional case in the (x, y) plane in some detail. Note that for the one-dimensional case the $\nabla \times (\mathbf{v}' \times \mathbf{B}_0)$ term identically vanishes. We set $\mathbf{B}' = (B_x, B_y, 0)$, $\mathbf{v}' = (u, v, 0)$, and $\mathbf{B}_0 = (a, b, 0)$. Then the two-dimensional representation of (7.66) and (7.68) becomes

$$
\rho_0 \omega u + \left(\frac{1}{4\pi}\right)(k_{0y} + ik_{1y})bB_x - \left(\frac{1}{4\pi}\right)(k_{0x} + ik_{1x})bB_y = 0,
$$

$$
\rho_0 \omega v - \left(\frac{1}{4\pi}\right)(k_{0y} + ik_{1y})aB_x + \left(\frac{1}{4\pi}\right)(k_{0x} + ik_{1x})aB_y = 0,
$$

$$
\tag{7.69}
$$

$$
(k_{0y} + ik_{1y})bu - (k_{0y} + ik_{1y})av + \left[\omega + i\left(\frac{c^2}{4\pi\sigma}\right)\mathbf{k} \cdot \mathbf{k}\right]B_x = 0,
$$

$$
-(k_{0x} + ik_{1x})bu + (k_{0x} + ik_{1x})av + \left[\omega + i\left(\frac{c^2}{4\pi\sigma}\right)\mathbf{k} \cdot \mathbf{k}\right]B_y = 0.
$$

The system (7.69) is a set of four homogeneous algebraic equations for (u, v, B_x, B_y), which are the real amplitudes of the corresponding components of $\mathbf{v}'$ or $\mathbf{B}'$. In order that the solution for these amplitudes be finite, we must have the condition that the determinant of the matrix of coefficients of these amplitudes is set equal to zero. This yields a fourth-degree algebraic equation in $\mathbf{k}$ relating the wave number to the frequency. Since, as mentioned above, $\mathbf{k}$

is complex we actually get two fourth-degree equations, one for the real part of $\mathbf{k}$ and one for the imaginary part. Rather than investigate the details of these algebraic equations we consider some special cases.

If we let $\mathbf{B}_0$ be in the x direction then $\mathbf{B}_0 = (a, 0, 0)$. Setting $b = 0$ in the first equation of (7.69) tells us that $u = 0$, which means that the flow field is normal to $\mathbf{B}_0$. The system (7.69) reduces to the following three algebraic equations for (v, B_x, B_y):

$$\rho_0 \omega v - \left(\frac{1}{4\pi}\right) k_y a B_x + \left(\frac{1}{4\pi}\right) k_x a B_y = 0,$$

$$-k_y a v + \left[\omega + i\left(\frac{c^2}{4\pi\sigma}\right) k^2\right] B_x = 0, \qquad (7.70)$$

$$k_x a v + \left[\omega + i\left(\frac{c^2}{4\pi\sigma}\right) k^2\right] B_y = 0.$$

In order that the amplitudes (v, B_x, B_y) be not identically zero the determinant of the coefficients of these amplitudes must be set equal to zero. This yields the following algebraic equation:

$$\left[\omega + i\left(\frac{c^2}{4\pi\sigma}\right) k^2\right]\left[-\left(\frac{1}{4\pi}\right) a^2 k^2 + \omega\rho_0\left(\omega + i\left(\frac{c^2}{4\pi\sigma}\right) k^2\right)\right] = 0. \quad (7.71)$$

If we set the first factor on the left-hand side equal to zero we get the root

$$\omega = -i\left(\frac{c^2}{4\pi\sigma}\right) k^2. \qquad (7.72)$$

Suppose $\mathbf{k}$ is real so that $\mathbf{k}_1 = 0$. Then we have a pure imaginary frequency. To see the meaning of this insert (7.72) into (7.46). We get

$$\mathbf{B}' = \mathbf{B}' \exp\left[i(k_{0x}x + k_{0y}y) - \left(\frac{c^2}{4\pi\sigma}\right) k^2 t\right]. \qquad (7.73)$$

The factor on the right-hand side involving x and y tells us that the spatial part of $\mathbf{B}'$ varies sinusoidally with x and y. The time-dependent factor shows an exponential decay in t where the time to decay to e^{-1} of its maximum value is τ, which is given by

$$\tau = \frac{4\pi\sigma L^2}{c^2}, \qquad L^2 = \frac{1}{k^2}.$$

This is essentially (7.18) which tells us that τ is the magnetic diffusion time for our two-dimensional model. We have used the fact that $\mathbf{k}$ is the wave number which is a characteristic inverse length. Thus, one of the roots of (7.71) yields the magnetic diffusion time.

If the factor $\omega + i(c^2/4\pi\sigma)k^2 \neq 0$, then the other roots of (7.71) can be obtained by setting the other factor equal to zero. We obtain

$$-\left(\frac{a^2}{4\pi}\right) k^2 + \rho_0 \omega\left[\omega + i\left(\omega + \left(\frac{c^2}{4\pi\sigma}\right) k^2\right)\right] = 0. \qquad (7.74)$$

Suppose we seek real roots of (7.74) which means that $\mathbf{k}$ is real. We then have

$$-\left(\frac{a^2}{4\pi}\right)k^2 + \rho_0\omega^2 = 0. \tag{7.75}$$

Using the definition of the Alfvèn wave speed v_A given by (7.28), we have

$$a = B_0 = \sqrt{(4\pi\rho_0)}\,v_A.$$

Then (7.75) becomes

$$v_A = \pm\frac{\omega}{|\mathbf{k}|}, \tag{7.76}$$

Thus, we see that (7.76) is the Alfvèn wave speed which represents a forward or backward propagating Alfvèn wave. In this special case we note that $\mathbf{k}$ can be in an arbitrary direction in the (x, y) plane, which means that the waves corresponding to this case are in an arbitrary direction.

We now consider the case where $\mathbf{B}_0$ is in the x direction (as above) but, in addition, we set $\mathbf{k} \cdot \mathbf{B}_0 = 0$. This means that $\mathbf{k} = \mathbf{j}k_y$ or the direction of wave propagation is normal to $\mathbf{B}_0$ (is in the y direction). The first equation of (7.69) is identically zero. The last equation tells us that $B_y = 0$ or that $\mathbf{B}'$ is in the same direction as $\mathbf{B}_0$. The remainder of (7.69) reduces to

$$\rho_0\omega v - \left(\frac{1}{4\pi}\right)k_y a B_x = 0,$$
$$-k_y a v + \left[\omega + i\left(\frac{c^2}{4\pi\sigma}\right)k^2\right]B_x = 0. \tag{7.77}$$

We set the determinant of the coefficients of (v, B_x) equal to zero and thereby obtain (7.76) for real $\mathbf{k}$. This tells us that this case also yields propagating Alfvèn waves.

CHAPTER 8

The Special Theory of Relativity

Introduction

In our development of electromagnetic theory the hypothesis of a Galilean transformation was assumed. This is a transformation from one coordinate system to another where the reference frame moves rigidly with a constant velocity with no rotation. Newton's laws of mechanics are invariant under such a transformation. But a Galilean transformation predicts that the velocity of light should be different in the two reference systems. Are not Newton's laws of mechanics therefore incorrect? The answer to this question led to a new and revolutionary system of mechanics, relativity theory, developed by Albert Einstein. Actually Einstein formulated two relativity theories: the special theory which does not account for gravitational effects, and the general theory which reformulates mechanics in the setting of a time–space whose curvature represents the effect of the gravitational field. The general theory is much more revolutionary and entirely a product of Einstein's genius, whereas other investigators such as Poincaré and Lorentz were skirting around the edges of the special theory. Indeed, Poincaré would have discovered the special theory if he had just made the leap of physical intuition that Einstein did. In this chapter we shall be concerned only with the special theory, since this deals with electromagnetic and optical phenomena.

We showed that Faraday's laws lead to Maxwell's equations which unified electricity and magnetism and optics. In a sense, Maxwell's laws forced Einstein to construct the special theory, for he observed that these equations were not invariant under a Galilean transformation, but were so under a Lorentz† transformation, as we shall see in the main body of this chapter.

At this point it is instructive to give a simple example showing that Maxwell's equations are not invariant with respect to a Galilean transformation. We take as our model, one-dimensional wave propagation in the x direction.

---

† Hendrik Antoon Lorentz was a Dutch theoretical physicist who, along with Pieter Zeeman, was awarded the Nobel prize in physics for his research into the influence of magnetism upon radiation.

Let f stand for the electric or magnetic vector. Then $f = f(x, t)$. In studying Maxwell's equations we noted that f satisfies the wave equation, so that we have

$$c^2 f_{xx} - f_{tt} = 0.$$

We now write the Galilean transformation from (x, t) to (x', t') space, taking into account that the coordinate x' translates with a constant velocity v with respect to x. Thus, we have

$$x' = x - vt, \qquad t' = t.$$

Note that t is invariant with respect to this transformation, which is the crux and also the weakness of Newtonian mechanics. From this transformation, $(x, t) \rightleftharpoons (x', t')$, we calculate the form of the wave equation for $f(x', t')$ (what the wave equation looks like in (x', t') space). From the above transformation we obtain

$$\frac{\partial}{\partial x} = \frac{\partial}{\partial x'}$$

$$\frac{\partial^2}{\partial x^2} = \frac{\partial^2}{\partial x'^2}$$

$$\frac{\partial}{\partial t} = (-v)\frac{\partial}{\partial x'} + \frac{\partial}{\partial t'},$$

$$\frac{\partial^2}{\partial t^2} = v^2\frac{\partial^2}{\partial x'^2} - 2v\frac{\partial^2}{\partial x'\partial t'} + \frac{\partial^2}{\partial t'^2}.$$

Using these transformations on the second-order operators, the wave equation for f in (x', t') space transforms into the following partial differential equation (PDE):

$$c^2 f_{x'x'} - f_{t't'} = v^2 f_{x'x'} - 2v f_{x't'}.$$

We note that, since $v \neq 0$, the right-hand side of this equation is not zero. From this we conclude that the Galilean transformation does not transform the wave equation into itself. Obviously, for the trivial case of $v = 0$, the Galilean transformation reduces to the identity transformation so that the wave equation is clearly invariant.

Thus far we have not studied the motion of high-speed charged particles. One of the fundamental parameters of the electron is the charge to mass ratio $r = e/m$. The charge e on an electron is a fundamental invariant of nature (independent of the coordinate system). The mass m, according to Newtonian mechanics, should also be an invariant. However, some crucial experiments done around 1906 by Kaufmann and others showed the rather disturbing fact that the ratio r for high-speed electrons (beta particles) emitted from radioactive substances is less than r for the rest mass. The decrease of r as the speed of the electron increased showed that the mass of the electron would increase with increasing speed and would tend asymptotically to infinity as the speed v of the electron approaches the speed of light c. The departure from the Newtonian idea of constant mass is one of the predictions of the special theory that Einstein formulated in 1905.

In essence, Einstein made a critical re-examination of the fundamental principles of mechanics and electromagnetic theory in the light of the crucial experiments of Michelson and Morley in 1881, who demonstrated that the velocity of light is a universal constant—it is invariant with respect to the rotation of the earth and the motion of any coordinate system. Actually, Michelson and Morley thought their experiment was a failure, because it failed to demonstrate the moving ether with respect to the earth. Anyhow, Einstein concluded that distances and time intervals between events would be given different values by observers moving in coordinate systems relative to each other. To do this he relied on the Lorentz transformations wherein time as well as space is transformed in such a manner that the velocity of light remains constant. It turns out that, in order to impose the constraint of the constancy of the velocity of light, time in the transformed system is different from that in the untransformed system. Therein lies the essential break with Newtonian mechanics. Thus, any observer will always find the same value of light $c = 3 \times 10^8$ m/sec relative to himself regardless of the direction of travel of the light beam. One of the advantages of the Lorentz transformations is to make Maxwell's equations invariant.

In the summer of 1905 Einstein wrote a paper entitled, "On the Electrodynamics of Moving Bodies", which was published in the *Annalen der Physik*. This seminal paper laid the foundation of the special theory of relativity.

It is not the purpose of this chapter to give an historical perspective of the special relativity theory. There are many interesting works on this aspect of relativity, including the more personal aspects of Einstein's life, which the interested reader may want to consult.† Einstein presented a semitechnical treatment of both the special and general theories in [17].

8.1. Collapse of the Ether Theory

Before Einstein came on the scene, investigators conceived of a massless fluid that permeated space and was the vehicle that allowed all waves to be transmitted through space, including electromagnetic waves which can propagate in a vacuum. This fluid was called the "ether". It was thought that this ether was a necessary vehicle for wave transmission in a vacuum. Many theories were proposed to account for the interaction between the ether and the traveling wave. At this point we observe that many first-rate mathematical

---

† See, for example, Clark, Ronald W., *Einstein, The Life and Times*, World Publishing, Singapore, 1971. Hoffmann, Banesh, *Albert Einstein, Creator and Rebel*, The Viking Press, New York, 1972.

techniques were developed to attempt to understand wave propagation in this setting. When the concept of the ether was abandoned, it was as if the scaffolding of the solid edifice of mathematical theorems was taken away, leaving the mathematics which now stands on its own feet without the artificial and false ether concept.

The earth moves around the sun with a velocity of around 20 miles/sec. If the ether is at rest with respect to the sun and the earth, then this fast motion of the ether with respect to the earth should make itself felt in a change of the velocity of light. This velocity should be different if the light were propagated in a direction parallel and perpendicular to the motion of the ether. Even if the ether does partly move with the earth, there should be some effect due to the "ether wind". A calculation of the expected effect showed that it should be very small, since it turns out to be proportional to the square of the ratio of the velocity of the earth to that of light, which is approximately 1.16×10^{-8}. One must carry out careful experiments on the interference of two light beams traveling parallel or perpendicular to the motion of the earth. This is exactly what Michelson and Morley did, as was indicated in the Introduction. It turned out that their first experiment was not sufficiently accurate. The experiment was repeated a number of times with more refined apparatus. Finally, the experiments of Morley and Miller in 1904 gave definite experimental proof that the effect of the expected order of magnitude did not exist. The inescapable conclusion was that the ether could not be detected, and the velocity of light was therefore a universal constant independent of its direction with respect to the motion of the earth.

It was also pointed out in the Introduction that Newton's laws of motion were invariant with respect to a Galilean transformation, which means that a uniform translational motion of a system does not produce any mechanical effects at all. We also showed that the laws of electromagnetic wave propagation as described by Maxwell are not invariant with respect to a Galilean transformation. This means that the effect of the moving ether on electromagnetic wave propagation should be felt. The negative results of Michelson, Morley, and Miller revived the idea that, like Newton's equations, Maxwell's equations should also be invariant with respect to a Galilean transformation. But, they were not! Again, this led to a serious flaw in the nature of time and space as described by Newton, where he postulated that time flows uniformly independent of the coordinate system.

On the other hand, there was an experiment done by Fizeau in 1851 that appeared to definitely contradict the Galilean principle of relativity. Fizeau had measured the velocity of light in a moving liquid. If the Galilean principle was correct, then the total velocity of light in the moving fluid should be the sum of the velocity of light in the fluid at rest and the velocity of the fluid. But this was not the case. Fizeau's experiment showed that the total velocity in the moving fluid was somewhat smaller.

Theoreticians were therefore inspired to look for other mathematical interpretations that would reconcile the wave equation for the propagation of light with perhaps another relativity principle. To this end, Lorentz suggested, in 1904, a mathematical transformation that would fulfill these requirements. He introduced the hypothesis that moving bodies are contracted in the direction of motion by a factor which depends on the velocity of the body. In different reference systems there are different "apparent" times which in many ways take the place of "real" time. His relativity principle showed that the "apparent" velocity of light was the same in every reference frame (which it should be according to the experiments of Michelson *et al.* Similar ideas were discussed by Poincaré, Fitzgerald, and other physicists.

The decisive step was taken by Einstein in his paper mentioned in the Introduction. In this paper he established Lorentz's "apparent" time as the "real" time and abolished what Lorentz called the "real" time. This concept of introducing "real" time, as a quantity that depends on the motion of a coordinate system, changed the very foundations of physics. This very radical and unexpected change took the courage of a young and revolutionary genius. To take this step, all Einstein needed to do was to consistently apply the Lorentz transformation in accordance with his new interpretation of the structure of time and space. This gave the physicists, not only a new and penetrating insight into wave propagation, but also showed them a new system of mechanics. For example, the substance ether, which gave theoreticians such trouble, could be abolished. Since all reference systems, in uniform translational motion with respect to each other, are equivalent for describing natural phenomena, there is no meaning to the statement that there is a substance (called ether) which is at rest in only one of these systems. Since the concept of ether can be abandoned it is much easier to say that electromagnetic waves can propagate in a vacuum without the need for any intervening medium. Thus EM waves have their own reality, solely by virtue of the electric and magnetic fields that are created.

8.2. The Lorentz Transformation

We now investigate in detail the Lorentz transformation alluded to above. We want the simplest transformation from the space–time (x, y, z, t) to the (x', y', z', t') space–time that preserves the constancy of the velocity of light c. The primed space–time reference system is translating with a constant velocity v with respect to the unprimed reference system. Let the origins of both systems coincide at the time $t = 0$, at which time a source of light fixed at the origin emits a signal in the form of a light pulse. The medium is assumed to be a vacuum. An observer in the unprimed system who is fixed at the origin will see a spreading spherical wave which propagates with the velocity c. The observed wave front satisfies the equation

$$x^2 + y^2 + z^2 = c^2t^2. \tag{8.1}$$

At $t = 0$ the radius of the front is zero since the observer is at the origin and the source begins to send out the pulse at $t = 0$. As t increases the radius of the spherical wave front increases as ct. The constraint of the constancy of c indicates that an observer in the primed reference system will also see the spherical wave front expanding at the velocity c. Thus the observer situated at the origin of the primed system will see that the wave front will satisfy the equation

$$x'^2 + y'^2 + z'^2 = c^2 t'^2. \tag{8.2}$$

Note again that $t' \neq t$ except for the trivial case where there is no relative motion between reference systems. Thus the desired transformation from the unprimed to the primed system must be such that the following equation is valid:

$$x^2 + y^2 + z^2 - c^2 t^2 = x'^2 + y'^2 + z'^2 - c^2 t'^2. \tag{8.3}$$

Equation (8.3) is reminiscent of an orthogonal transformation except for the fact that we have the $-c^2 t^2$ and $-c^2 t'^2$ terms.

Hermann Minkowski[†] took a decisive step in formulating what is called *world space* or *Minkowski space*. To get a formal correspondence with an orthogonal transformation in four-dimensional space, he invented the following symmetric notation:

$$x = x_1, \qquad y = x_2, \qquad z = x_3, \qquad ict = x_4.[‡] \tag{8.4}$$

The primed reference frame transforms to

$$x' = y_1, \qquad y' = y_2, \qquad z' = y_3, \qquad ict' = y_4. \tag{8.4a}$$

Using this notation (8.3) then becomes

$$\sum_{\nu=1}^{4} x_\nu^2 = \sum_{\nu=1}^{4} y_\nu^2. \tag{8.5}$$

Equation (8.5) is the condition of orthogonality. It tells us that the magnitude of the vector $\mathbf{x}$ is the same as that of $\mathbf{y}$. The Lorentz transformation (which we have yet to determine) is simply the orthogonal transformation from the $\mathbf{x}$ to the $\mathbf{y}$ reference system in Minkowski space. We may rotate these reference systems so that y_1 is contiguous to x_1, y_4 is contiguous to x_4, and the $\mathbf{y}$ system moves with a constant velocity v in the x_1 or y_1 direction. The orthogonal transformation from $\mathbf{x}$ to $\mathbf{y}$ space then becomes (in extended form):

$$y_1 = a_{11} x_1 + a_{14} x_4,$$

$$y_2 = x_2,$$

[†] Hermann Minkowski was a Russo-German mathematician who was a professor at the University of Göttingen. He presented an outstanding contribution to relativity in 1907.

[‡] Note the introduction of "imaginary time" given by x_4 so that some of the elements of the transformation matrix **A** are imaginary, as shown by (8.11).

$$y_3 = x_3,$$

$$y_4 = a_{41}x_1 + a_{44}x_4. \tag{8.6}$$

Note that y_2 and y_3 are invariant with respect to the transformation. The coefficients $a_{11}, a_{14}, a_{41},$ and a_{44} are to be determined from the orthogonality condition and the motion of y relative to x. The matrix representation of (8.6) is

$$\mathbf{y} = \mathbf{Ax}, \qquad \mathbf{x} = \mathbf{A}^{-1}\mathbf{y}, \tag{8.7}$$

where y is the column matrix whose elements are (y_1, y_2, y_3, y_4), x is the column matrix whose elements are (x_1, x_2, x_3, x_4), A is the orthogonal matrix given by (8.8) below, and $\mathbf{A}^{-1}$ is the inverse of A.

$$\mathbf{A} = \begin{pmatrix} a_{11} & 0 & 0 & a_{14} \\ 0 & 1 & 0 & 0 \\ 0 & 0 & 1 & 0 \\ a_{41} & 0 & 0 & a_{44} \end{pmatrix}. \tag{8.8}$$

The second equation of (8.7) is the inverse transformation which allows one to calculate x if A and y are known. Note that A is *nonsingular*, meaning that $\mathbf{A}^{-1}$ exists. We shall show that $\mathbf{A}^{-1}$ is equal to A* (A transpose) so that A is an orthogonal matrix. The matrix representation of the orthogonality condition (8.5) (invariance of length) is

$$\mathbf{y}^*\mathbf{y} = \mathbf{x}^*\mathbf{x}, \tag{8.9}$$

where y* is the transpose of the column matrix y, etc. Inserting (8.7) into (8.9) gives

$$\mathbf{y}^*\mathbf{y} = (\mathbf{Ax})^*\mathbf{Ax} = \mathbf{x}^*(\mathbf{A}^*\mathbf{A})\mathbf{x} = \mathbf{x}^*\mathbf{x},$$

which tells us that $\mathbf{A}^*\mathbf{A} = \mathbf{I}$ where I is the (4×4) identity matrix. This means that $\mathbf{A}^{-1} = \mathbf{A}^*$. This justifies the assertion made above that A is an orthogonal matrix. From (8.9) we obtain the orthogonality conditions on the elements of A which we write in the following extended form:

$$a_{11}^2 + a_{14}^2 = 1, \qquad\qquad a_{11}^2 + a_{41}^2 = 1,$$

$$a_{41}^2 + a_{44}^2 = 1, \qquad\qquad a_{14}^2 + a_{44}^2 = 1, \tag{8.10}$$

$$a_{11}a_{14} + a_{41}a_{44} = 0, \qquad a_{11}a_{41} + a_{14}a_{44} = 0, \qquad a_{14} = -a_{41}.$$

The system (8.10) tells us that A is not only orthogonal but is also *orthonormal*. The orthogonality condition means that the scalar product of each row of A with any other row is zero; and the same statement is true for any column. The "normality" condition arises from the fact that the magnitude of each row of A is unity, as is that of each column.

The system (8.10) does not allow us to determine all the elements of A. Therefore we need one more condition. This is obtained by recognizing that the y system is translated in the x_1 direction with a constant velocity v. At zero time (in each system) the origins of **y** and **x** are in coincidence. At time t the origin of **y** is at

$$x_1 = vt = -i\beta x_4, \quad \text{for} \quad y_1 = 0, \quad \text{where } \beta = \frac{v}{c},$$

upon using (8.4). Setting $y_1 = 0$ in (8.6) and using the above expression for x_1 gives

$$a_{14} = i\beta a_{11}.$$

This expression coupled with the orthogonality conditions yields the following results for the elements of **A**:

$$a_{11} = \frac{1}{\sqrt{1-\beta^2}}, \quad a_{14} = \frac{i\beta}{\sqrt{1-\beta^2}} = -a_{41}, \quad a_{44} = \frac{1}{\sqrt{1-\beta^2}},$$

$$a_{12} = a_{13} = a_{21} = a_{23} = a_{24} = a_{31} = a_{32} = a_{34} = a_{42} = a_{43} = 0, \quad a_{22} = a_{33} = 1.$$
$$(8.11)$$

Substituting these expressions into (8.6) gives the Lorentz transformation from the **x** to the **y** system (in Minkowski space)

$$y_1 = \left(\frac{1}{\sqrt{1-\beta^2}}\right)(x_1 + i\beta x_4),$$

$$y_2 = x_2,$$

$$y = x_3,$$

$$y_4 = \left(\frac{1}{\sqrt{1-\beta^2}}\right)(-i\beta x_1 + x_4).$$

$$(8.12)$$

Transforming the system (8.12) into physical variables, upon using (8.4), we obtain

$$x' = \left(\frac{1}{\sqrt{1-\beta^2}}\right)(x - vt),$$

$$y' = y,$$

$$z' = z,$$

$$t' = \left(\frac{1}{\sqrt{1-\beta^2}}\right)\left[t - \left(\frac{v}{c^2}\right)x\right].$$

$$(8.13)$$

The system (8.13) is the Lorentz transformation from (x, y, z, t) to (x', y', z', t') space. The first thing we observe is that for small particle motion $v \ll c$, so that we can neglect β^2 in the Lorentz transformation since (v/c) enters as a second-order term. We see immediately that (8.13) reduces to

$$x' = x - vt,$$

$$y' = y,$$

$$z' = z,$$

$$t' = t, \tag{8.14}$$

which is the Galilean transformation from the unprimed to the primed coordinate system, where time is invariant with respect to this transformation. This is the reason why Newton's laws of mechanics is so successful in treating systems involving small motion. Of course, Maxwell's equations are still not invariant with respect to a Galilean transformation. They are invariant with respect to a Lorentz transformation, which we shall now show.

8.3. Maxwell's Equations with Respect to a Lorentz Transformation

The following analysis, involving a Lorentz transformation applied to Maxwell's equations, is an generalization to three spatial dimensions of the Galilean transformation applied to the one-dimensional wave equation given in the Introduction. As before, we let f stand for any component of the electric intensity vector $\mathbf{E}$ or the magnetic flux density $\mathbf{B}$. Then f satisfies the three-dimensional wave equation in empty space

$$c^2(f_{xx} + f_{yy} + f_{zz}) - f_{tt} = 0. \tag{8.15}$$

We use the Lorentz transformation (8.13) to transform (8.15) to an analogous PDE in the primed coordinate system, and show that the resulting equation is again the wave equation in the primed coordinate system. For simplicity, we set

$$\gamma = \frac{1}{\sqrt{1 - \beta^2}}.$$

From (8.13) we obtain

$$\frac{\partial}{\partial x} = \frac{\gamma \partial}{\partial x'} - \left(\frac{\beta \gamma}{c}\right)\frac{\partial}{\partial t'},$$

$$\frac{\partial^2}{\partial x^2} = \gamma^2 \frac{\partial^2}{\partial x'^2} - \left(\frac{2\gamma^2 \beta}{c}\right)\frac{\partial^2}{\partial x'\partial t'} + \left(\frac{\beta^2\gamma^2}{c^2}\right)\frac{\partial^2}{\partial t'^2},$$

$$\frac{\partial}{\partial t} = -\gamma v \frac{\partial}{\partial x'} + \gamma \frac{\partial}{\partial t'}, \tag{8.16}$$

$$\frac{\partial^2}{\partial t^2} = \gamma^2 v^2 \frac{\partial^2}{\partial x^2} - 2\gamma^2 v \frac{\partial^2}{\partial x'\partial t} + \gamma^2 \frac{\partial^2}{\partial t'^2},$$

$$\frac{\partial^2}{\partial y^2} = \frac{\partial^2}{\partial y'^2}, \qquad \frac{\partial^2}{\partial z^2} = \frac{\partial^2}{\partial z'^2}.$$

Inserting the expressions for the second-order operators in (8.16) into (8.15), we obtain

$$c^2(f_{x'x'}) + f_{y'y'} + f_{z'z'}) - f_{t't'} = 0, \tag{8.17}$$

from which we conclude that Maxwell's equations are invariant with respect to the Lorentz transformation.

8.4. Contraction of Rods and Time Dilation

We first consider a rigid rod of length L in the unprimed coordinate system which lies along the x axis. This rod serves as a measuring instrument. The problem is to determine its length in the primed coordinate system as a result of the Lorentz transformation. If the end points of the rod are $(x_1, 0, 0, t)$ and $(x_2, 0, 0, t)$, we then obtain the expression

$$L = x_2 - x_1. \tag{8.18}$$

The inverse Lorentz transformation is obtained from (8.13) by merely reversing the sign of the velocity v. Clearly, it may be obtained by constructing $\mathbf{A}^*$, using (8.11) and transforming to the physical variables by using (8.4). We have

$$x = \left(\frac{1}{\sqrt{1 - \beta^2}}\right)(x' + vt'),$$

$$y = y',$$

$$z = z', \tag{8.13a}$$

$$t = \left(\frac{1}{\sqrt{1 - \beta^2}}\right)\left[t' + \left(\frac{v}{c^2}\right)x'\right].$$

Using this inverse transformation the length L' of the rod in the primed coordinate system becomes

$$x_2' - x_1' = L' = \frac{L}{\sqrt{1 - \beta^2}}. \tag{8.19}$$

Equation (8.19) tells us that the rod will appear contracted, by the factor $\sqrt{1 - \beta^2}$, to the observer moving with a velocity v in the x direction. (Note that it does not matter whether the observer is moving in the positive or negative x direction.) This result is essentially the famous Lorentz–Fitgerald contraction hypothesis, which was first used by Fitzgerald in an unsuccessful attempt to resurrect the ether hypothesis. Observe that it is not convenient to use the direct Lorentz transformation (8.13); for while both ends are measured at the same time t these are *not simultaneous events in the primed space*, since they are at different points x_1' and x_2'. Clearly for small velocity of the observer so that the second-order term $(v/c)^2$ term can be neglected $L' = L$, so that there is no contraction of the length. Also if $v = c$ then the length L' is contracted to a point, and if $v > c$ then L' is imaginary which again tells us that c is the maximum velocity that any mechanical system can have. We emphasize that the motion of the primed with respect to the unprimed coordinate systems is *relative*. Thus an observer in the unprimed system observes the same contraction of the length of the rod if it is fixed in the primed system. Therefore the motion of coordinate systems is relative in the sense that no one system is singled out as stationary and the other as moving. (The ratio v/c appears as a squared quantity.) This is the important principle of Einstein: *All uniformly moving systems are completely equivalent.*

We now investigate how a clock would measure time in the primed system if we prescribe a time interval in the unprimed system. Suppose the time interval Δt in the unprimed system is given by $\Delta t = t_2 - t_1$. Using the inverse transformation (8.13a) the corresponding time interval $\Delta t' = t'_2 - t'_1$ in the primed system becomes

$$t'_2 - t'_1 = \sqrt{1 - \beta^2}(t_2 - t_1).\dagger \qquad (8.20)$$

We will call the term $(1/\sqrt{1 - \beta^2})$ the *dilatation factor*. It is clearly greater than one for $\beta^2 > 0$. From (8.20) we deduce that, if the clock in the unprimed system (stationary with respect to the primed system) has a time interval of one hour, then the moving clock (in the primed system) has an interval of $(1/\sqrt{1 - \beta^2})$ hours, which clearly is greater than that of the stationary clock. Thus the moving observer will say that the stationary clock is slow or is losing time. This is the famous *time dilatation* principle which arises from the Lorentz transformation. By the principle of equivalence of uniformly moving coordinate systems, the observer in the unprimed system will also observe that the clock in the primed system is losing time with respect to his system.

One of the most fundamental of clocks that occurs in nature is due to the decay of radioactive particles. Each species of radioactive particle decays with a characteristic "half-life" that is independent of the external environment. If these particles travel with velocities that approach that of light, then dilatation can be clearly demonstrated. For example, consider high-energy mu mesons that occur in cosmic rays. These mesons are produced as secondary particles at heights of 10 or 20 km, and a large fraction of them reach the earth. But the mean lifetime of a mu meson is about $t_m = 2.2$ microsec. Therefore it could travel a distance no greater than $ct_m = 0.66$ km before decaying (actually it must be less than this distance because the particle has mass, and if it traveled with the velocity of light then its mass would be infinite, as we shall show later). How is it that it actually travels 10 to 20 km before decaying? The answer is that a time dilatation occurs which is of the order of ten or more since these mesons are high-energy particles and therefore travel with a velocity approaching that of light.

---

† Setting $x_1 = x_2$ in (8.13) yields this result.

Suppose the time dilatation factor is equal to γ which is a number greater than one. Then the time dilatation factor becomes

$$\frac{1}{\sqrt{1 - \beta^2}} = \gamma,$$

which means that the velocity of the meson is

$$v = \sqrt{(1 - \gamma^{-2})}c.$$

If $\gamma = 10$ then $v = \sqrt{99/100}$, $c = 9.9499c$, which is a high velocity associated with the high-energy meson.

8.5. Addition of Velocities

As mentioned several times, one of the fundamental principles of physics, as demonstrated by Michelson *et al.*, is that the velocity of light in a vacuum is independent of any coordinate system and is therefore an absolute constant of nature. Einstein as a young boy of sixteen puzzled over what would happen if he rode a light beam and shone another beam in front of it. According to Newton and Galileo this beam should travel with twice the velocity of light with respect to an observer fixed at a point on the earth. This was the result of their simply adding the velocity of the moving flashlight to the velocity of light. The experimental fact that light cannot travel greater than the velocity c led the young Einstein to puzzle over this paradox for a number of years. It was only when he became aware of the Lorentz transformation did he understand that this was the mathematical apparatus that would resolve this paradox. The Lorentz transformation taught him that velocities cannot be added in a simple manner as Newton and Galileo did. In discussing the derivation of the Lorentz transformation we showed that Lorentz used the principle that the velocity of light must be the same in any coordinate system moving uniformly with respect to another. This invariance of the velocity of light was the cornerstone of the Lorentz transformation.

Applying the Lorentz transformation successively we can show that a different law (different from that of Galileo and Newton) of adding velocities arises. By "applying the Lorentz transformation successively" we mean the following: Consider an observer fixed in the system $\mathbf{x} = (x, y, z, t)$. Let the $\mathbf{x}_1$ system move with respect to the $\mathbf{x}$ system with a velocity v_1. Consider another system $\mathbf{x}_2$ which moves with a velocity v_2 with respect to the $\mathbf{x}_1$ system. The problem is to find the velocity of $\mathbf{x}_2$ with respect to our observer in the $\mathbf{x}$ system. We have the following transformations: $\mathbf{x} \to \mathbf{x}_1 \to \mathbf{x}_2$, which are given by

$$\mathbf{x}_1 = \mathbf{A}_1\mathbf{x}, \qquad \mathbf{x}_2 = \mathbf{A}_2\mathbf{x}_1, \tag{8.21}$$

where the A_j's ($j = 1, 2$) are given by

$$
A_j = \begin{pmatrix} \gamma_j & 0 & 0 & -v_j\gamma_j \\ 0 & 1 & 0 & 0 \\ 0 & 0 & 1 & 0 \\ -\gamma_j v_j/c^2 & 0 & 0 & \gamma_j \end{pmatrix}, \tag{8.22}
$$

where

$$
\gamma_j = \frac{1}{\sqrt{1 - \beta_j^2}}, \qquad \beta_j = \frac{v_j}{c}, \qquad j = 1, 2. \tag{8.23}
$$

From (8.21) we obtain

$$
x_2 = A_2(A_1 x) = Cx, \qquad C = A_2 A_1. \tag{8.24}
$$

The matrix C which transforms x to x_2 is obtained by premultiplying A_1 by A_2. We obtain

$$
C = \begin{pmatrix} \gamma_1\gamma_2[1 + (v_1 v_2/c^2)] & 0 & 0 & -\gamma_1\gamma_2(v_1 + v_2) \\ 0 & 1 & 0 & 0 \\ 0 & 0 & 1 & 0 \\ -(\gamma_1\gamma_2)/c^2(v_1 + v_2) & 0 & 0 & \gamma_1\gamma_2[(v_1 v_2/c^2) + 1] \end{pmatrix}. \tag{8.25}
$$

Inserting (8.25) into (8.24) yields the transformation from x to x_2, which is given by

$$
\begin{aligned}
x_2 &= \gamma_1\gamma_2\left[1 + \frac{v_1 v_2}{c^2}\right]\left[\frac{x - (v_1 + v_2)t}{1 + (v_1 v_2/c^2)}\right], \\
y_2 &= y, \\
z_2 &= z, \\
t_2 &= \gamma_1\gamma_2\left[\frac{v_1 v_2}{c^2} + 1\right]\left[\frac{t - (1/c^2)(v_1 + v_2)x}{1 + (v_1 v_2/c^2)}\right].
\end{aligned} \tag{8.26}
$$

We now define v by the following equation:

$$
v = \frac{v_1 + v_2}{1 + (v_1 v_2/c^2)}. \tag{8.27}
$$

If we were to insert (8.27) into (8.26) we would note that the transformation (8.26) from x to x_2 is the same (apart from a multiplying factor) as (8.13), which is the transformation from the x to the x' system. Therefore v, as given by (8.27), is the sum of the velocities v_1 and v_2 with respect to the Lorentz transformation. Note that the term $v_1 v_2/c^2$ is the second-order effect that makes this addition different from that of Galileo and Newton if $v_1 v_2$ approach c^2. For small $v_1 v_2/c^2$ (8.27) reduces to Newton's addition $v = v_1 + v_2$ as an approximation. Suppose we have two positive constants k_1 and k_2 (each of which is not greater than one) such that

$$v_1 = k_1 c, \qquad v_2 = k_2 c.$$

Then (8.27) becomes

$$v = \left(\frac{k_1 + k_2}{1 + k_1 k_2} \right) c.$$

This expression tells us that, for the upper limit where $k_1 = k_2 = 1$, $v = c$, so that the sum of the velocities can never be greater than the velocity of light. For example, suppose $k_1 = k_2 = \frac{1}{2}$. Then we have $v = \frac{3}{4}c$, whereas Newton would tell us that $v = c/2$.

We now apply this rule of addition of velocities to measuring rods and clocks. First we note that, if a rod were contracted in going from x to x_1 and then contracted again in going from x_1 to x_2, the resulting contraction of the rod in going from x to x_2 is

$$L_2 = \left[\frac{\sqrt{(1 - \beta_1^2)(1 - \beta_2^2)}}{1 + (v_1 v_2/c^2)} \right] L.$$

This is seen from the inverse transformation corresponding to (8.26). And finally, the time dilatation for a clock in going from x to x_2 is given by

$$t_2 = \left[\frac{1 + (v_1 v_2/c^2)}{\sqrt{(1 - \beta_1^2)(1 - \beta_2^2)}} \right] t.$$

8.6. World Lines and Light Cones

We return to the world space or Minkowski space described in the section on "The Lorentz Transformation". At this point an historical note is appropriate. In 1908 at Cologne, Minkowski gave a famous lecture at the Eightieth Congress of German Scientists and Physicians. It was semitechnical and was noted for his sensational introductory words: "The views of space and time which I wish to lay before you have sprung from the soil of experimental physics, and therein lies their strength. They are radical. From now on space by itself, and time by itself, are destined to sink completely into shadows, and only a kind of union of the two will preserve an independent existence." This union is the *space–time* universe that Minkowski obtained from the Lorentz tranformation by introducing an "imaginary time" which allowed for the orthogonal tranformation that we previously discussed. But the important unification was that time and space merge with each other in Minkowski's space or the world space. We again point out that the Lorentz tranformation involves a time associated with each coordinate system. This clearly means that any event that occurs in nature must be described in this four-dimensional space–time continuum. The trajectory of a particle in this Minkowski space is called a *world line*.

For simplicity, we consider one spatial dimension so that the space $\mathbf{x} = (x, t)$ is mapped into $\mathbf{x}' = (x', t')$ by the Lorentz transformation, where $\mathbf{x}'$ is moving uniformly with a velocity v with respect to $\mathbf{x}$. The Lorentz transformation given by (8.13) and its inverse become

$$x' = \gamma(x - vt), \qquad x = \gamma(x' + vt'), \qquad \gamma = \frac{1}{\sqrt{1 - \beta^2}}, \qquad \beta = \frac{v}{c},$$

$$t' = \gamma\left[t - \left(\frac{v}{c^2}\right)x\right], \qquad t = \gamma\left[t' + \left(\frac{v}{c^2}\right)x'\right].$$

(8.28)

Consider two points in $\mathbf{x}$ space given by $P_1 : (x_1, t_1)$ and $P_2 : (x_2, t_2)$. The Lorentz transformation will map them into the corresponding points $P_1' : (x_1', t_1')$ and $P_2' : (x_2', t_2')$. Let $(\Delta s)^2$ and $(\Delta s')^2$ be defined by

$$(\Delta s)^2 = (x_2 - x_1)^2 - c^2(t_2 - t_1)^2,$$
$$(\Delta s')^2 = (x_2' - x_1')^2 - c^2(t_2' - t_1')^2.$$

(8.29)

$(\Delta s)^2$ is the *metric* associated with $\overline{P_1 P_2}$ (the square of the distance) and $(\Delta s')^2$ is the corresponding metric for $\overline{P_1' P_2'}$. The transformation (8.28) is the representation in one-dimensional space of the Lorentz transformation given in Minkowski space by (8.12). The correspondence is given by

$$x_1 = x, \qquad x_2 = 0, \qquad x_3 = 0, \qquad x_4 = ict,$$
$$y_1 = x', \qquad y_2 = 0, \qquad y_3 = 0, \qquad y_4 = ict'.$$

The metrics transform into

$$(\Delta s)^2 = (x_{12} - x_{11})^2 + (x_{42} - x_{41})^2, \qquad P_1 : (x_{1i}, x_{4i}), \qquad i = 1, 2,$$
$$(\Delta s')^2 = (y_{12} - y_{11})^2 + (y_{42} - y_{41})^2, \qquad P_i' : (y_{1i}, y_{4i}).$$

The orthogonality condition expressing the invariance of the metric or length is given by (8.9). Recall that this is the fundamental basis of the Lorentz transformation which rests upon the invariance of c with respect to the two coordinate systems. The orthogonality condition for our one spatial dimension model then becomes

$$(\Delta s)^2 = (\Delta s')^2.$$

(8.30)

Equation (8.30) expresses the invariance of the metric $\overline{P_1 P_2}$ as it maps into $\overline{P_1' P_2'}$. We may independently derive (8.30) by inserting the Lorentz transformation (8.28) into the second equation of (8.29), or the inverse into the first equation.

Suppose $(\Delta s)^2 = 0$. Then $(\Delta s')^2 = 0$, so that

$$\frac{x_2 - x_1}{t_2 - t_1} = \pm c,$$

$$\frac{x_2' - x_1'}{t_2' - t_1'} \pm c.$$

This means that P_2 is traveling with a speed c with respect to P_1, and similarly for P_2' with respect to P_1'. Let P_1 and P_1' be at their origins, and let $P_2 = P$ and $P_2' = P'$ be variable points. Then P and P' are generated by the following equations:

$$P: \quad x = \pm ct, \qquad P': \quad x' = \pm ct'. \tag{8.31}$$

These are the light paths that P follows in the x system and P' follows in the x' system. These trajectories are the two-dimensional space–time analogues of the *light cones* generated by P and P' if they satisfied the following equations in the four-dimensional space–time continuum:

$$P: \quad (\Delta s)^2 = x^2 + y^2 + z^2 - (ct)^2 = 0,$$
$$P': \quad (\Delta s')^2 = x'^2 + y'^2 + z'^2 - (ct')^2 = 0. \tag{8.32}$$

The first equation consists of two coaxial cones (t being the axis) in the four-dimensional x space whose vertex is at the origin and whose vertex angle is 90°. The origin represents the present time. Any point in the interior of the cone for $t > 0$ represents an event in the future and any point in the interior of the cone for $t < 0$ represents a point in the past. These points (both in the past and in the future) have trajectories (*world lines*) with a velocity $v < c$. Any point *on* each cone travels with a velocity c. Any point *outside* either cone is in the *elsewhere* region, meaning that it cannot be reached by a signal on or inside the cone, since it would require a velocity greater than that of light to reach that point. Similar statements hold for the light cones in the primed coordinate system.

For any two events P_1 and P_2 there are three possibilities:

(1) $(\Delta s)^2 = 0$, implying that $(\Delta s')^2 = 0$. P_2 travels on the light cone in the unprimed and primed systems with respect to P_1 with the velocity of light.

(2) $(\Delta s)^2 > 0$. Then the events are said to have a *space-like* separation, since we can always find a Lorentz transformation to the primed system such that $t_2' - t_1' = 0$ which yields

$$(\Delta s')^2 > 0,$$

$$(\Delta s)^2 = (1 - \beta^2)(x_2 - x_1)^2 + (y_2 - y_1)^2 + (z_2 - z_1)^2 > 0, \qquad \beta^2 < 1,$$

since $t = (v/c^2)x$. If the event P_1 is at the origin (the present) then the event P_2 must appear in the elsewhere region, both in the unprimed and primed systems. Clearly, this means that a signal sent by P_1 cannot be reached by P_2 in either system.

(3) $(\Delta s)^2 < 0$. The two events have a *time-like* separation. We can find a Lorentz transformation such that P_2 is coincident with P_1 in the primed system, meaning that $x_2' = x_1'$, $y_2' = y_1'$, $z_2' = z_1'$. This gives

$$(\Delta s')^2 = -c^2(t_2' - t_1')^2 < 0,$$

$$(\Delta s)^2 = \gamma^2(1 - \beta^2)(t_2' - t_1')^2 < 0, \qquad \beta^2 < 1.$$

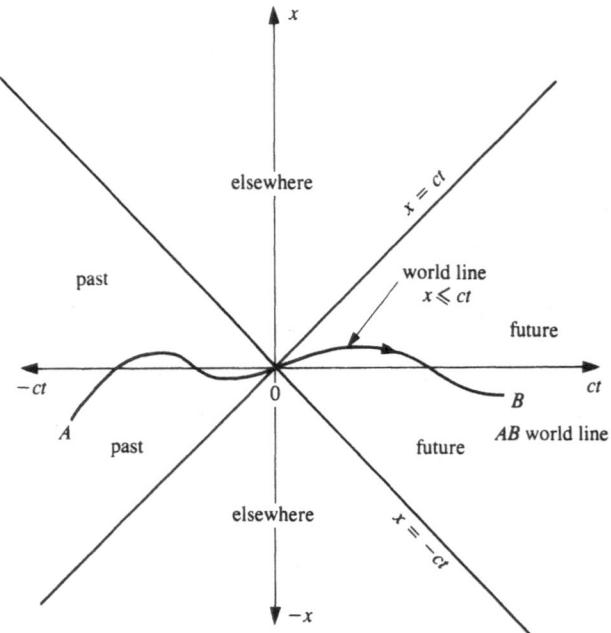

Fig. 8.1. Light cones and world line.

In the unprimed system we set P_1 at the origin and P_2 inside the light cone for $t_2 > 0$. This means that P_1 is a present event and P_2 is a future event which can sense a signal at time t_2 that is sent by P_1 at t_1. Thus we see that the separation of two events in space–time into the classes, space-like and time-like, is a Lorentz invariant division, meaning that the same class holds for the primed system as for the unprimed system.

Since we cannot get a graphical picture of these light cones in four-dimensional space–time, we revert to our two-dimensional space–time model. Figure 8.1 gives such a picture, where x is plotted versus ct (to preserve dimensions). The figure shows the various regions—past, present, future, and elsewhere. A world line for a particle traveling with velocity v is shown going from the past to the future. A similar diagram may be made for the primed system.

8.7. Covariant Formulation of the Laws of Physics in Minkowski Space

We have formulated the Lorentz transformation and cast it in the symmetric form given by Minkowski by introducing the "imaginary time" $ict = x_4$, thus yielding the Minkowski four-dimensional space $\mathbf{x} = (x_1, x_2, x_3, x_4)$. In this space $(\Delta s)^2 = \sum_{v=1}^{4}(x_{2v} - x_{1v})^2$ and $(\Delta s')^2 = \sum_{v=1}^{4}(y_{2v} - y_{1v})^2$, where $\mathbf{y} =$

Ax.† We saw that the Lorentz transformation is invariant with respect to whether events have space-like or time-like separation. But this invariance under the Lorentz transformation is not the only invariant property demanded by the laws of physics as envisaged by Einstein. The laws of electromagnetism as well as those of mechanics must have the same form in all uniformly moving coordinate systems. The task of investigating the laws of physics for invariance under the Lorentz transformation is greatly facilitated by writing them in terms of the four-dimensional world created by Minkowski (as described above).

Clearly, the physical content of any relationship cannot be affected by the particular orientation chosen for the spatial axes, which means that the laws of physics must be invariant with respect to a rigid rotation or a spatial orthogonal transformation. The Lorentz transformation in Minkowski space involves a rigid rotation in the (x_1, x_4) plane where the mapping into y space involves a translation along the x_1 axis. In part, a little reflection will convince the reader that the Lorentz transformation, given by the matrix **A** defined by (8.8) and (8.11), is an orthogonal transformation or a rigid rotation in the (x_1, x_4) plane through the imaginary angle ψ defined by

$$\tan \psi = i\left(\frac{v}{c}\right) = i\beta. \tag{8.33}$$

Figure 8.2 shows a rotation of axes through the angle ψ. The coordinates of the point P relative to the two sets of axes are related by

$$y_1 = \cos \psi x_1 + \sin \psi x_4,$$
$$y_4 = -\sin \psi x_1 + \cos \psi x_4. \tag{8.34}$$

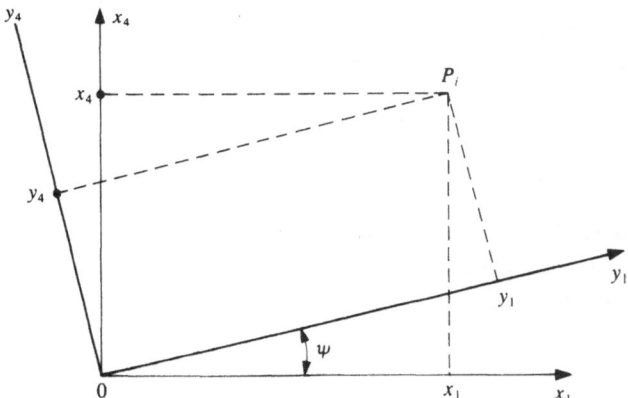

Fig. 8.2. Lorentz transformation as a rotation of axes.

---

† Refer to (8.7), (8.8), and (8.11).

Note that the coefficients of the right-hand side of (8.34) can be correlated with the appropriate (2×2) submatrix of (8.8), upon using the relation $\tan \psi = i\beta$.

The Lorentz transformation $\mathbf{x} \to \mathbf{y}$ given by (8.7) may be put in the following form:

$$y_\mu = a_{\mu\nu} x_\nu, \qquad \mu, \nu = 1, \ldots, 4. \tag{8.35}$$

The system (8.35) consists of four equations. The coefficients $a_{\mu\nu}$ are given by (8.11). We used the Einstein tensor summation convention where the right-hand side of each equation is summed over the double index ν from 1 to 4. (Note: It is customary to use Greek letters as subscripts for summation from 1 to 4, and latin letters for summation from 1 to 3.) The inverse transformation of (8.35) is

$$x_\mu = a_{\nu\mu} y_\nu, \tag{8.36}$$

since $\mathbf{A}^{-1} = \mathbf{A}^*$.

Any vector in Minkowski four-space has four components. It is called a *four-vector* or *world vector*. Every world vector transforms according to (8.35). Let $\bar{\mathbf{v}} = (\bar{v}_1, \bar{v}_2, \bar{v}_3, \bar{v}_4)$ be a world vector. To avoid confusion with spatial vectors a four-vector will be denoted by a typical one of its components $\bar{v}_\mu$ and an overbar. Then the transformation of $\mathbf{v}$ from $\mathbf{x}$ to $\mathbf{y}$ space is given by

$$\bar{v}_\mu(\mathbf{y}) = a_{\mu\nu} \bar{v}_\nu(\mathbf{x}) \qquad \text{(sum over } \nu). \tag{8.37}$$

Equation (8.37) tells us that each component of $\mathbf{v}$ in $\mathbf{y}$ space is a linear combination of all four components of $\mathbf{v}$ in $\mathbf{x}$ space.

As a particle moves in ordinary (x, y, z) space its corresponding point in Minkowski space will describe a trajectory known as the *world line*, as mentioned previously. In this four-space the vectors $d\mathbf{x}$ and $d\mathbf{y}$ are four-vectors. $d\mathbf{x}$ represents the change in the position four-vector $\mathbf{x}$ for a differential motion along the world line. (A similar statement holds for $\mathbf{y}$ space.) From the scalar product of $d\mathbf{x}$ with itself we can form a world scalar (and therefore a Lorentz invariant) denoted by $(d\tau)^2$ which is defined by

$$(d\tau)^2 = \left(\frac{1}{c^2}\right) dx_\mu \, dx_\mu \qquad \text{(sum over } \mu). \tag{8.38}$$

The significance of $d\tau$ can easily be seen by evaluating (8.38) in the $\mathbf{y}$ system where the particle is assumed to be at rest with respect to $\mathbf{y}$, which moves with the velocity $\mathbf{v}$ with respect to $\mathbf{x}$. In this system the components of $d\mathbf{y}$ are $d\mathbf{y} = (0, 0, 0, dy_4)$, where $dy_4 = ic \, dt'$. The invariant $(d\tau)^2$ is then given by

$$(d\tau)^2 = -\left(\frac{1}{c^2}\right) dx'_\mu \, dx'_\mu = (dt')^2, \tag{8.39}$$

where we use the primed system $\mathbf{x}'$ instead of the $\mathbf{y}$ system. Equation (8.39) tells us that $d\tau$ (which is the positive square root of (8.39)) is the differential time interval as measured by a clock traveling with the particle. (Note that it is also equal to dt'.) Therefore τ is called the particle's *proper time* or *world time.* Upon using (8.39) and transforming back to **x** space we obtain the following relation between $d\tau$ and dt:

$$\frac{d\tau}{\sqrt{1 - \beta^2}} = dt. \tag{8.40}$$

From (8.40) we see that a proper time interval $\tau_2 - \tau_1$ will transform to the time interval $t_2 - t_1$ given by

$$t_2 - t_1 = \int_{\tau_1}^{\tau_2} \frac{d\tau}{\sqrt{1 - (v^2(\tau)/c^2)}}. \tag{8.41}$$

Equation (8.41) is an illustration of the principle of time dilatation as discussed above.

We return to the more general four-vectors. Let **v** be such a vector. Then the components of **v** transform in the same manner as (8.35) from the **x** to the **y** system. We thus have $\mathbf{v}(\mathbf{y}) = A\mathbf{v}(\mathbf{x})$. Thus the components transform as

$$\bar{v}_\mu(\mathbf{y}) = a_{\mu\nu}\bar{v}_\nu(\mathbf{x}). \tag{8.42}$$

Suppose we have a scalar ϕ that is unchanged under a Lorentz transformation. Then ϕ is called a *Lorentz scalar*. The components of the four-vector **v** are Lorentz scalars. Let $\phi = \phi(\mathbf{y}(\mathbf{x}))$. Then

$$\frac{\partial \phi}{\partial y_\mu} = \frac{\partial \phi}{\partial x_\nu} \frac{\partial x_\nu}{\partial y_\mu} = a_{\mu\nu} \frac{\partial \phi}{\partial x_\nu}, \tag{8.43}$$

since

$$\frac{\partial x_\nu}{\partial y_\mu} = a_{\mu\nu}. \tag{8.44}$$

Equation (8.43) is the same form for the components of **grad** ϕ as (8.35) or (8.36), which means that ϕ is a Lorentz scalar. By using (8.42) it is easy to show that div **v** is Lorentz invariant:

$$\frac{\partial \bar{v}_\nu(\mathbf{y})}{\partial y_\mu} = \frac{\partial \bar{v}_\nu(\mathbf{x})}{\partial x_\nu}. \tag{8.45}$$

We now introduce the *D'Alembertian operator* $\square^2$ which is defined by

$$\square^2 = \frac{\partial}{\partial x_\nu} \frac{\partial}{\partial x_\nu} = \frac{\partial^2}{\partial x^2} + \frac{\partial^2}{\partial y^2} + \frac{\partial^2}{\partial z^2} - \left(\frac{1}{c^2}\right) \frac{\partial^2}{\partial t^2}. \tag{8.46}$$

Equation (8.45) shows that the div operator in Minkowski space is invariant with respect to the Lorentz transformation. Taking the scalar product of this four-dimensional divergence operator with respect to itself yields the important result:

The D'Alembertian operator is invariant with respect to the Lorentz transformation.

Equation (8.46) remains the same if the coordinates (x, y, z, ct) are replaced by the transformed coordinates (x', y', z', ct'). Clearly this means that the wave equation is invariant with respect to the Lorentz transformation. This may be formulated as

$$\Box^2 \phi(\mathbf{x}') = \Box^2 \phi(\mathbf{x}),\dagger \qquad (8.47)$$

where the left-hand side of (8.47) means that the D'Alembertian involves differentiation with respect to the components of $\mathbf{x}'$, and the right-hand side means differentiation with respect to $\mathbf{x}$. The scalar ϕ in (8.47) must be a Lorentz scalar. The wave equation (8.47) in four-space also holds for any four-vector or world vector.

8.8. Covariance of the Electromagnetic Equations

By demonstrating that the D'Alembertian is invariant with respect to the Lorentz transformation, we essentially showed that Maxwell's equations are similarly invariant, since both the electric vector $\mathbf{E}$ and the magnetic flux density vector satisfy the invariant wave equation in Minkowski space. In this section we shall investigate, in more detail, the covariance of the equations of electromagnetics, which is an example of the covariance of the laws of physics in Minkowski space discussed previously.

The invariance of the electromagnetic equations with respect to the Lorentz transformation was shown by both Lorentz and Poincaré before Einstein formulated his special relativity theory. One approach to this investigation is to take an experimentally proven fact such as the invariance of the charge on the electron and attempt to deduce that the resulting equations must be invariant. The other point of view is to demand that the equations be invariant in form (as an example of the invariance of the laws of physics), and thereby show that the Lorentz transformation properties of the various physical quantities such as the field strengths, charge, and current can be satisfactorily chosen to accomplish this. We shall adopt this latter approach. For simplicity, we shall consider the microscopic form of the electromagnetic equations.

As a prelude to this treatment we consider some preliminary ideas. The volume element in Minkowski space is defined as the real quantity

$$dV = dx_1 \, dx_2 \, dx_3 \, dx_0, \qquad (8.48)$$

where

$$dx_0 = -i \, dx_4 = d(ct). \qquad (8.49)$$

We now show that dV is a Lorentz invariant quantity. The transformation

---

† This is easily seen by remembering that $x_4 = ict$, so that $\partial^2/\partial t^2 = -c^2 \, \partial^2/\partial x_4^2$, and (8.46) transforms to $\phi_{x_{vv}} = \Box^2 \phi(\mathbf{x}) = \phi_{x'_{vv}} = \Box^2 \phi(\mathbf{x}')$

from the **x** to the **x'** system is given by

$$dV' = \left[\frac{\partial(x_1', x_2', x_3', x_4')}{\partial(x_1, x_2, x_3, x_4)}\right] dV = J\, dV, \tag{8.50}$$

where the Jacobian J is just the determinant of the matrix **A** defined by (8.8) and (8.11) and dV' is the volume element in the transformed space **x'**. Therefore dV is Lorentz invariant.

It was mentioned several times that the invariance principle of Einstein means that the laws of physics are covariant in form. This tells us that the equations which describe the physical laws (for example, Maxwell's equations) can be written so that both sides enjoy the same transformation properties under the Lorentz transformation. Therefore, the equations of physics must involve relations between Lorentz scalars, four-vectors, or in general four-tensors of the same rank. This necessarily follows from the fact that a physical relation valid in the **x** system will also hold in the **x'** system with respect to the Lorentz transformation. For example, it will be shown that Maxwell's equations can be cast in the following relativistic form:

$$\frac{\partial F_{\mu\nu}}{\partial x_\nu} = \left(\frac{4\pi}{c}\right) J_\mu, \qquad \mu = 1, 2, 3, 4, \tag{8.51}$$

where $F_{\mu\nu}$ is the field strength four-tensor and J_μ is a suitable current four-vector. Thus (8.51) consists of a set of relations between two four-vectors, since the divergence of a four-tensor is a four-vector. Clearly (8.51) refers to the **x** system. Since both sides of (8.51) are covariant, it is invariant with respect to the Lorentz transformation to the **x'** system. In this transformed system (8.51) can be written as

$$\frac{\partial F_{\mu\nu}'}{\partial x_\nu'} = \left(\frac{4\pi}{c}\right) J_\mu'. \tag{8.52}$$

We are now in a position to cast the electromagnetic equations into covariant form. We start with the continuity equation for the charge ρ and the current density $\mathbf{J} = (J_1, J_2, J_3)$:

$$\rho_t = -(J_{1,x} + J_{2,y} + J_{3,z}) = -\operatorname{div} \mathbf{J}. \tag{8.53}$$

(Note that in (8.53) **J** is a function of (x, y, z, t).) We may cast (8.53) in the covariant form by now letting $\bar{\mathbf{J}}$ be the *charge-current four-vector* defined by

$$\bar{\mathbf{J}} = (\bar{J}_1, \bar{J}_2, \bar{J}_3, ic\rho), \tag{8.54}$$

where $\bar{J}_\nu = \bar{J}_\nu(x_1, x_2, x_3, x_4)$, $\nu = 1, 2, 3, 4$. Recall that $x_4 = ict$. Inserting (8.54) intio (8.53) yields

$$\frac{\partial \bar{J}_\nu}{\partial x_\nu} = 0, \tag{8.55}$$

using the tensor summation convention. Equation (8.55) is the covariant form of the continuity equation. $\bar{\mathbf{J}}$ is the four-vector (defined in Minkowski space) as established from the experimentally known invariance of the electric charge. Note that $ic\rho$ transforms as the fourth component of $\bar{\mathbf{J}}$. Clearly, the transformation into the $\mathbf{x}'$ system yields the same equation as (8.55) for $\bar{\mathbf{J}}(\mathbf{x}')$, since $\bar{\mathbf{J}}$ is a four-vector.

We recall from Chapter 1 that the vector potential $\mathbf{A}$ and the scalar poential Φ satisfy the following wave equations in (x, y, z) space (in microscopic form):

$$\nabla^2 \mathbf{A} - \left(\frac{1}{c^2}\right)\mathbf{A}_{tt} = -\left(\frac{4\pi}{c}\right)\mathbf{J},$$

$$\nabla^2 \Phi - \left(\frac{1}{c^2}\right)\Phi_{tt} = -4\pi\rho. \tag{8.56}$$

The Lorentz condition (also in physical space) is

$$\text{div } \mathbf{A} + \left(\frac{1}{c}\right)\Phi_t = 0. \tag{8.57}$$

We now desire to construct a four-vector with which we can combine the two wave equations of (8.56). Our guide in this attempt is the D'Alembertian operator in Minkowski or $\mathbf{x}$ space defined by (8.46). We want this operator to operate on a single four-vector which would combine $\mathbf{A}$ and Φ. How do we do this? First of all we see from (8.56) that $\mathbf{J}$ and ρ are the nonhomogeneous terms. We also see from (8.54) that the J_4 component is $ic\rho$. We now let $\bar{\mathbf{A}}$ be the four-vector which we desire to construct. We do not worry about the first three components of $\bar{\mathbf{A}}$. However, the component A_4 in Minkowski space satisfies the first equation of (8.56), namely

$$\Box^2 \bar{A}_4 = -\left(\frac{4\pi}{c}\right)ic\rho.$$

If we multiply the second equation of (8.56) by i, we may then identify $\bar{A}_4$ with $i\Phi$, thus giving the correspondence which ties up $\bar{\mathbf{A}}$ and Φ. We thus obtain as the four-vector

$$\bar{\mathbf{A}} = (\bar{A}_1, \bar{A}_2, \bar{A}_3, i\Phi). \tag{8.58}$$

This definition of the four-vector $\bar{\mathbf{A}}$ allows us to rewrite the two equations of (8.56) into a single equation for $\bar{\mathbf{A}}$ in convariant form in Minkowski space. Thus each component of $\bar{\mathbf{A}}$ satisfies the covariant wave equation

$$\Box^2 \bar{A}_\nu = -\left(\frac{4\pi}{c}\right)\bar{J}_\nu, \qquad \nu = 1, 2, 3, 4, \tag{8.59}$$

with the definition of the $\bar{J}_\nu$'s given by (8.54). The covariant form for the Lorentz equation (8.57) then becomes

$$\frac{\partial A_\nu}{\partial x_\nu} = 0. \tag{8.60}$$

We now consider the electric and magnetic field strengths **E** and **B**, respectively. The relationship between **E** and the vector and scalar potentials **A** and Φ in physical coordinates is

$$\mathbf{E} = -\mathbf{grad}\ \Phi - \left(\frac{1}{c}\right)\frac{\partial \mathbf{A}}{\partial t}. \tag{8.61}$$

Recasting (8.61) into Minkowski space gives

$$\mathbf{E} = -\mathbf{grad}\ \Phi - i\frac{\partial \mathbf{A}}{\partial x_4}. \tag{8.62}$$

The relationship between **B** and **A** is

$$\mathbf{B} = \nabla \times \mathbf{A}. \tag{8.63}$$

Note that $\mathbf{E} = (E_1, E_2, E_3)$ and $\mathbf{B} = (B_1, B_2, B_3)$ have three components each so that neither of them is a four-vector. However, $\bar{\mathbf{A}}$ is a four-vector, recalling that $\bar{A}_4 = i\Phi$. Also, note that (8.63) is the same in Minkowski space and does not involve A_4. Multiplying (8.62) by i gives

$$i\mathbf{E} = -\mathbf{grad}\ \bar{A}_4 + \frac{\partial \bar{\mathbf{A}}}{\partial x_4}. \tag{8.64}$$

The three components of the vector equation (8.64) are represented by the following equations:

$$iE_j = -\frac{\partial \bar{A}_4}{\partial x_j} + \frac{\partial \bar{A}_j}{\partial x_4}, \qquad j = 1, 2, 3. \tag{8.65}$$

(The fourth component is trivially zero.) Writing out (8.63) in extended form gives

$$B_1 = \frac{\partial \bar{A}_3}{\partial x_2} - \frac{\partial \bar{A}_2}{\partial x_3}, \qquad B_2 = \frac{\partial \bar{A}_1}{\partial x_3} - \frac{\partial \bar{A}_3}{\partial x_1},$$

$$B_3 = \frac{\partial \bar{A}_2}{\partial x_1} - \frac{\partial \bar{A}_1}{\partial x_2}. \tag{8.66}$$

The right-hand sides of (8.65) and (8.66) involve the components of the divergence of the four-vector $\bar{\mathbf{A}}$, while the left-hand sides of these equations involve the E_j's and the B_j's, respectively. We now combine these six equations, thereby combining the components of **E** and **B**. To this end, we introduce the four-tensor $\bar{\mathbf{F}}$ whose $\mu\nu$th component is given by

$$\bar{F}_{\mu\nu} = \frac{\partial \bar{A}_\nu}{\partial x_\mu} - \frac{\partial \bar{A}_\mu}{\partial x_\nu}, \tag{8.67}$$

From this definition of $\bar{F}$ we observe that $\bar{\mathbf{F}}$ is an *antisymmetric tensor*, since $-\bar{F}_{\mu\nu} = \bar{F}_{\nu\mu}$. $\bar{\mathbf{F}}$ is called the *field-strength tensor*. For example,

$$\bar{F}_{14} = \frac{\partial \bar{A}_4}{\partial x_1} - \frac{\partial \bar{A}_1}{\partial x_4} = -iE_1.$$

We now write $\bar{\mathbf{F}}$ as the following (4×4) antisymmetric matrix:

$$\bar{\mathbf{F}} = \begin{pmatrix} 0 & B_3 & -B_2 & -iE_1 \\ -B_3 & 0 & B_1 & -iE_2 \\ B_2 & -B_1 & 0 & -iE_3 \\ iE_1 & iE_2 & iE_3 & 0 \end{pmatrix}. \tag{8.68}$$

We now need to recast Maxwell's equations in covariant form. The inhomogeneous pair are

$$\text{div } \mathbf{E} = 4\pi\rho, \tag{8.69}$$

$$\nabla \times \mathbf{B} - i\frac{\partial \mathbf{E}}{\partial x_4} = \left(\frac{4\pi}{c}\right)\bar{\mathbf{J}}.$$

The right-hand sides essentially form the components of the four-vector $(J_1, J_2, J_3, i\rho)$. Therefore the left-hand sides must also involve a four-vector. Now $\bar{\mathbf{F}}$ as defined by (8.67) is the appropriate four-tensor, which when used properly, will tie up these two Maxwell equations into one covariant form. We obtain

$$\frac{\partial \bar{F}_{\mu\nu}}{\partial x_\nu} = \left(\frac{4\pi}{c}\right)\bar{J}_\mu. \tag{8.70}$$

The system (8.70) is easily checked by referring to (8.65), (8.66), and (8.67).

The two homogeneous Maxwell equations are

$$\text{div } \mathbf{B} = 0, \qquad \nabla \times \mathbf{E} + i\frac{\partial \mathbf{B}}{\partial x_4} = 0. \tag{8.71}$$

Using the definition of $\bar{\mathbf{F}}$ it can be shown that (8.71) can be put into the covariant form represented by the following four equations:

$$\frac{\partial \bar{F}_{\mu\nu}}{\partial x_\lambda} + \frac{\partial \bar{F}_{\lambda\mu}}{\partial x_\nu} + \frac{\partial \bar{F}_{\nu\lambda}}{\partial x_\mu} = 0, \qquad \lambda, \mu, \nu = 1, 2, 3, 4. \tag{8.72}$$

8.9. Force and Energy Equations in Relativistic Mechanics

We saw that Newton's equations, which are invariant under a Galilean transformation, cannot be invariant under a Lorentz transformation. They must be suitably generalized to account for the basic equivalence principle of Einstein which involves the constancy of the velocity of light. Newton's equations of motion, we recall, states that the time rate of change of linear momentum is equal to the resultant of the external forces acting on the system. It is clear that the generalization we seek must reduce to the following familiar form when $\beta \ll 1$:

$$\frac{d}{dt}(m_0 v_i) = F_i, \qquad i = 1, 2, 3, \tag{8.73}$$

where v_i is the ith component of the particle velocity of mass m_0 and F_i is the ith component of the external force acting on the particle. It turns out that m_0 is the "rest mass" (the mass the particle has when its velocity is zero—the mass depends on the velocity, as will be shown below).

While (8.73) is itself not Lorentz invariant (as mentioned) we can expect that its relativistic generalization will be a four-vector equation whose spatial components reduce to (8.73) as $\beta \to 0$. The only velocity four-vector whose space part reduces to $\mathbf{v} = (v_1, v_2, v_3)$ for small β is the world vector $\bar{\mathbf{v}}$ whose components are $\bar{v}_\nu$. We know that m_0 can be taken as an invariant property of the particle. However, time is not a Lorentz invariant; but it must be replaced by the scalar proper time τ which approaches t as $\beta \to 0$. Therefore the νth component of the relativistic generalization of Newton's equations must be of the following form:

$$\left(\frac{d}{d\tau}\right)(m_0 \bar{v}_\nu) = \bar{K}_\nu, \qquad \nu = 1, \ldots, 4, \tag{8.74}$$

where $\bar{K}_\nu$ is the νth component of the four-vector $\bar{\mathbf{K}}$, and is called the *Minkowski force*. The spatial components of $\bar{\mathbf{K}}$ are not to be identified with the components of the external force. All that is required by (8.74) is that the spatial components of $\bar{\mathbf{K}}$ reduce to $\mathbf{F}$ as $\beta \to 0$.

One approach to relativistic mechanics is to avoid the necessity of using a physical theory beyond mechanics itself. This procedure simply applies Newton's law of motion by defining force as the time rate of change of linear momentum *in all Lorentz systems*, thus yielding

$$\frac{dp_i}{dt} = F_i, \qquad i = 1, 2, 3. \tag{8.75}$$

The ith component of the momentum p_i is not $m_0 v_i$, but rather some relativistic generalization which reduces to it in the limit of small velocities. The appropriate generalization is obtained by using the following three relations which arise from the Lorentz transformation:

(1) The spatial part of the relation between the ordinary external force $\mathbf{F}$ and the Minkowski force $\bar{\mathbf{K}}$ is

$$F_i = (\sqrt{1 - \beta^2})\bar{K}_i, \qquad i = 1, 2, 3, \tag{8.76}$$

(2) The relation between the time t and the proper time τ is given by (8.40):
$d\tau = (\sqrt{1 - \beta^2})\, dt$.

(3) The spatial part of the relation between the velocity four-vector $\bar{\mathbf{v}}$ and the particle velocity $\mathbf{v}$ is

$$\bar{v}_i = \left(\frac{1}{\sqrt{1 - \beta^2}}\right) v_i. \tag{8.77}$$

Equation (8.77) arises from the fact that

$$\bar{v}_i = \frac{dx_i}{d\tau} = \left(\frac{dx_i}{dt}\right)\left(\frac{dt}{d\tau}\right) = v_i\left(\frac{1}{\sqrt{1 - \beta^2}}\right).$$

Having accepted (8.74) as the form of the force law (conservation of momentum), the relativistic momentum and the meaning of the $\bar{K}_i$'s is easily obtained from the above relations. We obtain

$$\left(\frac{d}{d\tau}\right)(m_0\bar{v}_i) = \left(\frac{1}{\sqrt{1-\beta^2}}\right)\left(\frac{d}{dt}\right)\left(\frac{m_0v_i}{\sqrt{1-\beta^2}}\right) = \bar{K}_i,$$

which yields

$$\left(\frac{d}{dt}\right)\left(\frac{m_0v_i}{\sqrt{1-\beta^2}}\right) = \bar{K}_i\sqrt{1-\beta^2} = F_i, \tag{8.78}$$

which is the relativistic formulation of the spatial part of the equations of motion given by (8.75). Comparing (8.78) with (8.75) tells us that the spatial part of the momentum is

$$p_i = \frac{m_0v_i}{\sqrt{1-\beta^2}} = mv_i, \qquad i = 1, 2, 3. \tag{8.79}$$

Equation (8.79) tells us that each of the three spatial components of the momentum is equal to the mass m times the corresponding component of the particle velocity, providing m is defined by

$$m = \frac{m_0}{\sqrt{1-\beta^2}}. \tag{8.80}$$

This gives us the important relativistic result that the mass of a particle is a function of the ratio $(v/c)^2$ according to (8.80). Thus (8.80) is the analytical statement or the precise relation between the mass of a particle and its velocity v, and it gives the justification of the experimentally determined relationship between the ratio of charge to mass of a high-speed electron and its velocity as observed by Kaufmann, which was mentioned above. Equation (8.80) is usually expressed as the ratio of the mass m of the particle traveling at a velocity v to its rest mass m_0. Expanding this ratio we get

$$\frac{m}{m_0} = \frac{1}{\sqrt{1-\beta^2}} = 1 + \frac{1}{2}\left(\frac{v}{c}\right)^2 + \frac{3}{8}\left(\frac{v}{c}\right)^4 + \cdots.$$

Our first observation is that as $v \to c, m \to \infty$, or we have a singularity at $v = c$. This tells us that no particle can travel as fast as the speed of light. Moreover, if $v > c$ then m is imaginary. Clearly a photon which travels at the speed of light (by the definition of a photon) cannot have a mass. The above expansion tells us that if $v/c \ll 1$, so that we can neglect the second- and higher-order terms, then $m = m_0$ to a good approximation and (8.79) reduces to the Newtonian definition of momentum.

We now derive the fourth component of the relativistic equations of motion. We first need the full velocity four-vector $\bar{v}$. The metric vector in Minkowski space is

$$d\mathbf{s} = \mathbf{i}\, dx + \mathbf{j}\, dy + \mathbf{k}\, dz + \mathbf{l}ic\, dt,$$

where $(\mathbf{i}, \mathbf{j}, \mathbf{k})$ are unit vectors in the (x, y, z) directions and $\mathbf{l}$ is the unit vector in the ict direction. The velocity four-vector is given by

$$\bar{\mathbf{v}} = \frac{d\mathbf{s}}{d\tau} = \left(\frac{1}{\sqrt{1 - \beta^2}}\right)(\mathbf{i}u + \mathbf{j}v + \mathbf{k}w + \mathbf{l}ic),$$

where $u = dx/dt$, etc. It follows that

$$\bar{v}^2 = \left(\frac{1}{1 - \beta^2}\right)(u^2 + v^2 + w^2 - c^2) = -c^2,$$

since $\beta^2 = (u^2 + v^2 + w^2)/c^2$. To get the fourth component of the Minkowski force $\bar{\mathbf{F}}$ we take the scalar product of the relativistic equation of motion (whose vth component is given by (8.74)) with the four-vector $\bar{\mathbf{v}}$ and obtain

$$\bar{\mathbf{v}} \cdot \left(\frac{d}{d\tau}\right)(m_0\mathbf{v}) = \left(\frac{d}{d\tau}\right)\left(\left(\frac{m_0}{2}\right)(\bar{\mathbf{v}} \cdot \bar{\mathbf{v}})\right) = \bar{\mathbf{v}} \cdot \bar{\mathbf{K}}.$$

Since $\bar{\mathbf{v}} \cdot \bar{\mathbf{v}} = -c^2 = $ const. it follows that

$$\bar{\mathbf{v}} \cdot \mathbf{K} = \frac{\bar{v}_i \bar{F}_i}{1 - \beta^2} + \left(\frac{1}{\sqrt{1 - \beta^2}}\right)(icK_4) = 0 \qquad \text{(sum over } i \text{ from 1 to 3).}$$

This gives the fourth component of the Minkowski force as

$$\bar{K}_4 = \left(\frac{1}{ic}\right)\frac{\bar{F}_i \bar{v}_i}{\sqrt{1 - \beta^2}}. \qquad (8.80)$$

Since fourth component of the velocity four-vector is ic, it follows that the fourth component of relativistic equations of motion (8.74) becomes (upon transforming to the operator d/dt):

$$\left(\frac{d}{dt}\right)\left(\frac{m_0 c^2}{\sqrt{1 - \beta^2}}\right) = F_i v_i = \mathbf{F} \cdot \mathbf{v}. \qquad (8.81)$$

We now turn to the relativistic formulation of the kinetic energy. In classical mechanics the element of work dW done by a force $\mathbf{F}$ in moving a body an element of distance ds along a prescribed path is

$$dW = \mathbf{F} \cdot d\mathbf{s} = \mathbf{F} \cdot \mathbf{v}\, dt, \qquad \text{or} \qquad \frac{dW}{dt} = \mathbf{F} \cdot \mathbf{v}.$$

We now identify the kinetic energy T with W and obtain, upon using (8.81):

$$T = \frac{m_0 c^2}{\sqrt{1 - \beta^2}}. \qquad (8.82)$$

We set the constant of integration equal to zero, as explained below. Expanding the right-hand side of (8.82) yields

$$T = m_0 c^2 + \tfrac{1}{2} m_0 v^2 + \tfrac{3}{8} m_0 \left(\frac{v}{c}\right)^4 + \cdots \to m_0 c^2 + \tfrac{1}{2} m_0 v^2, \qquad (8.83)$$

This limiting value, which is obtained by neglecting terms of fourth order and higher in β, is equal to the classical expression for T plus the $m_0 c^2$ term—which has enormous physical significance. In fact, this term is obtained by setting $v = 0$ in (8.83). We might ask the question: Why take the constant of integration to be zero when it is arbitrary? The answer is that it is preferable to define T by (8.82); for then the momentum p_i as defined by (8.79) and iT/c form a four-vector or world momentum vector $\mathbf{p}$, so that

$$\bar{\mathbf{p}} = \frac{1}{c}(m_0 u, m_0 v, m_0 w, iT).\dagger \tag{8.84}$$

Equation (8.84) tells us that, if the momentum is conserved (is constant) then T is also conserved. Thus we have an important unifying principle of special relativity which states that the conservation laws for momentum and kinetic energy are no longer separate; they appear as different aspects of the unifying conservation law for the four-vector given by (8.84). This is in contrast to classical or nonrelativistic mechanics where the conservation of momentum and conservation of energy are two separate laws.

As stated above $T(0) = m_0 c^2$ so that $m_0 c^2$ is aptly called the *rest mass*. Its physical importance occurs when we consider a change in the rest mass $m_0 c^2$. Associating this with the corresponding change in energy ΔE as represented by ΔT, we have the famous Einstein relation, *the equivalence of mass and energy*:

$$\Delta E = \Delta m_0 c^2. \tag{8.85}$$

Many examples of the use of (8.85) abound in the literature. One of the first examples proposed was the case of an inelastic collision of two bodies traveling at nonrelativistic velocities. We would say ordinarily that the energy lost due to the collision is converted into heat. But if the relativistic kinetic energy is to be conserved as part of the momentum four-vector, then there must be an increase in the rest mass of the system in proportion to the amount of heat produced. This change in rest mass is clearly very small since, for example, a joule of energy corresponds to a mass of only about 1×10^{-14} gm. However, in modern physics there is a rich variety of examples where relativistic mass changes due to energy changes are of much greater consequence, for example, pair creation in which two particles of finite mass are created from the energy of a massless photon. The most striking demonstration of (8.85) is the conversion of mass into energy in an atomic bomb explosion. Momentum is conserved but a tremendous amount of kinetic energy is produced. But despite the fantastic energies produced the loss in rest mass can at most amount to only about 0.1% of the original mass. In general, in a nuclear transformation process which results from the bombardment of elements by α-particles, protons, neutrons, etc., the equivalence of mass and energy, as expressed by

---

† Formally, the connection between the momentum and kinetic energy is expressed in the statement that the magnitude of this momentum four-vector is constant. This gives $\bar{\mathbf{p}} \cdot \bar{\mathbf{p}} = -m_0^2 c^2 = p^2 - T^2/c^2$, or $T^2 = p^2 c^2 + m_0^2 c^4$.

(8.85), has been amply confirmed. The law is:

The sum of the reacting masses, together with the mass equivalent of the kinetic energy of the bombarding particles (or photons), is always greater than the sum of the resulting masses. This difference is the equivalent mass of the kinetic energy of the particles generated, or the release of the electromagnetic energy (γ photons).

As in the atomic bomb, the mass of a spontaneously disintegrating radio-active atom is always greater than the sum of the masses of the resulting products by the mass equivalent of the kinetic energy of the particles generated (or of the photon energy). Precise measurements of the energy of the products of nuclear reactions in conjunction with the appropriate equations describing these reactions make it possible to calculate atomic weights to a high degree of accuracy.

8.10. Lagrangian Formulation of Equations of Motion in Relativistic Mechanics

We now investigate the Lagrangian formulation of the relativistic equations of motion. It is a little difficult to determine a Lagrangian which will furnish the correct relativistic equations of motion, by starting with D'Alembert's principle (as described in Chapter 4) which is still valid. However, the momentum vector is a four-vector, and this would lead to complications. Therefore we approach the problem of determining the Lagrangian formulation by way of Hamilton's variational principle, described by (4.17) of Chapter 4, which we repeat

$$\delta I = \delta \int_{t_1}^{t_2} L \, dt, \tag{8.86}$$

where $L = L(\mathbf{q}, \dot{\mathbf{q}})$, and we have reverted to the generalized coordinates $\mathbf{q}$ and generalized velocities $\dot{\mathbf{q}}$ (not the velocity four-vector).

A simple example is the case of a single particle acted upon by conservative forces independent of the velocity. A suitable relativistic Lagrangian would be

$$L = -m_0 c^2 \sqrt{1 - \beta_i^2} - V(\mathbf{q}), \qquad \beta_i^2 = \frac{\dot{q}_i^2}{c^2}, \tag{8.87}$$

where V is the potential independent of the velocity. Lagrange's equations for a single particle in a conservative force field are

$$\left(\frac{d}{dt}\right)\left(\frac{\partial L}{\partial \dot{q}_i}\right) - \frac{\partial L}{\partial q_i} = 0, \qquad i = 1, 2, 3. \tag{8.88}$$

From (8.87) we obtain

$$\frac{\partial L}{\partial \dot{q}_i} = \frac{m_0 \dot{q}_i}{\sqrt{1 - \beta_i^2}} = p_i. \tag{8.89}$$

The right-hand side of (8.89) is the spatial part of the relativistic momentum vector **p**. Therefore the relativistic equations of motion become

$$\left(\frac{d}{dt}\right)\left(\frac{m_0 \dot{q}_i}{\sqrt{1-\beta^2}}\right) = -\frac{\partial V}{\partial q_i} = F_i, \tag{8.90}$$

which is the same as (8.78) for a force derivable from a velocity-independent potential. We note that the Lagrangian is no longer given by $L = T - V$ (as in nonrelativistic mechanics). However, the partial derivative of L with respect to the velocity is still the (spatial part of) momentum. We also note that we are still using generalized coordinates. The canonical momenta are still defined by

$$p_i = \frac{\partial L}{\partial \dot{q}_i}, \qquad i = 1, 2, 3,$$

so that the correspondence between cyclic coordinates and conservation of the spatial part of the corresponding momentum remains valid in relativistic theory.

When we investigate the conservation of energy for the relativistic case we must alter the derivation somewhat. We recall in Chapter 4 that if L does not contain the time explicitly, the Hamiltonian may be defined as the constant of motion H such that

$$H = \sum_i \dot{q}_i p_i - L. \tag{8.91}$$

Equation (8.91) is also valid in relativistic mechanics since its derivation only involves the general form of L and the definition of canonical momentum. The proof that the Hamiltonian H is indeed the total energy for the relativistic case must be changed, since L (as mentioned above) is no longer given by $T - V$. Inserting (8.87) and (8.89) into (8.91) for a system of particles we obtain

$$H = \sum_i \frac{m_0 \dot{q}_i^2}{\sqrt{1-\beta^2}} + m_0 c^2 \sqrt{1-\beta^2} + V$$

$$= \frac{m_0 c^2}{\sqrt{1-\beta^2}} + V = T + V = E, \tag{8.92}$$

where E is the total energy and $\beta^2 = \sum_i (\dot{q}_i/c)^2$. The Hamiltonian H is thus seen to be the total energy E which is a constant.

Extending the relativistic theory to a velocity-dependent potential does not introduce any difficulties. It can be treated in the same manner as shown in Section 4.8 of Chapter 4. For example, the Lagrangian for a system of particles in an electromagnetic field is given in terms of the scalar potential Φ and vector potential **A**

$$L = -m_0 c^2 \sqrt{1-\beta^2} - q\Phi + \left(\frac{q}{c}\right)\mathbf{A}\cdot\mathbf{q}. \tag{8.93}$$

8.11. Covariant Lagrangian

Up to now we have made no effort to keep to the covariant form for the Lagrangian system in Minkowski space. Thus time has been treated as a parameter entirely distinct from the spatial coordinates. A covariant formulation would require (as previously mentioned) use of Minkowski space which unifies space and time into the space–time continuum. If we desire to trace the progress of the system point in configuration space it is clear that we must use the proper time τ instead of t, since τ is invariant with respect to the Lorentz transformation, thus satisfying the condition of covariance. Note that the Lagrange functions discussed above have no particular invariance properties with respect to the Lorentz transformation. For a covariance formulation the Lagrangian should be an invariant property of the system only, independent of the particular coordinate system used. Therefore, we must expect that L must be a world scalar or Lorentz scalar, since we want it to be invariant with respect to all Lorentz transformations. Instead of L being a function $\mathbf{q}$ and $\dot{\mathbf{q}}$ with t as a parameter, the Lagrangian expressed correctly in four-space should be a function of the components of $\bar{\mathbf{q}}$: the $\bar{q}_v$'s and the components of $\bar{\mathbf{q}}$: $\bar{q}_v = d\bar{q}/d\tau$, with τ as a parameter. (Note that we use a bar over the q's to represent the components of a four-vector, a dash means differentiation with respect to τ, and v goes from 1 to 4.)

Coming back to Hamilton's variational principle, a covariant formulation should have the form

$$\delta \bar{I} = \delta \int_{\tau_1}^{\tau_2} \bar{L}(\bar{q}_v, \bar{q}_v', \tau)\, d\tau = 0, \qquad (8.94)$$

where $\bar{I}$ and $\bar{L}$ represent the covariant formulation of the functional and the Lagrangian, respectively.

We now give a few comments on the limitations of the covariant representation of the Lagrangian. It is not always possible to set up a completely covariant formulation in the manner presented above. Clearly, the covariant formulation of the relativistic Lagrange equations of motion depends on formulating the external forces in a covariant form. We know that in Newtonian mechanics the gravitational force is derivable from a potential. We also know that this gravitational potential cannot be represented as a Lorentz scalar so that the gravitational force, as represented by Newton, is not covariant (is not a four-vector). Therefore it cannot be incorporated in the relativistic equations of motion. The physical reason for this is easy to see. The gravitational force is static in the sense that it is an example of the "action at a distance" principle, which means that the effect of a gravitational field produced by a body at a point in space is felt instantaneously at any place in space; meaning that the gravitational source produces an effect that travels with infinite speed! This makes no sense in relativity theory. In Einstein's mechanics no source can propagate a signal that travels faster than the velocity of light. This point is at the very heart of why the Newtonian gravitational force is incorrect and must be replaced by Einstein's general relativity theory which takes into account the gravitational field. It is based on an entirely different principle from Newton: that a gravitational field produces a curvature in the four-dimensional space–time continuum. Therefore the covariant formulation of Lagrange's equations of motion are restricted in the sense of not allowing a gravitational field. The general theory of relativity is beyond the scope of this book.

The covariant form of the Euler–Lagrange equations of motion for a single particle deduced from the covariant formulation of Hamilton's principle are

$$\left(\frac{d}{d\tau}\right)\left(\frac{\partial \bar{L}}{\partial \bar{q}_v}\right) - \frac{\partial \bar{L}}{\partial \bar{q}_v} = 0, \qquad v = 1, 2, 3, 4. \tag{8.95}$$

The left-hand sides of the four equations of (8.95) transform as the components of a four-vector. For a free particle (no external force acting on it) the covariant Lagrangian obviously must be such that (8.95) reduces to

$$\left(\frac{d}{d\tau}\right)\left(\frac{\partial \bar{L}}{\partial \bar{q}_v}\right) = \left(\frac{d}{d\tau}\right)(m_0 \bar{v}_v)\dagger = 0, \tag{8.96}$$

which is similar in form to the conservation of momentum for the nonrelativistic case, namely, $(d/dt)(mv_i) = 0$. This suggests that we construct $\bar{L}$ by replacing v^2 in the nonrelativistic case with the square of the velocity four-vector and obtain

$$\bar{L} = \tfrac{1}{2} m_0 \bar{v}_v \bar{v}_v = -m_0 c^2. \tag{8.97}$$

This is indeed the correct choice as is easily verified.

Suppose electromagnetic forces act on a particle. Then a suitable covariant Lagrangian is

$$\bar{L} = \tfrac{1}{2} m_0 \bar{v}_v \bar{v}_v + \left(\frac{q}{c}\right) \bar{v}_v A_v. \tag{8.98}$$

The vth component of the canonical momentum four-vector is

$$p_v = \frac{\partial \bar{L}}{\partial \bar{v}_v} = m_0 v_v + \left(\frac{q}{c}\right) A_v. \tag{8.99}$$

Using (8.99) the relativistic Lagrange equations (8.95) become

$$\left(\frac{d}{d\tau}\right)(m_0 \bar{v}_v) = \frac{\partial}{\partial \bar{q}_v}\left(\left(\frac{q}{c}\right)\bar{v}_\mu A_\mu\right) - \left(\frac{q}{c}\right)\frac{dA_v}{d\tau} = K_v, \tag{8.100}$$

where the right-hand side of (8.100) is set equal to the vth component of the Minkowski force, thus relating (8.100) to (8.74) which is the vth component of the relativistic equation of motion.

---

† Note that we revert to setting the vth component of the velocity four-vector equal to $\bar{v}_v$ instead of $\dot{\bar{q}}$.

The p_4th component of the canonical momentum four-vector (8.99), for the case of an electromagnetic field, becomes

$$p_4 = \frac{iT}{c} + \frac{iq\Phi}{c} = \frac{iE}{c},\qquad(8.101)$$

where the total energy is $E = T + q\Phi$. Therefore (8.101) tells us that the momentum canonical to the ict coordinate is proportional to the total energy.

From (8.99) we obtain

$$m_0^2 v_\nu v_\nu = -m_0^2 c^2 = \left(\mathbf{p} - \frac{q\mathbf{A}}{c}\right)^2 - \frac{T^2}{c^2}.$$

This gives

$$T^2 = \left(\mathbf{p} - \frac{q\mathbf{A}}{c}\right)^2 + m_0^2 c^4,\qquad(8.102)$$

which relates the kinetic energy to the momentum for the case where electromagnetic forces act on the particle.

We close this section with another remark on cases in physics which do not have a relativistic analogue. We observed above that the gravitational field does not fit into the covariant setting of Lagrange's equations, and thus does not have an analogue in the special theory of relativity. The Lorentz transformation is concerned only with uniformly moving coordinate systems and thus is not applicable to accelerating or rotating systems. Therefore those problems in physics which involve accelerating or rotating coordinate systems cannot be put into a relativistic framework (within the constraints of the special theory of relativity). Also, the whole subject of rigid body mechanics has no analogue in special relativity theory. The constraints are only concerned with the spatial parts of the position four-vector and are therefore not covariant.

Bibliography

Note: This bibliography represents the multidisciplined nature of wave propagation in electromagnetic media. Thus, it contains references to various disciplines of applied mathematics, especially in the area of partial differential equations, and mathematical physics related to this subject, and the related fields of quantum mechanics and cosmology. Note that a selected sampling of the pertinent standard works is given rather than references to the numerous papers in these areas, which would only tend to confuse the nonspecialist reader. Therefore, this bibliography is not complete.

1. Ames, W.F., *Nonlinear Partial Differential Equations in Engineering*, Academic Press, 1966.
2. Bateman, H., *Partial Differential Equations of Mathematical Physics*, Cambridge University Press, 1959.
3. Bergmann, P.G., *Introduction to the Theory of Relativity*, Prentice-Hall, 1942.
4. Birkhoff, G. and S. MacLane, *A Survey of Modern Algebra*, 4th ed., Macmillan, 1977.
5. Bliss, G.A. *Lectures on the Calculus of Variations*, University of Chicago Press, 1947.
6. Chandrasekhar, S., *Plasma Physics*, University of Chicago Press, 1960.
7. Chester, C.R., *Techniques in Partial Differential Equations*, McGraw-Hill, 1971.
8. Churchill, R.V., *Complex Variables and Applications*, 2nd ed., McGraw-Hill, 1960.
9. Churchill, R.V., *Fourier Series and Boundary Value Problems*, 2nd ed., McGraw-Hill, 1963.
10. Churchill, R.V., *Operational Mathematics*, 3rd ed. McGraw-Hill, 1972.
11. Courant, R. and D. Hilbert, *Methods of Mathematical Physics*, Vol. I—Linear Algebra, Eigenfunction Theory, and Theory of Vibrations, Vol. II—Wave Propagation in Solids and Fluids, Interscience, 1962.
12. Courant, R. and K.O. Friedrichs, *Supersonic Flow and Shock Waves*, Interscience, 1948.
13. Davis, J.L. *Finite Difference Methods in Dynamics of Continuous Media*, Macmillan (marketed by McGraw-Hill), 1986.
14. Davis, J.L., *Introduction to Dynamics of Continuous Media*, Macmillan (marketed by McGraw-Hill), 1987.
15. Davis, J.L. *Wave Propagation in Solids and Fluids*, Springer-Verlag, 1988.

16. Eddington, A.S., *The Mathematical Theory of Relativity*, Cambridge University Press, 1922.
17. Einstein, A. *Relativity, the Special and General Theory*, 17th ed., Crown, 1961.
18. Feynman, R.P., R.B. Leighton, and M. Sands, *The Feynman Lectures on Physics*, Vol. II—Electromagnetism and Matter, Addison-Wesley, 1964.
19. Feynman, R.P., R.B. Leighton, and M. Sands, *The Feynman Lectures on Physics*, Vol. III—Quantum Mechanics, Addison-Wesley, 1965.
20. Friedman, B., *Principles and Techniques of Applied Mathematics*, Wiley, 1956.
21. Gamow, G., *Biography of Physics*, Harper & Bros., 1961.
22. Goldstein H., *Classical Mechanics*, Addison-Wesley, 1959.
23. Hadamard, J., *Lectures on Cauchy's Problem*, Dover, 1952.
24. Hawking, S.W., *A Brief History of Time from the Big Bang to Black Holes*, Bantam, 1988.
25. Hildebrand, F.B., *Advanced Calculus for Applications*, 2nd ed., Prentice-Hall, 1976.
26. Jackson, J.D. *Classical Electrodynamics*, Wiley, 1962.
27. Jeans, J.H., *Mathematical Theory of Electricity and Magnetism*, 5th ed., Cambridge University Press, 1948.
28. Jeffreys, H. and B. Jeffreys, *Methods of Mathematical Physics*, 2nd ed., Cambridge University Press, 1950.
29. Lamb, H., *Hydrodynamics*, 1st American ed., Dover, 1945.
30. Landau, L.D. and E.M. Lifshitz, *Classical Theory of Fields*, Addison-Wesley, 1951.
31. Landau, L.D. and E.M. Lifshitz, *Electrodynamics of Continuous Media*, Addison-Wesley, 1960.
32. Maxwell, J.C., *Treatise on Electricity and Magnetism*, 3rd ed. (2 vols.), reprint by Dover, 1954.
33. McLachlin, N.W., *Theory of Vibrations*, Dover, 1951.
34. Morse, P.M. and H. Feshbach, *Methods of Theoretical Physics* (2 parts), McGraw-Hill, 1953.
35. Noble, B. and J.W. Daniels, *Applied Linear Algebra*, 2nd ed., Prentice-Hall, 1977.
36. Panofsky, W.K.H. and M. Phillips, *Classical Electricity and Magnetism*, Addison-Wesley, 1955.
37. Rayleigh, J.W.S., *The Theory of Sound* (2 vols.), 1st American ed., Dover, 1945.
38. Shapiro, S.L. and S.A. Teukolsky, *Black Holes, White-Dwarfs, and Neutron Stars, the Physics of Compact Bodies*, Wiley, 1983.
39. Sokolnikoff, I.S., *Tensor Analysis*, 2nd ed., Wiley, 1964.
40. Spitzer, L., *Physics of Fully Ionized Gases*, 2nd ed., Interscience, 1962.
41. Stratton, J.A., *Electromagnetic Theory*, McGraw-Hill, 1941.
42. Winch, R.P., *Electricity and Magnetism*, 2nd ed., Prentice-Hall, 1963.
43. Whittaker, E.T., *History of the Theories of the Aether* (2 vols.), Nelson, 1951, 1953.
44. Whittaker, E.T., *A Treatise on the Analytical Dynamics of Particles and Rigid Bodies*, 4th ed., Dover, 1944.

Index

Action integral   171
Almost plane waves   201
Ampere's law   3
Anomalous dispersion   16
Antimatter   221
Atomic energy, development of   216

Bicharacteristic equations   117, 121
Bicharacteristics   114f
Big bang theory   221f
   three-dimensional model   224f
Big crunch   226
Black-body radiation   206f
Black holes   227f
Bohr's atom   209f
Bose–Einstein statistics   217
Brachistochrone problem   141f
Brewster's angle   27

Calculus of variations   135, 137f, 144f
Canonical coordinates   167, 182
Canonical form of a PDE (see under Normal form)
Canonical momentum   167
Canonical transformations   181f
Cauchy initial value problem   35f, 93f
Chandrasekhar limit   228
Characteristic condition in n dimensions   103f, 113
Characteristic cone   121
Characteristic coordinates   12, 34, 70
Characteristic curves   34
Characteristic determinant   106
Characteristic equation   33, 41, 49
   parametric form   45

Characteristic equation for the eigenvalues   131
Characteristic equations, quasilinear case   92
Characteristic ODEs for fully nonlinear case   85
Characteristic strip   95
Characteristic surfaces   99f
Characteristic theory, applied to magnetohydrodynamics   239f
Characteristics
   geometry of for second-order PDEs   112f
   method of   40f
   method of for first-order system   48f
Charge-current four-vector   275
Charge density   2
Charge separation   232
Coaxial cable (see under Telegraph equation)
Conductivity   2
Configuration space   146f
Conjugate momentum   167
Conservation
   of charge (continuity equation)   3
   of energy   7
   of mass   3, 233
   of momentum   3f, 4, 233
Conservation laws   4f
   generalized   63, 66f
   physical interpretation   65f
Conservative force system   156, 158
Constitutive equations   2
Constraints   152f
Contact transformations (see under Canonical transformations)
Continuity equation (see under Conservation of mass)

Contraction of rods   263
Cosmology   221f
Covariance of electromagnetic equations
    274f
Covariant formulation of laws of physics
    270f
Covariant Lagrangian   285f
Cyclic coordinates   169f

D'Alembert's principle   151, 154
D'Alembert's solution of wave equation
    35f
D'Alembertian operator   273, 276
De Broglie waves   210f
Degrees of freedom   146
Depth of penetration   31
Diagonalization of symmetric matrix
    133f
Diffusion equation   238
Dilatation factor   264
Direction field   42
Directional derivative   12, 148f
    in n dimensions   95f
Dispersion   14, 124f
Displacement vector   2
Dissipation   30, 124
Domain of dependence and deter-
    minacy   55f
Duality of ray cone and normal cone
    118f

Eiconal equation   121, 202
Eigenvalues   131
Eigenvectors   131
Electric field   2
Electromagnetic field,   3
Elliptically polarized wave   20f
Equation of state, adiabatic   235
Equipartition theorem   207
Equivalence of mass and energy   282,
    283
Ether theory, collapse of   256f
Euler's equations   139
    n dimensions   145
    for continuum   177
Extremum (*see under* Stationary value)

Faraday's law   3
Fermi–Dirac statistics   217
Field-strength tensor   277
Finite difference method   82f

Fizeau's experiment   257
Fluid dynamics, general treatment   232f
Four-vector   272
Functionals   135f
    nonexistence of   137f
Fundamental forces in particle physics
    218

Galilean transformation   255
Gauge transformations   10
Generalized coordinates   146
Generalized force   155
Generalized potential   159
Generalized velocity   147
Geometric optics, example   82f, 86f
Group velocity   14
GUT   222

Hamilton density function   179
Hamiltonian   167
Hamilton's (canonical) equations of
    motion   167f
Hamilton's characteristic function   191
Hamilton's equations of motion for
    continuum   178f
Hamilton's principle for continuum   176
Hamilton's principle function   187f
Hamilton's variational principle   146,
    147, 148, 162
Hamilton–Jacobi equations   188, 193,
    194
Hamilton–Jacobi theory   187f
    and quantum mechanics   203f
    and wave propagation   197f
    related to characteristic theory   193f
Heisenberg's uncertainty principle   211f
Hubble's law   223f
Hyperbolic PDEs in more than two
    independent variables   Chap. 3

Ill-posed problem   58
Index of refraction   15, 121
Inner derivative   96, 98
Integral surfaces   42

Jacobi's form of principle of least action
    174

Kinematics   234

Lagrange density function   175
Lagrange's equations of motion   148f
   for continuum   174f
Lagrangian   148, 163
   formulation in relativistic mechanics
      283f
Light cones   269, 270
Limiting wave speed   127
Linear vector function   131, 133
Linearization of the field equations   243f
Linearly polarized wave   19
Lorentz condition   10
Lorentz force   4f, 159
Lorentz scalar   273
Lorentz transformation   258f, 261, 262,
   271

Magnetic diffusion   237f
   time   238
Magnetic field   2
Magnetic-flux density   2
Magnetic forces (magnetohydrodynamic
   effect)   249f
Magnetohydrodynamic domain   231
Magnetohydrodynamic field equations
   235f
Magnetosonic waves   246
Mass defect   215f (*see also under*
   Equivalence of mass and energy)
Maxwell's equations   1f, 105f, 262
Maxwell's stress tensor   6
Method of characteristics for first-order
   equation   40f
Metric, relativistic   268
Minimum principle   24
Minkowski space   259, 274
Momentum density function   179
Monge axis   42
Monge cone   84, 94, 118
Monge equation   88
Monge pencil   42

Nonconservative force field   158
Nonlinear hyperbolic PDEs   82f
Nonsingular matrix   128
Normal cone   118f
Normal dispersion   16
Normal ellipsoid   119
Normal forms for second-order PDEs
   69f
Nuclear particles   217f
Nucleosynthesis   223

Orthogonal transformations   130f, 186
Orthonormal matrix   110, 133

Parametric coordinates   43f
Particle accelerators   213
Particle physics   221f
Pauli's exclusion principle   213f
PDEs
   classification of second order   49f
   n space   108, 111f
   second-order quasilinear   50f
   two-dimensions   Chap. 2
Permeability   2
Phase for plane waves   123
Phase velocity   14
Photoelectric effect   207f
Plane electromagnetic waves in a non-
      conducting medium   10f
Plane wave propagation   28f, 244f, 247
Plane waves   122f
Plasma   231
   oscillations   232
   physics domain   231
Polarization   17f
   vector   2
Potential, velocity-dependent   158f
Poynting's theorem   7
Poynting's vector   7
Principle of least action   171f
Principle of least time   136f
Principle of virtual work   153
Propagation of discontinuities   60f

Quadratic forms   109f, 129, 133
Quantum of action   210
Quasilinear PDEs   40
   first order in *n* variables   91f
   second order in *n* variables   107f

Radioactive decay   214f
Range of influence   55f
Ray cone   118f
Ray ellipsoid   120
Rayleigh–Jeans law   207
Rayleigh's dissipation function   160f
Rays (*see under* Bicharacteristics)
Reflection coefficient   26
Reflection of electromagnetic waves   21
Refraction coefficient   26
Refraction of electromagnetic waves   21f
Relativistic mass   280

Relativistic mechanics, force and energy equations 278f
Riemann function 75, 79f
Riemann's method of solving PDEs 74f
Rutherford's atom 208f

Scalar potential 8f, 159
Schrödinger's equation 205, 211
Similarity transformations 110, 127f, 132f
Skin effect 31
Snell's law 24f
Spin 220f
Stationary value 137, 138
Steady-state theory 223f

Tangential derivative (see under Inner derivative)
Telegraph equation 124f
Time dilation 263
Time-harmonic waves 13
TOE 222
Trace 132
Transformation to generalized coordinates 154f
Transmission line (see under Telegraph equation)
Tricomi's equation 73

Ultraviolet catastrophe 207
Undetermined multipliers 165f

Variational methods Chap. 4
Vector potential 8f, 159
Velocities, addition of 265
Virtual displacement 151, 163
Virtual work, principle of 153, 163
Viscosity effects 248

Wave equation
  general solution 33f
  n dimensions 120f
Wave fronts 115f
Wave function 205
Wave number 17
  complex 29
Wave propagation in electromagnetic media by Hamilton–Jacobi theory 199f
Wave propagation
  one-dimensional 10f
  three-dimensional 16f
Weak solutions of PDEs 63f
World lines 267, 269
World space (see under Minkowski space)
World vector (see under Four-vector)